T0139004

Resource Proportional Software Design for Emerging Systems

Resource Proportional Software Design for Emerging Systems

Suparna Bhattacharya

Kanchi Gopinath

Doug Voigt

CRC Press
Taylor & Francis Group
Boca Raton London New York

CRC Press is an imprint of the
Taylor & Francis Group, an **informa** business

A CHAPMAN & HALL BOOK

CRC Press
Taylor & Francis Group
6000 Broken Sound Parkway NW, Suite 300
Boca Raton, FL 33487-2742

© 2020 by Taylor & Francis Group, LLC
CRC Press is an imprint of Taylor & Francis Group, an Informa business

No claim to original U.S. Government works

Printed on acid-free paper

International Standard Book Number-13: 978-1-138-05354-0 (Hardback)

**Visit the Taylor & Francis Web site at
http://www.taylorandfrancis.com**

**and the CRC Press Web site at
http://www.crcpress.com**

Contents

II The Antidote: Resource Proportional Software Design   65

4 Resource Proportional Software Design Principles to Reduce Propensity for Bloat   67

Contents

III   Responding to Emerging Technologies: Designing Resource Proportional Systems   155

Authors

Suparna Bhattacharya is a Distinguished Technologist at Hewlett Packard Enterprise. She has spent most of her career in systems software development and research (5 years at HPE preceded by 21 years at IBM), including several enjoyable years as a well-recognized open source contributor to the Linux kernel. Her recent work advances the use of nonvolatile memory technologies and cross-layer optimization in storage and hyper-converged systems for edge to core data services, containers, machine learning, and artificial intelligence. Suparna is an ACM India eminent speaker and has served on program committees for ASP-LOS, OOPSLA, MASCOTS, ECOOP, HotStorage, and USENIX FAST. She holds a B.Tech from IIT Kharagpur (1993) and a (late-in-life) PhD with a best thesis award from the Indian Institute of Science (2013).

Kanchi Gopinath is a professor at Indian Institute of Science in the Computer Science and Automation Department. His research interests are primarily in the computer systems area (Operating Systems, Storage Systems, Systems Security, and Systems Verification). He is currently an associate editor of IEEE Computer Society Letters and was earlier an associate editor of ACM Transactions on Storage (2009-2018). His education has been at IIT-Madras (B.Tech'77), University of Wisconsin, Madison (MS'80) and Stanford University (PhD'88). He has also worked at AMD (Sunnyvale) ('80-'82), and as a PostDoc ('88-'89) at Stanford.

**Doug Voigt** is a retired Distinguished Technologist who worked for HP and HPE storage for his entire 40 year career. He has developed firmware and software for disk controllers, disk arrays, and storage management. He has led HP and HPE virtual array advanced development projects and strategy since 1990. For the last 10 years his focus has been on non-volatile memory systems. Throughout his career Doug was a strong proponent of industry standards. He has been a member the Storage Network Industry Association (SNIA) since 2009. He served on the SNIA board of directors, technical council, and as co-chair of the NVM Programming Technical Working Group. Doug has over 50 patents, mostly in the areas of virtual arrays and persistent memory. Doug's hobbies include music, photography, and reading science fiction/fantasy.

Preface

This book is an attempt at distilling what good designers instinctively do when confronted with a complex design challenge at the software and systems levels. While it is difficult to state what good designers do concisely, there is one heuristic among the many that stands out and captures this well: the resource proportionality argument ("pay as you go" or "pay for what you get").[1] C++ design has been stated to be as such by B. Stroustrup and a similar principle appears in many other recent designs such as the concept of energy proportional computing introduced by Barroso and Hölzle. However heuristics can also lead us astray: note the effectiveness of the non-intuitive non-work conserving strategies[2] that have been employed in some contexts. Thus even when talented designers design systems, multiple heuristics can interact in unusual ways and system behavior can still surprise us all. One such issue is the problem of software bloat, another topic of discussion in this book.

Throughout our careers the authors of this book have been both observers and participants of the incredible progress in computer hardware and software for the past many decades. Collectively we have been involved in many research and industry projects in operating systems, embedded systems, analytics software, software architecture, as well as teaching and mentoring students and younger designers. Given the criticality of the digital world in our everyday lives, we have also observed or participated in interesting social concerns or broad technical community endeavours such as open source development and free software models for the digital economy (as in GNU/Linux systems and many others) as well as sustainable models of digital economy especially from the energy point of view. One example of the rich interplay between technical and social domains is how the Linux kernel evolved in flexibility to support diverse needs because the community of developers and users involved was very diverse in terms of the scenarios they cared about.

Even though the digital economy is constrained by resources in the long term and the advancement of technology generally leads to greater density and higher throughput per unit of energy, the rate of increase in demand in the digital economy has generally continued to exceed the rate of efficiency

---

[1]Some others include the "end to end argument," "strength of the whole system is that of the weakest link," "a design is best when nothing can be removed from it," etc.

[2]Even when work is available, the system idles for some time to wait for an opportunity to arise to attempt to amortize some cost.

improvements. This is clearly non-sustainable in the long term. Furthermore, the larger economy has often rewarded rapid deployment of inefficient implementations ("roll out updates fast and let customers debug the code").

There is also the tendency of complexity to increase as the computer industry drives technology to its limits. We have also seen, for example, the advent of persistent memory, the proliferation of machine learning, and the increased importance of distributed data centric solutions. We have repeatedly heard that technology streams in memory and storage have reached their limits only to find that scientists and engineers find ways to circumvent them. Rarely does this result in a net simplification of whole systems. For example, flash technology demands wear leveling which introduces the complexity of managing deferred work such as garbage collection.

We find that Resource Proportional Design (RPD) is an interesting perspective that straddles both the technical and social domains of computing. In this book, we have chosen to use RP as a guiding principle to understand the software and systems problem space with suggestions on ways to navigate this design space. Resource proportionality is an efficiency metric describing how system resources are being used on functionality that is actually needed by the current consumers of the system. Our approach to the book involves the following objectives.

- Encourage the use of Resource Proportional Design (RPD) as a methodology

- Expand on nuances of quantitative RP metrics

- Codify RPD as comprising specific methods for improving RP

- Illustrate the role of RPD in use cases across a wide range of software and hardware scenarios.

The creation of long lasting solutions in the RP design space is primarily the work of researchers and developers who are also the primary audience for this book:

- Software Architecture Researchers and Solution Architects - Evaluate RPD approaches and lay the groundwork for tools and practices that work.

- System Engineers (developers and architects of system software including runtime libraries, data access stacks such as drivers, file systems, databases, and caches, OS's, and optimizers as well as the underlying hardware) - Create systems that embed RPD and support application RPD through co-design with runtime libraries, data access stacks, OS's optimizers, and hardware.

- Tool Developers - Create and maintain RPD tools and frameworks that enable developers to efficiently incorporate RP considerations into their designs and implementations.

- Application Developers - Use RPD tools and practices to add this important missing software architecture dimension to their work to make sure that every code feature they develop as programmers is RP by construction.

In addition to researchers and developers, readers may also be interested in specific technology areas such as upper level software, system related software, and distributed applications. Various chapters in the book highlight different technology areas.

The book is divided into four parts with three or four chapters in each. We begin with an overall problem statement and basic description of RPD in Chapter 1. Chapter 2 dives deeper into RPD in the area of software bloat which occurs when resource usage and complexity is out of proportion with utility. Chapter 3 examines the relationship between software bloat and power which is important because power efficiency is the primary goal of RPD. This part of the book should be of general interest to all types of readers. Chapter 1 in particular serves as an important introduction regardless of which subset of the book is of primary interest.

Part 2 focuses on RPD in software development. Chapter 4 elaborates further on RPD principles and definitions while Chapter 5 describes RPD strategies for tool and application developers. Chapters 4 and 5 should be of interest to most readers as they are referenced frequently in subsequent chapters. Chapter 6 speaks to developers responsible for existing (as opposed to new) applications. Chapter 7 follows through with implications of RPD deployments on non-functional requirements such as security.

Part 3 highlights RPD in the context of emerging technologies. Chapter 8 focuses on Persistent Memory while Chapter 9 addresses the related topic of memory interconnects. Chapter 10 addresses RPD in the context of complex, deeply layered software stacks. Chapter 11 applies RPD to data (as opposed to "just" compute) intensive workloads. Reader interest in these chapters is technology specific.

Finally, Part 4 looks forward to managing radically non-uniform systems (Chapter 12), open challenges (Chapter 13), and conclusions (Chapter 14). Part 4 should be of interest to researchers and advanced developers.

We sincerely hope that you, dear reader, take the notion of RPD to heart as an important and interesting perspective. We realize that the current state of RPD tools and frameworks is fragmented, and does not address many of the future looking or advanced concepts described here. Our goal is to further new work in this important area to avoid design bottlenecks and to enable socially responsible and sustainable progress in the digital economy across academia and industry.

Acknowledgments

The perspectives expressed in this book have evolved over decades of lively and enriching discussions with colleagues each of us has worked with throughout our careers. We are deeply grateful to them for shaping our thought process.

Some portions of the book are derived from Suparna's doctoral dissertation and have appeared in the following papers, which are reused here with permission from the publishers IEEE, ACM, and Springer as appropriate. The thesis was guided by Prof. Kanchi Gopinath from the Indian Institute of Science and Dr. Manish Gupta from IBM. We would like to specially acknowledge them and collaborators Dr. Kathick Rajamani and Dr. Mangala Gowri Nanda for their contributions to these papers:

- Suparna Bhattacharya, Kanchi Gopinath, Karthick Rajamani, and Manish Gupta. Software Bloat and Wasted Joules: Is Modularity a Hurdle to Green Software? *in IEEE Computer, vol. 44, no. 9, pp. 97-101, Sept. 2011* (the discussion on software bloat in parts of Section 2.1, 2.5)

- Suparna Bhattacharya, Karthick Rajamani, Kanchi Gopinath, and Manish Gupta. 2012. Does lean imply green?: A study of the power performance implications of Java runtime bloat. *In Proceedings of the 12th ACM SIGMETRICS/PERFORMANCE Joint International Conference on Measurement and Modeling of Computer Systems (SIGMETRICS '12). ACM, New York, NY, USA, 259-270* (the experimental study and analytical model described in Chapter 3)

- Suparna Bhattacharya, Kanchi Gopinath, and Mangala Gowri Nanda. 2013. Combining concern input with program analysis for bloat detection. *In Proceedings of the 2013 ACM SIGPLAN International Conference on Object Oriented Programming Systems Languages and Applications (OOPSLA '13). ACM, New York, NY, USA, 745-764* (the Concern Oriented Program Analysis technique described in Section 5.1.3)

- Suparna Bhattacharya, Mangala Gowri Nanda, Kanchi Gopinath, and Manish Gupta. 2011. Reuse, Recycle to De-bloat Software. *In: Mezini, M. (eds) ECOOP 2011 – Object-Oriented Programming. ECOOP 2011. Lecture Notes in Computer Science, vol 6813. Springer, Berlin, Heidelberg* (the introduction to object reuse in parts of Section 5.2.1)

The authors would also like to thank:

- The Indian Institute of Science for providing a good research ambiance

- Milan Shetti, Mark Potter, Joel Lilienkamp, Peter Corbett, and Naresh Shah from Hewlett Packard Enterprise for their support and encouragement in writing this book

- The SNIA for providing a rich and inspiring vendor neutral context for storage technology discussion in pursuit of standardization and education

- Gary Sevitsky, Nick Mitchell, Matt Arnold, Edith Schonberg, and Harry Xu for their work on Java runtime bloat and regular discussions that helped sharpen our understanding of the deep systemic issues that lead to bloat. Also Bob Blainey and members of the IBM Academy of Technology study on "Software Bloat - Causes and Consequences" (2009) for their contributions and rich insights on the problem of software bloat in the context of hardware and software technological trends

- Parthasarathy Ranganathan for insights on requirements-aware energy scale-down and for feedback on our early ideas about resource proportionality of software features

- LakshmiNarasimhan Naveenam for detailed review of Chapter 8

- Muthukumar Murugan and Kalapriya Kannan for their detailed reviews of Chapter 11. Also Sam Adams, Mukthukumar Murugan, Kalapriya Kannan, Madhumita Bharde, and Annmary Justine for collaboration on externally published papers referenced in this chapter

- Aravinda Prasad for feedback on Chapters 7, 10, and 13

- Kushal Arun at the Computer Architecture and Systems Lab, IISc for her help at various times

- Our families for inspiring us and putting up with our absences through the many nights, weekends, holidays, and precious vacation time consumed in writing this book

इदं नमु ऋषिभ्यः पूर्वजेभ्युः पूर्वेभ्यः पथिकृद्भ्यः ॥

This is our homage to the Rishi-s (seers) of old, the pioneers and the path makers... Rig Veda 10.14.15.2

Part I

Software Bloat, Lost Throughput, and Wasted Joules

Chapter 1

Introduction

A smartphone today has a million times the computing power of the systems
that NASA used to land on the moon[356] and one hundred thousand times
the amount of memory in personal micro-computers we used 30 years ago.

Yet, it does not seem enough to satisfy our needs for long. In cloud based environments, there is a remarkable growth in the pace of software delivery using continuous integration models where new software is now released every few minutes instead of years. As new applications and data sources proliferate rapidly, we also find that every software update requires more resources than its previous version. Is this as efficient as we can get without slowing down the rate of application innovation or is the resource usage for computing growing out of proportion over the years?

Radical improvements in the ease of gathering data and the artificial intelligence revolution continue to raise the demand for computing resources. The cost of sequencing a human genome has reduced by a factor of a million in less than decade, out-shadowing Moore's law[401]. How do we tell if the computing resources we have in any given system are being utilized appropriately for the purpose it serves? If not, then how do we bring in this sense of proportion when designing next generation software and systems solutions?

In this chapter we cover an overview of these questions, why they matter now, our approach to address them, and how the rest of this book will unfold.

1.1 Green Software for the Expanding Digital Universe: Designing with a Sense of Proportion

"Anything that is produced by evolution is bound to be a bit of a mess"

Sydney Brenner, Nobel laureate
In Roger Lewin: Why is Development So Illogical?

Beauty is more important in computing than anywhere else in technology....

Beauty is important in engineering terms because software is so complicated; beauty is the ultimate defense against complexity.

David Gelernter
Machine Beauty: Elegance and the Heart of Technology

Cost, energy efficiency, and performance remain crucial considerations across the entire spectrum of computing systems from the edge to the cloud that collectively serve an ever expanding digital universe. Billions of smartphones, personal devices, and IoT[1] have become indispensable at the edge while data centers gravitate to countries with cold climates[2] and even dive underwater.[3]

The digital expansion is not just a matter of quantity. It is representative of a natural desire for progressively richer and more sophisticated capabilities, as we entrust more and more responsibilities of our lives to the power of software genies residing in the devices we wear or hold in our palms, our homes, our vehicles, our hospitals, our schools, our factories, our farms, spanning both remote edges and the mighty clouds. This unprecedented scale of digitization propels a demand for efficient software and computing systems. Fueled by the constant influx of data to learn from, software is now central to innovation. To sustain the progressive expansion of software capabilities, the functionality supported by a software system needs to be designed and reused with a sense of proportion about the appropriateness of its resource usage.

However, the inevitable complexity in software stacks (a reflection of the complexity in real world needs) is a barrier to reasoning about efficiency and assessing "appropriateness" with respect to utility. The mass of detail in any software system makes it hard to understand what happens at runtime and how to reuse it proportionally for a given scenario. Building efficient solutions requires some form of elegant co-design across different layers of a stack that contain multiple levels of information transfer (e.g., hardware-software co-optimization in smartphones).[4]

---

[1] Internet of Things.

[2] E.g., Facebook's data center in Lulea, Sweden and Google's Hamina data centre in Finland (since 2011) or the White data center snowcooling experiment in Japan[11].

[3] E.g., Microsoft Project Natick[16].

[4] A radical alternative, discussed later in this book, would be to use a deep neural network to infer behavior functions and learn a specialized stack (model) from execution traces of a generalized reusable (but potentially inefficient) implementation. This approach has the drawback of being fragile as it suffers from a loss of explainability, which is problematic from an architect's point of view.

Analog vs. Digital Design

Analog computing in the '50s and '60's provided results at the "speed" of light in some cases but accuracy or controllability was poor. Digital representations and computations based on such representations became necessary to provide some functional guarantees of repeatability and accuracy. However, there are situations where computational or energy requirements are still an issue. Moving from analog to digital fully may be a few orders of magnitude costlier in terms of energy and sometimes also orders of magnitude slower. For example, Spice simulation for circuits is orders of magnitude slower than live electronic speeds.

A system designed with a better sense of proportion about appropriate use of resources may therefore be partly "analog" (for example, to conserve energy) and partly digital (for accuracy). Analog here could refer to systems without quantization, whether electronic, optical, chemical (such as DNA-based computing or storage) or, in the even longer timeframe, quantum-based systems. In communication systems, storage systems, and the like, it is now a widespread practice to use fiber optics that use light ("fast" photons) for transfer of information ("optical") while using electrons ("slow") for computation as it needs bistable states ("electronic"). In a recent hybrid design, certain complex computations needed in a digital design are done "simply" optically and without using power, e.g., preprocessing image data with optical filters, resulting in a substantial speedup, after which the rest of the computation is performed electronically[134].

1.2 The Challenge Posed by Emerging Systems: Why Hardware Advancements Are Not Enough

For decades, we have witnessed a near exponential pace of advancements in hardware technologies and systems that has powered this expansion in software capabilities. These advancements have made framework based software development paradigms practical, resulting in order of magnitude improvements in software development productivity through reuse and decoupling from hardware. The cumulative dividend of these hardware developments has made the smartphone a reality and cloud computing affordable. Software defined storage is replacing traditional storage arrays which were built using specialized hardware (ASICs and battery backed NVRAM[5]). The rise of general purpose GPUs has helped usher in the deep learning revolution

---

[5]Non-volatile random access memory.

by unlocking the computational capabilities needed for training multi-layer neural networks.

Any recommendations geared at specific emerging systems, even those based on technologies that are cutting-edge at the time of writing this book, are likely to become obsolete.[6] However, the nature of hardware advancements in recent times is driving a need for changes at many levels in software to unlock the benefits. It is getting increasingly difficult to simply rely on powerful hardware to compensate for inefficiencies that have accumulated in software stacks over previous generations. Adapting to the tension between shifting tradeoffs and opportunities provided by emerging systems is a recurring challenge in software systems design.

This tension is exacerbated by several technology trends embodied in emerging memory centric system architectures [392], [40] which promise new levels of performance and flexibility with lower power consumption, e.g., by leveraging the speed of persistent memory, hardware acceleration, and low-latency fabrics. Such hardware technology trends and operational challenges are now forcing us to pay greater attention to mitigating the impact of any disproportionate resource usage by software on energy consumption and system performance while preserving the undeniable benefits of flexible abstractions.

At the same time it is becoming more and more practical to use machine learning (ML) techniques such as neural networks and alternatives such as approximate computing in response to these challenges. These techniques expend analysis resources to gain runtime efficiency improvements. When is this a good trade-off? Perhaps there are some lessons to be learned from the early days of computation where both analog and digital approaches were in play.

Here are a few examples of the tensions and opportunities we explore in this book:

1.2.1 Runtime bloat in framework based software

The emergence of powerful frameworks for redeployable software has transformed the pace of large scale software development. Framework based software systems are provisioned with a high degree of built-in excess flexibility to support standardization and cope with evolving requirements. An unintended side effect of this trend is that software systems tend to accumulate "runtime bloat" - the resource overhead of excess functions and objects. According to Mitchell, Schonberg and Sevitsky [264], bloat can manifest as a pervasive pattern of excessive memory, processing, and IO overheads in framework based applications. These overheads can cause applications to "miss their scalability goals by an order of magnitude," resulting in the deployment of additional hardware and complex system infrastructure to meet service level objectives.

---

[6]We cannot (thankfully) anticipate the future.

Runtime bloat often results from a pile-up of unnecessary consumption of memory and processing resources caused by a lack of effective coupling between layers in a software stack. Specific sources of this, as revealed through research and case studies, include duplication and transformation of data structures across software layers and execution of features in one software layer that are not needed by the layers above. This ineffective coupling is a byproduct of software developer training cornerstones such as encapsulation and code reuse. Thus software stacks grow resource heavy as a side effect of the very paradigm of framework based reuse that revolutionized the pace of information technology. The same flexibility that enables applications to rapidly evolve in scale and function to handle unanticipated changes in demanding environments also tends to encourage the accumulation of excess functionality and data.

Bloat imposes a large and pervasive infrastructure tax[7] which eludes even sophisticated JIT optimizers. This tax consumes excess CPU cycles, clogs memory bandwidth (refer to "Allocation wall" discussion in Chapter 3), increases lock contention, and amplifies memory footprint, IO, and storage costs. The cumulative effect of these power-performance inefficiencies can significantly limit the benefits of emerging systems, especially when adapting to technology advancements that shift tradeoffs.

1.2.2   Software interaction with systems

In general there are multiple ways of implementing a software function, some of which are better suited to particular underlying systems. In many cases, underlying system or technology improvements change the optimal relationship between resources consumed and work accomplished. This raises questions about how software deals with change that originates with underlying systems or technologies in addition to those that arise from functionality required by an application. Often software behavior must change in order to fully realize the potential improvement of a system or technology change.

As an ordinary everyday life example, consider the scenario of walking in the rain to an important meeting. The requirements are to arrive on time and dry. The functional choice to be made is that of rain gear. The system is represented by weather in this example where the unexpected variable is wind. Given the requirements one might choose an umbrella as the solution, ignoring the possibility of wind, that imposes the least transit time overhead. Obviously if wind is a factor the umbrella might be worse than useless relative to both requirements. A raincoat would have been a better choice.

In this example, wind represents a system condition that may not have been accounted for by an otherwise efficient software solution. The ability to adapt the choice of rain gear to wind represents a resource proportional real time response to a system condition. The reader can imagine similar exam-

---

[7]In contrast with localized hotspots.

ples where choices made with awareness of available technology (e.g., trench coat vs. poncho) could be relevant depending on additional non-functional requirements such as appearance.

More to the point, trade-offs with implications that are inconsequential on one system at one point in time may have dire consequences in a different system or temporal context. Stated differently, a choice that results in an acceptable perturbation from a desired system behavior target in one environment may result in an unacceptable perturbation in another. System behavior, in this case, includes aspects of both performance and resources consumed.

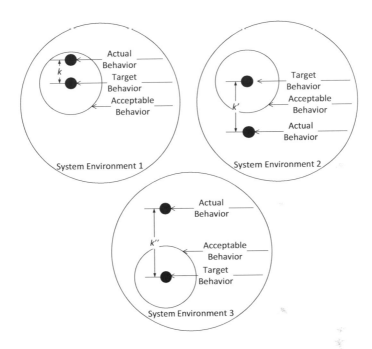

FIGURE 1.1: Perturbation within or beyond acceptable operational bounds.

In Figure 1.1, system environment 1 represents a scenario in which a piece of software runs within acceptable behavior bounds, perhaps represented by a set of service level objectives and a target resource footprint that is nearly optimal. At some point within a conceptual behavioral range there is a target behavior for the function in system environment 1, and an actual behavior point at which a software implementation of a function currently operates. Often, resources consumed or performance are metrics of interest that can be used to create a ratio representing actual behavior relative to target behavior. The acceptable behavior ring represents a range of metric values that is deemed acceptable, perhaps because it is within some bounds of efficiency that accounts for a service level objective that the function must meet. The ratio of actual to target metric values is represented by a distance k.

In system environment 2, a change has occurred that causes the software function to consume more resources or perform more poorly than it did before, even though the implementation has not changed. Thus the behavior of the function in that system has moved a distance of k' from the target. Examples of system environment changes that could cause this might include the following:

(a) Increase in memory pressure causing additional swapping

(b) Degradation in a subsystem such as a network that changes the effectiveness of a retry policy

In the scenario illustrated for system environment 2 the target and acceptable behaviors may not have changed, but the function implementation necessary to achieve them is different. For example, more aggressive garbage collection may be needed to shrink memory footprint, or the choice of network access point may need to change. In the absence of an implementation change, k' in system environment 2 falls outside of acceptable behavior.

In system environment 3, the target behavior and acceptable behavior have changed, perhaps due to new expectations resulting from some underlying change in system implementation or provisioning. If the function implementation is not able to take advantage of this change then the distance from the behavior of the function to the target behavior in that system has changed to k''. Examples of system environment changes that could cause this might include the following:

(a) Increase in the amount of memory or the number of cores in the system

(b) Introduction of new types of components such as Persistent Memory (PM) into the system that could assist the function

The scenario of system environment 3 represents the reverse of system environment 2 because the "cheese moved." The target behavior changed, and moved the acceptable behavior so that k'' no longer meets expectations.

These stories illustrate behavior perturbation that originates in system changes rather than variations in the functional requirements of a piece of software. System interaction can create inefficiencies analogous to bloat except that the problem originates outside of the software itself. Let's look deeper into some specific factors that can perturb system behavior.

1.2.3   Impact of non-volatile memory and low-latency fabrics

The advent of non-volatile memory technologies (also known as storage class memory and henceforth referred to as persistent memory in this book) is driving one such dramatic shift in emerging systems. Traditionally, data used by applications have persisted in high capacity storage systems (based on hard disk technology) that operate at latencies about five orders of magnitude slower than expensive (and volatile) memory systems. Software components at multiple levels of the stack have therefore typically been designed to operate with the assumption of access to limited amounts of fast memory and large amounts of slow storage.

Figure 1.2 illustrates how this trade-off has shifted over the years. The stacked lines depict the relative latencies of storage, memory interconnect, and software between 2000 and 2020. Solid-state storage technologies such as flash based SSDs reduced this gap by two orders of magnitude. With emerging persistent memory technologies, the gap has shrunk so much that it has created an inflection point where the latencies are close enough to the speed of memory to be accessible using memory semantics (Chapter 8). Further, with the availability of low latency memory interconnect (MI) technology (Chapter 9) it is even possible to access persistent memory attached to a remote node using memory semantics. At these low latencies, path length overheads induced by traditional software stacks may no longer be affordable.

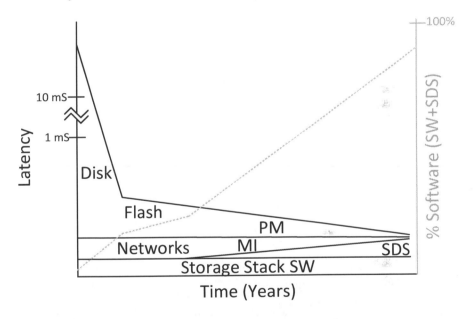

FIGURE 1.2: Stacked line graph of storage technology latencies relative to storage software overhead. The lowest line shows the storage software overhead, which used to be a small fraction of the overall latency compared with hard disk access times but now represents a very high percentage compared to persistent memory (PM).

The figure illustrates how software overheads begin to dominate as storage latencies reduce (see "Software considered harmful" discussion by Swanson and Caulfield in [374]). While both storage and network latencies have dropped substantially over the years, software path lengths have actually increased with the rising number of layers present in a typical software defined storage (SDS) stack. The percentage of software contribution is now so high that it can overshadow the benefits anticipated from switching to persistent memory and low latency interconnects, limiting performance gains. The figure does not include the additional number of application layers beyond the core storage stack that are involved in typical data access operation such as databases and file systems.

Ensuring proportional gains from such significant technology advancements is non-trivial because it requires rethinking assumptions that are deeply embedded in different layers of the software stack.

1.2.4   Large scale connected architectures from edge to cloud

Operating systems and runtimes enable applications to be de-coupled from systems trade-offs. However, emerging computing systems are extending beyond individual (single node) servers to large scale connected infrastructure, where solutions are architected to utilize processing resources distributed in many different ways, e.g., across a cluster, a data center, from mobile elements, across a WAN between IoT edges and a core or across multiple clouds. Examples of such (existing and emerging) systems are very diverse:

- Data analytics clusters (Hadoop, Spark [363])

- High performance computing clusters

- Datacenter as a computer (warehouse scale computers, cloud computing)

- Mobile first internet services

- Scale out storage, object storage

- Edge computing and edge to core analytics

- Distributed deep learning

- Hybrid cloud solutions

In all these cases, it is no longer sufficient to design software stacks with a purely "local" or "single system" scope. The impact of inefficiencies as well as opportunities typically involve a collective view of trade-offs spanning resources from multiple (and potentially heterogeneous) systems and locations. This has necessitated a diffusion of systems software functionality from traditional operating systems (such as resource provisioning, scheduling, fault tolerance) into frameworks that operate at higher levels of the stack. Much innovation in this space has been realized via layers of middleware that step around the gaps between the needs of this emerging application computing environment and the scenarios that established operating system mechanisms such as virtual memory and storage architectures were traditionally designed for.

One implication of this diffusion is duplication of functionality across layers (e.g., a local filesystem and HDFS) and a need to rethink assumptions and trade-offs about where certain types of functions should be implemented to avoid disproportionate resource usage. For example, in their white paper, "Disks for datacenter"[80] Brewer et al. advocate simplifying individual disk drives by sacrificing a small amount of reliability (e.g., by reducing retries) and instead letting higher level software ensure reliability across a collection of drives, resulting in more consistent performance (reduced tail latency).

In an edge to core analytics workflow, costs and constraints on data that can be moved to the core for analysis and analysis functionality that can be shipped to the edge must be factored in to ensure appropriate resource usage based on an assessment of the utility of computation at each end. Besides data transfer overheads, real time constraints, and security restrictions at the edge, computing resource costs may not be uniform across all the nodes. To compound the situation, software code that is traversed from edge to core to cloud comprises many independently developed components and frameworks with their own CI/CD (continuous integration and deployment) pipelines.

1.2.5 Emerging software models and data centricity

The needs of emerging applications are also driving a shift in software development and deployment models (e.g., cloud native, mobile first, artificial intelligence (AI) driven, anything-as-a-service). Containers and serverless computing (Chapter 2) simplify how software functionality is delivered, deployed, managed, and moved across diverse infrastructure environments. Containers offer the convenience and efficiencies of lightweight virtualization and separation of application data from images which in turn are built as reusable layers. Further, as applications become more data centric and insight driven, building, training, deploying, and updating machine learning models in production across large scale infrastructure is becoming integral to solution workflows and continuous integration and deployment pipelines. Transfer learning techniques that allow pre-trained models to be reused for related tasks are becoming popular especially in deep learning where the amount of training data and computation resources needed for training a model from scratch is huge[2].

Despite the improved efficiencies, designing with a sense of proportion is still essential with emerging software models, which also introduce fresh challenges. Containers can get bloated, microservices have data exchange costs, and much data collected for analysis is often bloated with low value information which can overwhelm storage and processing resources. Proportionality in resource consumption with respect to software utility translates into better economics in terms of lower cloud costs, reduced edge to core constraints, and greater data tiering savings.

1.3   The Heart of the Matter: Why a Plea for Lean Software Is Not Enough

"Software girth has surpassed its functionality largely because hardware advances make this possible.

About 25 years ago, an interactive text editor could be designed with as little as 8000 bytes of storage (Modern editors request 100 times that much!). An operating system had to manage with 8000 bytes and a compiler had to fit into 32 Kbytes, whereas their modern descendants require megabytes. Were it not for a thousand times faster hardware, modern software would be utterly unusable."

Niklaus Wirth
"A Plea for Lean Software" [410]

Concern about software code and data bloat is not a new issue - yet for over two decades since the publication of Niklaus Wirth's article "A Plea for Lean Software" [410], the incidence of bloat has mostly continued unabated. Instead as we observed in the previous section, there are many examples where its proportion has even grown as a consequence of prevalent software and hardware trends.

1.3.1   The flexibility, productivity, and efficiency trade-off

The deep systemic cause of this persistent issue is not one of "bad programmers," but practical cost-benefit trade-offs that have emerged in reaction to the inherent difficulty in meeting the pressures of simultaneously addressing the need for flexibility, productivity and efficiency[8] of software solutions. Any two of these considerations can typically be achieved at the price of compromising on the third consideration (Figure 1.3). In the case of framework based software the trade-offs have leaned in favor of flexibility and productivity at the cost of efficiency [264].

For example, one explanation for why the problem of software bloat has received very little sustained attention until now is that there have been dramatic improvements in hardware performance over successive CMOS technology generations. Such improvements would tend to obscure the significance

---

[8] Although there are other important non-functional considerations such as security, standardization, RAS (reliability, availability and serviceability), for simplifying the discussion we include those considerations either under the category of flexibility or productivity, as appropriate.

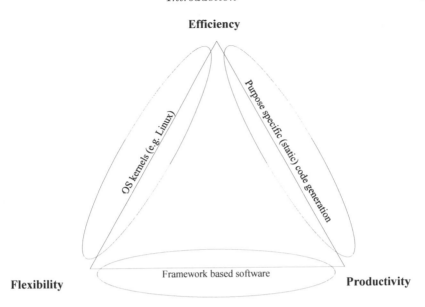

FIGURE 1.3: Trade-off between software engineering considerations of solution flexibility, runtime efficiency, and development productivity.

of inefficiencies caused by bloat in comparison with the huge gains in software flexibility and development productivity. Efficiency in software is now regaining its importance with the shifting technology trade-offs we discussed in the last section, the gradual reduction in energy efficiency improvements over successive generations of hardware, and the rising operational costs of maintaining overly complex and bulky software stacks. At the same time, both flexibility and productivity continue to be important software engineering considerations.

1.3.2 Unsustainability of tightly coupled hardware-software abstractions

Hence, traditional solutions that tackle code bloat by relying on detailed attention to programming, finely crafted abstractions, tightly coupled hardware software development, or carefully customized feature configuration are unsuitable for the scale and pace of software development required in large flexible IT solutions.

For example, operating system code such as Linux can support a very diverse range of requirements and conditions very efficiently, but require a lot of careful programming and deep expertise, which is a barrier to development productivity as depicted in Figure 1.3. On the other hand, one way to decouple hardware software abstractions from a programmer's perspective is the use of domain-specific languages or higher level models with automatic code generators (compilers and runtimes) optimized for a desired hardware system, e.g., using a DSL for graph processing to generate optimized code for GPUs. As illustrated in Figure 1.3, this approach preserves efficiency and productivity at the cost of flexibility.

1.3.3   Traditional performance optimization is not enough

In "Profiling a Warehouse Scale Computer," Kanev et al. found that data-center workloads show no single hot application to optimize for and no major hotspots within each application. A similar observation has been made in studies of sofware bloat which finds large applications to be "relatively free of hotspot methods."[264] In both cases the overheads (bloat or "datacenter tax") appear to be diffused more pervasively, cutting across methods in an application and across applications in a data center or cutting across machines in a datacenter. Thus traditional performance optimization approaches which are geared at optimizing algorithms in critical methods fail to be useful even though the collective overhead across all methods can lead to disproportionate costs.

1.3.4   Difficulty in quantifying the opportunity and impact of software bloat reduction

Designing software with a sense of proportion requires conducting early cost-benefit assessments to determine where opportunities exist for reducing disproportionately high resource usage. This requires one to have line of sight from software features in code to their relative impact on resource consumption and system power-performance. Distinguishing excess resource utilization from resource utilization due to essential function is non-trivial, as it typically requires a deep knowledge of intended semantics of complex software stacks. Hence software designers only have access to a limited understanding of the circumstances under which bloat arises and the extent to which it affects runtime resource usage and system power-performance. As a result it is difficult to perform a systematic optimization of software to address disproportionate resource consumption.

For example, in order to establish the relationship between bloat and energy efficient design, there is a need to quantitatively relate the overall power-performance impact of bloat in a system to specific sources of bloat and to reason about energy savings expected from alternative strategies for bloat reduction. However, this is a very broad and challenging problem that raises several interesting questions which we attempt to address in this book:

- How can the extent of bloat attributed to a given source be determined?

- How much does bloat matter for power-performance?

- What can be done to mitigate the impact of bloat once we have identified it?

- How do we assess and reduce the propensity for bloat of a given software implementation?

1.4   The Resource Proportional Software Design Principle

Given that simply imploring developers to write efficient code is not particularly effective, is there a principled way to design software with a sense of proportion, in the absence of a free ride on hardware system advancements? In this book, we encourage a shift in software engineering that counters the overhead of deeply layered code by making its resource consumption proportional to situational software utility, defined here as "resource proportional software design." Resource proportional software design (RPD) introduces a principled approach for developing software components in a way that consciously mitigates the effects of excess resource usage from that contributing to essential function, which would otherwise be highly non-trivial without a deep knowledge of intended semantics of very complex software.

The RPD approach systematically tackles some of the challenges described in previous sections:

1.4.1   How to assess the propensity for bloat in a software component?

If software components are designed to be RP, then it would be easier to extend the methodology to then compose RP stacks using those components. In order to build resource proportional components, we need a way to first assess and quantify whether a given component implementation has propensity for bloat that could cause disproportionate resource consumption with respect to its utility in certain situations. The term "utility" and what is appropriately proportional is subjective and can vary based on deployment environment constraints. We typically assess it specifically for an aspect of interest, e.g.:

1. proportion of code features (functionality) in a program (component) actually utilized or essential for a given scenario

2. proportion of software stack utilized by the underlying hardware operation (e.g., media access latency compared to software level access time observed for data reads and writes)

3. proportion of input data actually utilized or essential for contributing to a useful analysis outcome in data centric software

A general measure for RP can be devised by comparing the actual resource consumption of a given component under each utilization scenario with the resource consumption of a design of a component that is optimized just for the specific utilized proportion in that scenario. If this ratio is less than k across all utilization scenarios and inputs, we say that the given component is k-RPD.

The k-RPD metric represents an ideal because, in practice, it is unlikely that component designs that are optimized for each utilization level would be available for direct comparison. Further, such comparison may not be directly useful in identifying ways to improve RP of a given component. A more practical approach to assess if the resource usage of a given component is appropriately proportional is to develop line of sight into the resource overhead and power performance cost incurred due to unutilized or non-essential features, capabilities, and data. This overhead is a source of non-resource proportionality. Comparing the overhead with actual resource usage provides an estimate of the percentage bloat, which can then be used to derive estimates of bloat propensity and k. Notice that the overhead and bloat propensity is zero in the ideal perfectly resource proportional component, where $k = 1$. If $k = 2$, on the other hand, the component is consuming twice the resources of an implementation optimized for the utilization scenario in question.

RP analysis can be broken down into the following questions which will be explored in Chapters 2, 3, 4.

- Why do excess features accumulate in a software stack? (Chapter 2)

- When do excess features lead to runtime bloat? (Chapter 4)

- When is the resource overhead due to bloat most pronounced? (Chapter 4)

- How much does bloat matter for power-performance? (Chapter 3)

- What information is necessary to automatically estimate the extent of bloat? (Chapter 4)

1.4.2   How to broaden supported features without a runtime overhead?

A common practice adopted by developers to reduce the burden of excess features in generalized components is to reduce the features or create specialized versions of the component for different scenarios. However, such an approach compromises flexibility, e.g., reduces the situations where the component may be used/reused. This could be visualized as gaps in the RPD utilization scenario space which are not addressed by the component. Such a solution no longer qualifies as a k-RPD design, even though k would be much lower for the supported scenarios. Overspecialization also affects productivity by limiting reuse or forcing developers or operators to anticipate exactly what features would be relevant in each deployment scenario where the component is leveraged. Instead, the goal of RPD is to (a) broaden the features supported (cover the utilization space) for flexibility, (b) maximize reuse for productivity and efficient utilization of features and, at the same time, (c) minimize the runtime overhead incurred due to unutilized features.

To contain bloat when unifying code features and broadening them to cover more scenarios, two key questions need to be tackled:

- What information is necessary to automatically de-bloat software without sacrificing reuse?

- Can systems be redesigned to enable software to avoid propensity for bloat without losing flexibility or productivity?

Chapter 4 addresses these questions by showing that the presence of excess features, in itself, may not lead to runtime bloat, unless these features have some structural interaction with essential features. Structural interactions usually arise in order to enable efficiencies from reuse when these features are used together. Further, the overheads due to such structural interactions, in turn, may not cause substantial runtime resource bloat in a long running (server) application unless they are incurred repeatedly during program execution. Finally, even such resource bloat has a pronounced impact on power-performance only if it affects a system bottleneck or a hardware resource that has a high relative energy proportionality and consumes a high fraction of system power compared to the other system resources (Chapter 3). Chapter 4 integrates these insights into a high-level cause-effect flow diagram that provides a foundation to enable the development of systematic strategies for constructing RP components.

1.4.3 Can software be designed to cope with changing trade-offs as technology evolves?

Section 1.2.2 introduced the notion of behavior perturbation originating in system changes which we now can describe and measure using resource proportionality concepts. The metric k in Figure 1.1 can be analyzed using the k-RPD definition in Section 1.4.1. Even though the scenarios in play are driven by system change instead of software functionality, similar techniques for orchestrating responses are applicable. One significant difference is that responses to system change are guided by information about the system itself in addition to software feature requirements and usage.

System changes may also occur more rapidly than those that originate with software context. Even if software has a designed-in ability to adapt to system conditions, there is a question of how disruptive re-optimization of the software is to the progress of the application. For some changes it may be acceptable for re-optimization to require that software be rebuilt and restarted. More contained, less disruptive responses may occur quickly based on algorithms that are already engaged. For rapid or erratic changes, questions of system stability become relevant.

A disruptive re-optimization mandates more planning based on knowledge of ongoing or concurrent system changes. If a system administrator knows that a group of changes are planned, re-optimization may be delayed until all have occurred. On the other hand if a condition arises that has an unknown but protracted duration, more immediate re-optimization may be in order.

Chapters 8, 9, and 12 further explore responses to behavior changes due to system and technology perturbation.

1.4.4   Can we anticipate what proportion of data processed by application is truly useful?

Besides code features and perturbations due to technology changes, another potential source of disproportionate resource consumption is excess input data processed by applications.

Applications are becoming more data and insight driven and the amount of data generated and stored is increasing at an exponential rate. As the cost of collecting data goes down, there is a natural tendency to generate more and more data - such data could come from the physical world through sensors and cameras, from people, and from machines or computer systems (e.g., telemetry from data centers). This trend has the upside of providing us a high resolution view of real world events allowing increasingly sophisticated decisions to be derived from data, but also has the downside of overwhelming human attention and computer system resources with a flood of not so informative data. One example is the huge rise in the number of photographs taken on average every day with the advent of smartphone cameras, only a fraction of which are of lasting value and quality, as compared to the previous digital camera era and the era of non-digital photography.

While it is desirable to spend computation and storage resources proportionately on the useful data and reduce resources spent processing and storing less relevant data, making this distinction is non-trivial work and typically requires some computation resources as well. In a multi-stage workflow, the earlier we can discriminate between data that would prove to be useful (high utility) for an analysis decision and data that are not worth processing further in a given deployment scenario, the more resource proportional the solution. Chapter 11 discusses the concept of data centric resource proportionality and such trade-offs in systematically tackling this data bloat problem.

1.5 Dimensions of Resource Proportional Design

Figure 1.4 depicts a high level RPD workflow that addresses the three dimensions of resource proportionality introduced in the previous section: (a) Code, (b) Technology (System), and (c) Data. The remaining chapters of this book systematically address each of these dimensions.

Chapter 2 and Part II primarily deal with challenges of software bloat and approaches to tackle it by designing resource proportional code features, with tips for software developers (components, frameworks, applications), architects, and tool developers (compilers, runtimes, static and dynamic analysis). Chapter 7 extends this perspective to non-functional implications of RPD and the challenging issues that arise as a result. Chapter 3 and Part III are more heavily focused on designing response proportional responses to technology and systems evolution, with tips for system software developers and architects. Part III also includes cases studies of complex emerging application stacks (Chapter 10) with tips for full stack developers, end to end system architects, and data centric solution architects. In particular, Chapter 11 deals with opportunities and challenges in designing stacks that make resource proportional use of data for machine learning, AI, and analytics.

Building on this foundation, Part IV develops a more advanced architectural perspective and future directions for RPD.

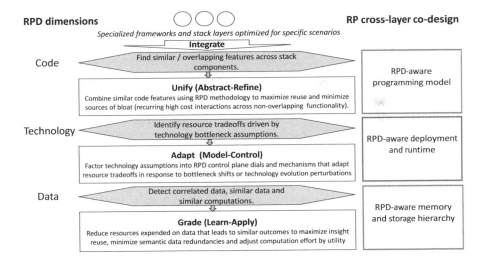

FIGURE 1.4: RPD dimensions.

1.5.1   Resource proportional code features

Based on the observations made previously in Section 1.4.2, a code feature would be perfectly resource proportional (1-RPD)

- if the addition of that feature does not induce any runtime cost when the feature is not essential or logically unutilized in a given situation (minimal interaction overhead when unused)

- if the utilization of the feature incurs only an incremental resource consumption for the functionality added by the feature, i.e., it efficiently reuses and is reusable by existing features efficiently (minimal interaction overhead when used)

In this situation resource consumption of the software (e.g., CPU, memory, IO) would scale proportionally as additional features are utilized.

Optimizing for both these conditions is a key challenge in designing resource proportional code features. If a feature is heavily utilized (reused) then we need to reduce the interaction overhead to use the feature as well as the interaction overhead induced on it by features that are not useful in that deployment scenario. Notice that there is no inherent constraint for the code that implements the feature to be localized to a function or module or micro-service; each of these approaches involves a different trade-off. Separating features using indirections can reduce the overhead of unutilized features; however, directly embedding code may be more efficient when features are utilized together. In spite of modularization attempts, some components of the state of one subsystem can spill into another and make it difficult or costly to keep such a state consistent across the whole system.

Chapter 4 systematically introduces different levels of intervention that address this trade-off as part of the RPD methodology. As summarized in the top pane of Figure 1.4, the first step is to unify commonly used functionality (instead of duplicating it across layers or across specialized components) and increasing variability through efficient reuse to then support a wide feature set. Efficient reuse includes minimizing the interaction cost to utilize a common essential feature. The next step is reduce the interaction overhead (typically caused by the presence of structurally entwined code and data structures) induced on frequently executed code by all other features when they are not used.

In Chapter 5 we describe a series of strategies and practices that could be used by developers to realize these steps in order to construct resource proportional software components or add resource proportional features to an existing component or application. If software is not already written to be resource proportional, it is difficult to optimize it for RP. In Chapter 6 we suggest strategies and practices that could be used to optimize or refactor an existing application to make it more resource proportional.

1.5.2 Resource proportional response to technology and system evolution

A number of system factors may affect resource proportionality in positive or negative ways, especially if they are not accommodated by system or application software design. The Figure 1.5 illustrates a taxonomy of several relevant examples.

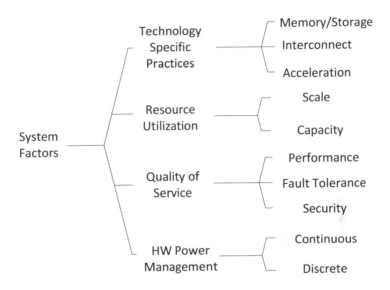

FIGURE 1.5: System factors that impact resource proportionality.

Some new technologies create opportunities to improve system behavior if software can be modified to take advantage of them. In many cases modifications may be optional, but they lead to greater resource efficiency if they are implemented. A change in efficiency such as code path length reduction does not necessarily imply a change in resource proportionality in and of itself. Perhaps not using the capability of the new technology or using it in a backward compatible way is just a missed opportunity. In other cases a technology substitution may radically change system dynamics causing new resource proportionality considerations to emerge, or existing ones to change significance.

Technology-specific practices may arise in any fundamental system element including memory or storage, interconnect, or computational acceleration. The emergence of persistent memory and new memory interconnects described in Section 1.2.3 are prime examples of this. In Chapters 8 and 9 we will review RP considerations that result from these system level changes.

Resource utilization is a perennial source of non-RP behavior that may interact with emerging technologies in new ways. In this taxonomy, resource or system scale refers to the amount of resource available such as memory capacity or network connectivity as with scale out or memory centric architectures. Capacity refers to space for storing information and to the rate at which the information can be accessed such as effective bandwidth or operations per second. Existing systems exhibit non-RP behavior related to both of these characteristics. For example, as memory allocation increases and available free space decreases, an effect described as the "allocation wall" may cause non-linear system behavior due to increasing fragmentation and garbage collection (Chapter 3). Likewise, as network demand approaches saturation, non-linear delays due to congestion are likely. These remain relevant in the analysis of RP behavior in emerging technologies.

Another important aspect of resource utilization has to do with system bottlenecks. Important non-linearities often occur when workloads cause the most heavily utilized (bottleneck) resource to shift from one system component to another (Chapter 3).

The imposition of Quality of Service (QoS) objectives on system workloads tends to exacerbate RP issues by imposing new non-functional requirements such as performance constraints, fault tolerance, and security. As a performance example, one can view maximum latency objectives as placing a hard limit on acceptable deviation of system behavior from a target operational state (Figure 1.1). As a result the system may be viewed as defective if its RP is not sufficient. Fault tolerance tends to drive additional load into systems for redundancy and recovery. As a result system health degradation is more likely to lead to non-RP behavior before the system is deemed unusable due to failure. In Chapter 12 we suggest a response to these utilization and QoS challenges. Security generally demands additional functionality and resources over and above the work output of the system. Security is discussed in Chapters 7 and 13, in particular in Section 7.3.1 on RP remediation of security features.

Finally, some computer systems have sophisticated automatic power management. These include control systems applied to continuous system parameters such as varying clock speed to decrease power consumption while still meeting performance goals. Still other systems shift work onto smaller numbers of components so as to power down parts of a system. Ironically these hardware approaches tend to amplify the observable effects of non-RP software because they convert inefficient resource usage into potentially rapid variation in power consumption (Chapter 3). Also, some data centers are managed using power capping so that whole systems are not allowed to consume more than a designated amount of power. In such environments non-RP software may trigger resource caps that quickly drain resources from other software running on the same system, thus impacting broad system behavior in ways that might otherwise have been localized.

1.5.3   Resource proportional data processing

To obtain a measure for data resource proportionality, the actual resource usage of an application that uses the entire data set can be compared to the resources expended by an application that uses only the portion of data that is sufficient for deriving the insight required, under each deployment scenario. If this ratio is less than k across all deployment scenarios, we say that the given component is k-dataRPD.

Based on the observations made previously in Section 1.4.4, data processing would be perfectly resource proportional (1-dataRPD)

- if the addition of data does not induce any runtime cost when the data are not essential or logically unutilized in a given situation (minimal interaction overhead when unused)

- if data that are essential (in a given situation) get utilized and incur only an incremental resource consumption for the insight added by the data, i.e., it efficiently supplements existing data and can be efficiently supplemented by more data, for example, by reusing results from similar computations on similar data [206, 170] (minimal interaction overhead when used)

Under these conditions resource consumption of the software (e.g., CPU, memory, IO) would scale proportionally as additional data are utilized for insight.

The key challenge in designing resource proportional data processing is how to distinguish data that are essential for deriving insights from data that do not add insight in any situation. Techniques for detecting correlations and data similarity can be applied to minimize semantic redundancies in the data. This data reduction step adds some computation overhead. The solution is resource proportional if the reduction obtained is substantial and later stages in the workflow are far more resource intensive than the data reduction step.

In Chapter 11 we describe a few case studies to illustrate resource proportional design considerations for data centric applications and how to design systems software stacks that enable data resource proportionality. We illustrate how a system that remembers what data it found essential and what was non-essential can reuse that learning across analytics tasks, thus amortizing the cost of the data reduction step. Generalizing the concept further we explore the possibility of designing RPD aware virtual memory and storage tiers which can learn to grade data from most essential to least informative. For example, data that results in analysis outcomes very similar to previous data are less informative. The bottom pane in Figure 1.4 summarizes this process of data gradation and how it can be used to adjust the amount of data and computation costs in proportion to utility.

1.6   Approach in This Book

The rest of this book develops a foundation for these underlying principles, in the form of a whole systems perspective derived from a survey of recent research on the problem of runtime bloat and resource amplification in large framework based applications and studying its implications for efficient design in emerging systems. This forms the basis for suggestions to developers on sustainable ways to improve performance, resource footprint, and power consumption of software and systems using RPD.

Applying our methodology typically involves a four step approach:

1. **Identify** potential sources of resource amplification in terms of code, systems technology, or data

2. **Quantify** resource proportionality in terms of code, systems technology, or data

3. **Transform** the problem to make RP trade-offs visible as familiar measures and solution opportunities

4. **Apply** known tools and techniques to address these opportunities and manage the trade-offs

For example, in Chapter 8 we analyze the application journey from Disks to PM as a technology transition scenario similar to the transition from IPV4 to IPV6[9] in order to **Identify** sources of resource amplification. We then use a performance model to **Quantify** amplification impact. In Chapter 9 we add modeling of memory interconnects and **Transform** the problem into a tabular structure that enables us to **Apply** pre-existing tools such as ML to make placement decisions. In Chapter 12 we use models and metrics from chapters 8 and 9 to formulate a management framework that **Transforms** data and function decisions into a memory pool hierarchy that **Applies** parametric, rule based, or ML tools in a combination of background and real time placement decision making.

Chapters 5 and 6 apply the same pattern by using metrics to decompose RPD into a series of actions as summarized in figures 5.12 and 6.1.

Both software and system level changes needed to achieve this are explored in a series of case studies throughout the book. RPD strategies include a new set of best practices for application software developers and a new set of constructs and optimization techniques for developers of software and system development tools. The practices and tools described enable vertical optimization within a software stack to eliminate bloat and resource amplifica-

---

[9]To support the IPV4 to IPV6 transition, interoperable network stacks have been implemented with both technologies at the same time in the same system.

tion. These include automated code analysis and language aids for tracking a component's candidate non-resource proportional concerns.

The RP tension recurs in emerging systems where memory hierarchies and specialized processing create non-uniformity. To manage complexity, non-uniformity is purposefully hidden from applications and developers using resource allocation policies and system specific plug-ins. The resource proportional software development principles used to address bloat are extended to resolve analogous problems in managing this emerging system complexity.

The book concludes with a summary and discussion for practitioners and researchers in the field, including some interesting open questions to tackle along the transition to wider adoption of resource proportional design principles.

Figure 1.6 shows chapter order recommendations for selective reading of this book once the reader has identified chapters of interest based on subject matter as described above. These are just suggestions for people who are only interested in some specific topics and have a specific background. In general as chapter numbers increase, subject matter becomes more advanced and forward looking.

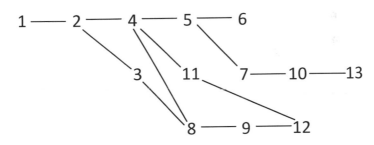

FIGURE 1.6: Chapter recommendations based on area of interest.

To recap, Chapters 1, 2, and 4 are important to the reader's understanding of RPD overall and should be viewed as prerequisites for the remainder of the book. Chapter 3 should be of interest to readers who desire background in the relationship between RPD and hardware.

Chapters 5 and 6 are primarily for those interested in digging deeper into programming and tools for RPD. Chapters 7, 10, and 13 are for those interested in broader considerations about cross stack interactions and full system architecture. Chapter 11 focuses on data centric RPD.

Finally, Chapters 8, 9, and 12 elaborate on the implications of emerging hardware technologies including systems software and application software programming models to exploit them.

Chapter 2

The Problem of Software Bloat

"Software is getting slower more rapidly than hardware becomes faster"

Wirth's Law
Niklaus Wirth attributes this quote to Martin Reiser

"From 1963 to 1966, I worked for the Pennsylvania State Drivers License Division, programming an IBM 1401 computer. The first widely used transistor-based computer, our 1401 handled all six million drivers in the state.

The machine had 12K of memory. The point of this story is that our department never added the extra 4K memory module, upgrading the machine to its whopping 16K maximum... it would have cost several thousand dollars, and we never found it necessary. Our programs, written in IBM assembly language, first on paper and then transcribed to punch cards, were extremely compact. We condensed program steps wherever we could. There was no graphical interface, because there was no screen. We debugged our programs with memory dump printouts, and we didn't even have an operating system. What for? We just wrote our own input/output routines.

But, we processed an entire state!

Know anyone these days processing a state on his or her desktop computer with a million times as much memory as we had in 1963? Software bloat. Better believe it. Happy computing!"

Alan Freedman

[147]

Over the past few decades, software design paradigms have evolved to prioritize programmer productivity over runtime efficiency. In contrast to programs tuned for a specific use, large software systems are standardized around deeply layered frameworks that facilitate rapid development. Each (component) layer is designed to ensure composability of its functions to support a high degree of flexibility for reuse and interoperability. In a typical execution scenario, the system uses only a small subset of the functionality but still pays the overhead for supporting the full functionality. With more layers, the number of potential function combinations grows exponentially, compounding the hidden burden of largely unused combinations.[1]

As a result, although functionally richer and more flexible, newer software packages often incur a larger resource overhead in typical execution scenarios than their older editions [410].

In this chapter we review the problem of software bloat with an emphasis on runtime bloat in framework based applications.

---

[1]Several paragraphs in this section, Section 2.1 and Section 2.5 contain material ©2011 IEEE. Reprinted, with permission, from [S. Bhattacharya, K. Gopinath, K. Rajamani and M. Gupta. Software Bloat and Wasted Joules: Is Modularity a Hurdle to Green Software? in Computer, vol. 44, no. 9, pp. 97–101, Sept. 2011.]

2.1   The Notion of Software Bloat

Both in the popular press and in published research, the term "software bloat" is commonly used in a very broad sense to refer to a number of related issues

- Growth in the size and resource requirements of software across product generations

- Cost incurred by software (both in terms of resource usage and complexity) that seems out of proportion compared to its utility (benefits) in a given situation

- Overheads of excess functionality in a reusable software component

Framework based software can exhibit many forms of runtime bloat (runtime resource consumption disproportionate to the actual function being delivered) such as the execution of excess function calls, the generation of excess objects, and the creation of excessively large data structures. For example, Mitchell, Schonberg, and Sevitsky [264] mention cases involving (1) the creation of hundreds of thousands of method calls and objects to service a single web request that retrieves and formats a few database records and (2) the consumption of a gigabyte of memory per hundred users in an application that needs to scale to millions of users. They document 15 anecdotes drawn from their experience in analyzing real world applications to illustrate how current software engineering trends can lead to bloat.

Our own earlier experience and review of several case studies support similar findings: such as a document-exchange gateway that creates six copies or transformations for each input document processed, a telecom application intended to support high transaction rates that generates over a megabyte of temporary objects per transaction, and an object cache with 90% data structure overhead when used for storing small objects[61]. More recently, Bu et al. have shown that Big Data applications can be even more vulnerable to runtime bloat than regular Java applications; they mention that even a moderate-size application on Giraph[155], a big data graph processing framework, with 1GB input data can easily run out of memory on a 12 GB heap[87].

In the past, the benefits of flexibility have outweighed the perceptible cost of overheads incurred, given the pace of improvements in hardware performance enabled through CMOS technology. However, shifting hardware technology trends and operational challenges prompt a need for paying greater attention to these inefficiencies, as the overheads tend to negate gains from increases in computing hardware capacity and can result in higher power consumption and wasted energy.

2.2   Software Bloat: Causes and Consequences

2.2.1   Principal aspects of bloat

At a high level, we can categorize principal aspects of bloat in terms of the nature of perceived effects as well as the nature of perceived sources (Figure 2.1).

- **(a) Perceived Effects:**

  1. Excessive complexity (and its impact on consumability, manageability, and security)

  2. Resource inefficiency (and its impact on system performance)

- **(b) Perceived Sources:**

  1. Overprovisioning (excess or largely unnecessary) features

  2. Poor implementation choices (sub-optimal coding)

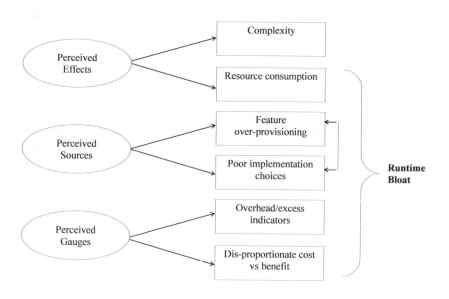

FIGURE 2.1: Different aspects of bloat.

These aspects are not entirely independent, however. It is this characteristic of bloat which makes it difficult to arrive at precise definitions that clearly demarcate a particular source-effect combination from the others. For example, resource inefficiency leads to higher demand for hardware capacity scaling which can translate into more complex configurations, installation, and administration overheads. Likewise the more complex and obscure the software becomes, the harder it is to write efficient code and optimize resource usage. Poor implementation choices made when designing without context is a source of over-general framework components with excess features. On the other hand, the presence of unnecessary features and abstractions can obscure the overheads of implementation decisions and thus cause poor choices to be made.

Note that in our use of the term "poor implementation choices," we distinguish implementation choices that lead to runtime bloat[2] from choices related to inefficient algorithms or traditional performance issues such as contention, scheduling, and concurrency, which also impact the effective utilization of resources by software.

2.2.2 Definitions of software runtime bloat relevant for this book

RPD is focused on the perceived effects on runtime resource inefficiency rather than the perceived complexity effects of software bloat. We use the following different variations of the definition of runtime bloat to qualify the nature of perceived sources (from 2.2.1(b)) that are relevant for a given analysis:

- **Definition I** *(emphasizes both perceived sources)* :– <u>Bloat</u> is the runtime resource overhead induced by the presence of excess functionality and objects.

- **Definition II** *(emphasizes poor implementation choices as the perceived source)* :– <u>Bloat</u> is runtime resource consumption disproportionate to actual function delivered.

- **Definition III** *(emphasizes overprovisioning features as the perceived source)* :– <u>Bloat</u> is the runtime execution overhead induced by the presence of excess features or program concerns.

A *feature* is defined as "a unit of functionality of a software system that satisfies a requirement, represents a design decision, and provides a potential configuration option" [34].

Concerns are features, design idioms, or other conceptual considerations that can impact the implementation of a program.

---

[2]Such as a combination of disparate local decisions and component reuse choices that can result in global inefficiencies, e.g., by inducing excessive data copies.

2.2.3 Systemic practices in framework based development attributed as causes of bloat

In large applications, bloat results from a pileup of inefficiencies stemming from a combination of disparate decisions across different development groups.

1. *Inadvertent reuse of costly abstractions* Frameworks can have high resource costs which are compounded when they are easy to misuse, when they induce frequent data transformations, and when API boundaries introduced by additional layers limit effectiveness of JIT (just-in-time compiler) optimizations and cause global efficiencies. For example, Mitchell and Sevitsky describe a case where developers estimated that they were using only about 2KB of state per session while actual measurements showed 200KB of per-session state, which explained why the application was scaling poorly and running out of memory at much lower loads than expected [267]. In this case the developers were two orders of magnitude off in their estimates.

2. *Feature creep, technology evolution, and backward compatibility* New features keep getting added to meet evolving standards, data formats, technology constraints and changing requirements, but once the code for a feature is introduced in a reusable component it is very hard to remove it even in environments when it is no longer useful. As Joel Spolsky describes in an article on the "Law of Leaky Abstractions"[364], it is hard for any non-trivial abstraction to fully hide implementation detail or isolate low level dependencies from the rest of the stack.

3. *Systemic bloat of managed language platforms* Java's data modeling limitations[3] combined with high data representation and communication costs can force applications into resource heavy designs that are hard to optimize [264]. Automated optimizations performed by runtime systems are mostly local in scope and ineffective in tackling problems of bloat which typically tend to span components.

4. *Bloated specifications and protocols* Standard specifications and protocols built for high flexibility, interoperability, and extensibility tend to result in bloat for common use cases. Excessive use of XML[4] and SOAP[5] has been known to introduce processing and communication overheads that reduce achievable peak transaction rates. For example, a legacy financial transaction message (ISO8583) of 122 bytes turns into a 15.3KB

---

[3]For example, composition is often preferred over inheritance but composition in Java must be implemented as delegation, resulting in extra indirections and the use of several objects to represent even basic data structures.

[4]Extensible Markup Language.

[5]Simple Object Access Protocol.

message when represented in XML. Accessing a NOAA[6] weather service using a WSDL[7] API was observed to generate 30X temporary Java objects (churn) and invoke 20X more method calls compared to a REST[8] based API. Hence the use of these protocols causes bloat in situations which do not require the power of interoperability and flexibility provided.

Such examples illustrate that applications tend to accumulate excess function and objects as a side effect of the same software engineering paradigms that enable support for ever more complex business logic, integration requirements and operational considerations. Mitchell, Schonberg and Sevitsky discuss four software development trends that lead to Java runtime bloat [264] in large framework based applications:

1. A culture of using objects (both long lived and temporary) to represent even the simplest of concerns without consideration of costs

   - Example: Fine grained data modeling with high per-object overheads
   - Effect: Memory bloat, object delegation costs, excess temporaries

2. A sea of (costly) abstractions and excessive layering

   - Example: Designing without context and inappropriate reuse
   - Effect: Barrier to optimization (both for humans and JITs)

3. Computers as communicators integrating disparate services and data sources

   - Example: Impedance mismatch between constituent services and data sources
   - Effect: Excessive transformations and data duplication

4. "Just-in-case" programming and anticipatory flexibility

   - Example: Insulation abstractions and (self-describing) dynamic types
   - Effect: Memory bloat (due to embedded strings and collections used for associating structural information description such as field names with data values) and execution bloat due to the overheads of data-driven dynamic dispatch

These trends can be collectively viewed as emphasizing a sound general design principle: opting for a safe (but heavy) overapproximation (superset)

---

[6]National Oceanic and Atmospheric Administration.
[7]Web Service Definition Language.
[8]Representational State Transfer.

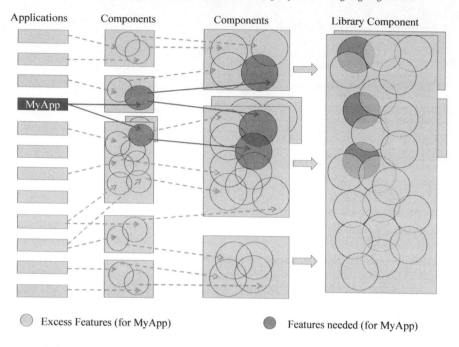

Applications   Components   Components   Library Component

MyApp

○ Excess Features (for MyApp)   ● Features needed (for MyApp)

FIGURE 2.2: Accumulation of excess functionality across layers of redeployable components at many levels of granularity. The circles represent features implemented by a component; intersecting circles signify feature interactions. The arrows connect features in a component with features it uses from other components. Dark circles represent features that are needed by a particular application, MyApp. Light circles represent features that are currently in excess for MyApp but exist to support other components or future needs. Not all excess features induce runtime bloat, however.

of requirements when choosing or implementing standardized reusable functionality and data structures at all granularities from low-level libraries to high level solution composition.

This view highlights two related systemic sources of excess functionality (Figure 2.2). First, it is safer to overprovision features supported by components built for widespread deployment, where it is difficult to anticipate all potential usage situations. This emphasis on flexibility encourages the creation of overgeneral component and data structure implementations. Second, when choosing components and data structures to construct solutions from, software developers tend to rely on simpler and less detailed mental models that overapproximate desired requirements in a sound manner. This is similar to the manner in which abstract interpretation techniques aid the construction of computable yet sound static analysis tools. They rely on an overapproximation of possible behaviors of a program by a less detailed and hence simpler to verify model.

The emphasis on productivity encourages the selection of concrete components that satisfy more features than necessary in a given situation. The combination of these two effects (productivity pressure and flexibility pressure) at every layer leads to the presence of excess features in software.

Although this simplified systemic view illustrates why excess features can accumulate in redeployable software components (Figure 2.2), as we show later in this book the presence of excess features does not necessarily explain the presence or intensity of runtime bloat even though it increases code size. In other words, runtime bloat need not be an inevitable consequence of over-provisioning features, which is often an appropriate design strategy for such applications. Runtime overheads can occur due to excess state that must be traversed or processed at runtime even when the corresponding feature is not relevant. This includes excess intermediate objects and transformations that are generated in order to reuse APIs that support some excess functionality. The challenge is to figure out how such side effects could be mitigated to enable flexibility without incurring bloat.

2.3 Bloat in Containerized Software

2.3.1 Application containers vs. virtual machines

Application container technologies such as Docker and Kubernetes[4] have become popular as a convenient application delivery vehicle and software deployment environment that can be hosted on both cloud and private infrastructure. Containers provide a way to compose a self-contained software stack which is deployed using a lightweight virtualization mechanism. Unlike virtual machines, applications running in different containers on a host share the same operating system and thus incur far lower resource overheads.

Containers are typically light and stateless so that moving a container is as simple as pulling and starting in on a different host. Persistent data is saved in one or more persistent volumes which are attached to containers (cloned or replicated as needed) and managed separately from ephemeral runtime data and image files. Docker allows container images to built as layers that can reuse other base container images, allowing for futher savings across multiple containers that share a parent image.

2.3.2 Container image bloat

Even so, containers are prone to getting bloated with excess software just the same way that excess features accumulate across redeployable components, resulting in bulky images[118, 133] and slow startup times[177]. As containers are intended to be self-contained, stateless, and easy to move (not dependent on a specific host OS environment), it is safer and easier to overprovision software packaged in a container using standardized image layers, but these layers may contain a lot of programs and files that are unnecessary for running the application and make it more prone to security exposures[327].

Several best practices have been described for building optimized containers, such as splitting discrete functionality into separate containers or microservices, the use of minimal shareable parent images (e.g., using Alpine Linux instead of a full Ubuntu Linux image as the base distribution), multi-stage builds, and optimized dockerfile instructions (to avoid pulling in container build dependencies and intermediate stages into the runtime environment).

2.3.3   Runtime bloat in serverless computing

A different model uses standard containers as an underlying substrate for serverless computing which supports a function-as-a-service (FaaS) (e.g., AWS Lambda, Apache OpenWhisk, Google Cloud Functions, Microsoft Azure Functions) approach to cloud native application delivery and deployment. This model frees developers from maintaining the operational stack and ensuring scalability (sometimes at the cost of potential vendor lock-in) but can introduce other potential sources of runtime bloat, which can impact efficiency and latency. For example, as the functions are stateless, data is saved, restored, and transfered between functions using external data services resulting in costs of message serialization-deserialization and data copies [190, 47].

2.4   Different Forms of Software Runtime Bloat

Let us explore some instances of software runtime bloat, the kinds of software bloat that directly result in inefficient usage of runtime resources and can thus impact system power-performance.

2.4.1   Runtime bloat categories in Java applications

As a case study, we focus on managed language frameworks, in particular Java based software where certain types of runtime bloat have been documented or analyzed by previous researchers. The following are some examples:

- Framework bloat

- Protocol bloat

- Feature bloat

- Data structure bloat [266]

- Temporary object churn bloat [128]

- Transformation bloat [268]

- Java heap bloat

- Java native memory bloat [289] (or internal bloat of the runtime system)

- Code size bloat

- Execution pathlength bloat

The above types are interrelated and sometimes overlapping. For example, both data structure bloat and temporary object churn lead to bloat in the Java heap. Framework and protocol bloat can cause unnecessary (bloated) transformations. Transformation bloat generates excess objects and bloats execution pathlength. Feature and framework bloat lead to increased code size and pathlengths. Protocol bloat can lead to higher communication payloads. Code size bloat and communication bloat can lead to internal bloat in Java native memory usage especially as a side effect of JVM optimizations such as the use of direct byte buffers and bytecode generated at runtime to optimize reflective invocations.

2.4.2 Relationship between the various runtime manifestations of bloat

We characterize the different forms in which bloat is manifested at runtime according to the following major categories:

- Excess Work (Code execution)

- Excess State (Data generated)

- Excess Communication (I/O transfer)

Figure 2.3 illustrates the interrelated nature of the categories. Data structure bloat (which can affect the size of both long lived and short lived objects), temporary object churn bloat (excessive generation of short lived objects), and Java native memory bloat are examples of bloat involving *excess state (data)*. The presence of excess data is typically associated with excess work to maintain the state (e.g., to create, lookup, update, copy, parse, format, and transform data). Likewise, *excess work* typically results in the generation of excess objects as vehicles for processing relevant state, dynamic dispatch of relevant methods, and ferrying data across API boundaries. Bloat at a higher level of abstraction, such as framework, protocol, and feature bloat, may be viewed as the inclusion of *excess or incidental software concerns* and can thus result in excess work as well as excess state and *excess communication*.

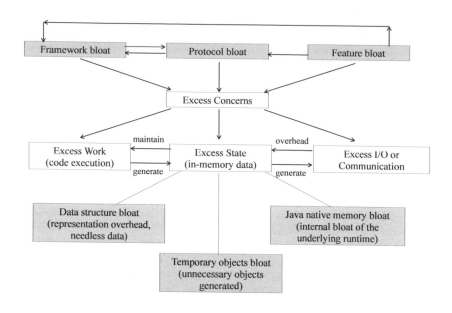

FIGURE 2.3: Relationship between different types of bloat and their runtime manifestations.

2.5   Progress in Bloat Characterization, Measurement, and Mitigation

2.5.1   Modeling and measuring bloat

Software does not come with built-in labels that indicate which portions of computation are necessary for a given application and which lead to bloat. Estimating the amount of resources a non-bloated implementation would have consumed for a specific execution is also difficult. Understanding the nature and sources of different types of software bloat is the first step to addressing the issue. The second is to quantify the magnitude of excess resource consumption attributed to each type of bloat. While the latter enables an estimate of how much room for improvement exists, the former provides insights on how to fix the problem.

In their study of large framework-based applications, Mitchell et al. observed that information-flow patterns for a data record involved a sequence of expensive transformations with multiple levels of nesting [268]. For example, moving a single date field from a SOAP message to a Java object in a stock-

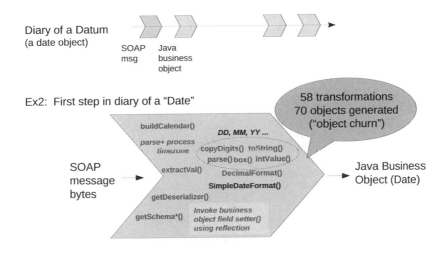

Ex2: First step in diary of a "Date"

FIGURE 2.4: Excessive transformations in framework based software. A sequence of expensive transformations with nested transformations [268].

brokerage benchmark involved 58 transformations and generated 70 objects (Figure 2.4). Many of these were facilitative transformations for reusing existing parsers, serializers, and formatters. The observations highlight the considerable overhead expended in supplying data to the application's core business logic. The study also proposed metrics based on modeling runtime information flow to classify and characterize the nature and volume of data transformations executed. However, these measures have not been automated to date.

Automated measures of bloat rely on a variety of heuristics to distinguish incidental overhead due to a specific category of bloat from strictly necessary resource usage. When analyzing Java heap snapshots, for example, it is possible to automatically differentiate data structure representation overhead such as bytes expended by the JVM object headers, pointers (object references), collection glue, and book keeping fields from the actual application data contained in the structures. Mitchell and Sevitsky introduced the notion of a *data structure health signature*, a relative measure of memory consumed by actual data versus associated representational memory overhead (Figure 2.5), which reveals that data-structure bloat can increase the memory footprint of long-lived heap data structures by anywhere between a factor of two and five [266]. Their work was the first to automate the diagnosis and quantitative

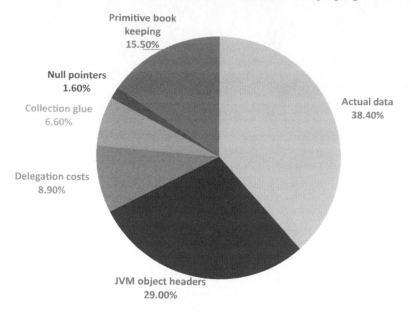

FIGURE 2.5: Data structure bloat: Due to a high percentage of representation overhead (60% of bytes in this example), this Java application's heap uses 2.5X the memory of the actual data size for long lived data structures[266].

measurement of a type of bloat and to show how to generate formulas to predict asymptotic behavior of the health of a given data structure design at high data scales.

From a power-consumption standpoint, execution bloat and associated excess temporary-object generation are even more interesting. However, such overheads are more difficult to distinguish and characterize automatically. Xu et al. and other researchers have made good progress in diagnosing some signs of potential bloat by focusing on data flow patterns[9] in the use of temporary objects – for example, excessive or expensive data copies [417] and the creation of expensive data structures with low utility [418].

Another automated measure of code bloat is the extent of extraneous code in libraries or runtime layers (managed language environments and operating systems) pulled in by applications. For example, Quach et al. conducted a study of code bloat across user programs using C/C++ libraries, managed language environments (Java and Python), and OS kernels (Linux and FreeBSD)[319] using both static and dynamic measurement approaches. They observed that only 21% of instructions were executed on average across user level programs and only about 32% code is executed in the managed runtime environments.

---

[9]In contrast with compiler optimizations and performance analysis tools that focus on control flow, such as expensive or frequently executed methods.

Table 2.1 categorizes existing diagnostic approaches in terms of whether they focus on identifying specific overheads[10] or on identifying cost-benefit imbalances.[11]

TABLE 2.1: State-of-the-art techniques for measuring different indicators of bloat categorized according to the underlying heuristics and extent of reliance on intrusive profiling

Bloat characterization approach	Diagnostic measure	Analysis type	Profiling expense
Overhead indicators	Data structure health [266]	(Dynamic) heap snapshot	Low
	Loop invariant fields(hoistability) [422]	Static	Low
	Copy profile [417]	Dynamic slicing	High
	Temporary object usage [128]	Blended (dynamic+static)	High
	Percentage of unused code [319]	Static and Dynamic	Medium
Out-of-proportion costs	Inefficient collection usage [421]	Static (or dynamic)	Low
	Low utility data structures [418]	Dynamic slicing	High

Characterizing overhead indicators Data structure representation overhead [266] and chains of data copies [417] are both indicative of overheads with respect to actual data. Detecting the former requires only a heap snapshot [263] to be captured at runtime. Detecting the latter requires JVM instrumentation that incurs a high profiling overhead as it traces detailed information flows. Another indication of bloat is the repeated construction of data that is actually invariant across loop iterations and can potentially be hoisted outside the loop [422] – these indicators can be detected using a static analysis. Unused code overhead can be detected using conservative static analysis (lower bound) or using dynamic analysis for concrete execution scenarios[319, 320, 200].

Characterizing out-of-proportion costs Inefficiently used collections and low utility data structures which are expensive to create are indicative of cost vs. benefit imbalances. Xu et al. show how the former can be detected using a static analysis [421] and the latter can be identified by using abstract thin dynamic slicing [418] to compute relative object construction costs. These costs are computed in terms of the number of

---

[10]To characterize what is *in excess.*
[11]To characterize what is *excessive.*

bytecode instructions required to produce a value from already available fields; the analysis relies on JVM instrumentation with a high profiling overhead.

2.5.2   Mitigating and avoiding bloat

2.5.2.1   Semi-automated approaches

Tackling the source of bloat usually involves manual source-code fixes and assumes some domain knowledge about the application. In a few cases, such as when the bloat originates in inefficient data structures, advisory tools can minimize manual effort. For example, tools such as *Chameleon* [137] for Java and *Perflint* [243] for C++ have adopted rule based approaches for recommending improvements in the choice of collection data structures used by an application. *Chameleon* has an option to apply changes automatically by enabling an adaptive selection of appropriate efficient collections at runtime. Chis et al. [106] have developed an analysis framework for discovering high impact patterns of memory inefficiency from heap snapshots along with suggested remedies. Their tool covers eleven commonly occurring patterns of memory misuse observed in Java applications.

Detection of performance anti-patterns in component based applications for enabling manual or even automatic performance optimization is an active area of research. Parsons and Murphy have developed monitoring techniques for capturing runtime paths and inter-component communication patterns and mining these to automatically detect several J2EE performance design and deployment anti-patterns [300]. A few of these anti-patterns are likely to be related to bloat (e.g., the unused data object pattern, bloated session anti-pattern, fine grained remote calls) while others reflect high level performance problems, e.g., inefficient configuration of container pools or inappropriate transaction sizes. Wang et al. explore a pattern based performance optimization approach using automatic refactoring to replace such anti-patterns with more optimal patterns at runtime [399] by exploiting a reflective middleware framework. Such approaches are suitable only for bloat patterns that can be detected and fixed at a coarse level of introspection. Bu et al. proposed a bloat-aware design for developing efficient and scalable big data applications without giving up the benefits of an object oriented GC enabled managed language environment[87].

2.5.2.2   Automated code optimization

Traditionally, software development paradigms have evolved with the expectation that it is the responsibility of system tools, runtime optimizers, and underlying infrastructure to ensure efficient program execution. However, runtime bloat occurs despite the best efforts of modern compiler and runtime optimizers to automatically minimize inefficiencies in software code. There are two key limitations that such optimizers face when it comes to address-ing bloat. First, being inherently agnostic to the higher level purpose of the code (e.g., software features provided), they cannot distinguish function and objects that are truly essential from those that are unnecessary (in excess) but still happen to be invoked or referenced during a possible program execution. Second, although some redundancies induced by bloat could in principle be detected purely by observing low level code, the problems of bloat typically span several API boundaries and deep call chains - situations where existing analysis approaches are too conservative and/or too expensive in practice to exploit optimization opportunities effectively.

Even so, automatic code-optimization techniques can mitigate finer symp-toms of bloat. In particular, storage representation optimizations such as object fusing [408] (and inlining) or space-optimized object headers can reduce some of the overheads of fine-grained and highly delegated[12] data models. Automated JVM level techniques have been developed for reducing string memory inefficiencies [211], an important source of memory bloat. JVM level approaches have also been explored for tackling the incidence of excessive temporary object generation in component based applications. For example, Shanker, Arnold and Bodik applied escape analysis after inlining method call chains in high object churn regions identified using a lightweight dynamic analysis [345]. Language extensions such as additional modeling options and specification of design intent (relationships, transformations in type systems) have also been proposed to enable more effective automated optimization and efficient storage representation.

A runtime optimizer could exploit speculative optimization by monitoring and intercepting dynamic data flow behavior to realize de-bloating oppor-tunities spanning several layers of API boundaries without becoming overly conservative. However, many of the dynamic analysis techniques for diagnos-ing bloat that are described in Table 2.1 incur a heavy runtime overhead - they can cause a tenfold slowdown and are only intended to help developers perform offline optimization. To achieve continuous object access profiling of sufficient accuracy while incurring a low enough runtime overhead to be used as a basis for new online object optimizations, Odaira and Nakatani propose a memory protection based technique called pointer barrierization [288]. Their object access profiles capture dynamic behavior properties such as write-only objects, immutable objects, and non-accessed bytes. Write-only object access profiles are used as basis for speculative compression of character arrays as an online optimization which employs pointer barrierization as a recovery mech-anism. The profile of non-accessed bytes is used for another optimization - the dynamic adjustment of initial sizes of containers to optimal sizes to avoid memory wastage. Profiling immutable objects can enable copy-on-write opti-mizations. Although such online optimizations result in performance gains that well exceed the profiling overheads in some situations according to the

---

[12]The pointer indirection costs of using mutiple levels of object delegation for representing composition.

paper, there are other situations where the overheads are too high for online profiling (the technique is still useful for offline profiling in these situations). Further, the energy consumption overheads of using continuous object profiling have not been studied.

Automatic techniques for de-bloating programs by trimming unused code in libraries have gained attention in the light of the risk potential security vulnerabilities exposed by extraneous code. Quach et al. found that, on average, only 5% of libc functions are used across two thousand programs in Ubuntu Desktop 16.04 and observe that debloating the remaining code could significantly reduce the potential attack surface[320]. They propose a late stage debloating framework that utilizes piecewise compilation and loading, combining compile time (static) and load time (dynamic) approaches[320]. JRed[200] is a static analysis tool built using the Soot[12] framework for trimming unused code from both Java applications and Java runtime environment (JRE).

Trimmer[346] is a tool implemented in the LLVM framework to reduce code bloat by specializing C/C++ applications for a given deployment context and user configuration. CHISEL[182] generates a minimized version of a program that satisfies a given set of properties or test cases (covering a desired subset of functionality), using reinforcement learning to accelerate the search for the reduced version of the program.

De-bloating application containers automatically is another area that has received much research interest given its implications for both cloud and edge computing. Cimplifier[327] is a tool that automatically splits a container into minimal containers (or microcontainers[367]) and proposes further research to specialize the stack for each minimal piece, including kernel specialization to tackle OS bloat[199]. Slacker[177] adopts a different approach which focuses on just reducing runtime overheads by implementing a Docker driver with a lazy fetching optimization based on the observation that pulling packages accounted for 76% of container start-up time, even though only 6.4% of that data is actually read.

2.5.2.3   Can runtime bloat be avoided by construction?

Modularity is fundamental to the composability of software packages and to their rapid development and deployment. However, the prevalent approach to achieving it can lead to significant software bloat, which is detrimental to power, performance, and energy efficiency. The real issue isn't modularity itself but rather that programmers inadvertently introduce superfluous processing and data overhead for reuse because of the difficulty in modularizing functions exactly as needed.

The traditional maxim for creating lean software, as advocated by David Parnas [299] and Niklaus Wirth [410], among others, is to engineer it right by adopting minimalist design principles that avoid bloat. Such software is built in a series of stepwise refinements carefully crafted to provision each potential use case without sacrificing extensibility or reuse. The Linux kernel illustrates the successful adoption of this principle to efficiently satisfy diverse environments and requirements [62]. In framework-based environments, however, this approach is impractical. Many redeployable components must be dynamically programmable by business analysts and integrate with dozens of heterogeneous systems and information sources. Thus, it is not easy to anticipate usage of a component at design time, nor is it feasible to incrementally change intermediate interfaces later. Programmers are often unaware of the overhead that systems might incur during actual deployment, and a low-level runtime optimizer cannot "know" their intentions. Improving cross-layer line of sight into high-level functional intent and interoperability overhead (including data-supply inefficiencies such as transformations and copies to facilitate reuse) will help both programmers and runtime systems deliver better energy-optimization solutions by minimizing bloat automatically. In the next part of the book, we discuss systematic approaches that programmers and tool developers could adopt to address this challenge.

2.6   Conclusions

We find that the first challenge in addressing the problem of bloat is that the term "software bloat" lacks a consistent definition. It has several connotations depending on the perceived effects of bloat (complexity or resource inefficiency), its perceived sources (excess features or poor implementation), and how the presence of bloat is gauged in terms of trends over successive software versions, out-of-proportion costs compared to useful work achieved, or indicators of excess functionality and overheads. For the purpose of this book, we are interested in understanding potential systemic characteristics that influence the origin of software runtime bloat and its direct impact on system power-performance. Thus, we will focus mainly on bloat's perceived effect on resource inefficiency gauged in terms of indicators of overheads due to the combination of perceived sources of excess features and poor implementation choices.

Techniques for measuring, managing, and mitigating runtime bloat face significant challenges but there has also been progress in these areas during the past few years. Most of this work has been focused on inefficiencies observable from data structure and data flow patterns in the form of specific overhead indicators or out-of-proportion costs. Although this has been a very fruitful line of research, mitigating these inefficiencies requires substantial manual effort due to the limited scope of automated de-bloating techniques available. Thus, such efforts need to be complemented by a higher level analysis perspective to characterize the implications of bloat for system power-performance and to relate the popular (user level) view of software bloat [406] in terms of optional features to the deeper forms of inefficiencies identified as runtime bloat by previous researchers. For instance, it would be useful to qualify program characteristics that determine when feature bloat (or more generally, program concerns that are in excess for a given deployment situation) can result in runtime execution bloat. There is also a need for more automated techniques for mitigating bloat in existing software. Finally, this leads us to the key objective of this book: understand how a principled approach could be applied to guide the construction of new systems that possess a lower propensity for runtime bloat without compromising on flexibility or development productivity.

Chapter 3

Does Lean Imply Green? How Bloat in Software Impacts System Power Performance

As mentioned in earlier chapters, excess resource usage from bloat[1] could lead to energy inefficiency and reduced performance [420]. Energy savings from reducing bloat are anticipated via both direct and indirect effects (e.g., server

---

[1]In this chapter we will use **Definition I** from Chapter 2, Section 2.2.2: *"Bloat" is the resource overhead induced by the presence of excess functionality and objects*, typically as a result of using highly general components standardized around deeply layered frameworks.

consolidation opportunity) of reducing excess resources.[2] However, reducing bloat is non-trivial, particularly because its origin is linked to the same software development trends [264] that have been extremely successful in fueling the growth and widespread impact of redeployable software. Hence, there is a need to develop approaches to assess cost-benefit implications of reducing bloat and focus these efforts where they matter the most.

While reduction of bloat is expected to (obviously) have an impact on power-performance, we find that the inherent slack in large scale IT solution architectures and the presence of elements in the system that are not *energy proportional*[3] may require this intuition to be qualified.

Is the effect of bloat more pronounced or less pronounced in the presence of energy proportional hardware? Should we expect an increase or a decrease in peak power consumption with lower bloat? In two different experiments on the same hardware platform, but involving different kinds of bloat, one with Apache DayTrader and another with an object creation microbenchmark, results indicated opposite answers[68]. Even with the same workload and same de-bloating optimization, an empirical study we discuss later in this chapter has shown wide variations in the effects with different hardware and software characteristics. These examples illustrate why it is important to validate our intuitions and gain a deeper understanding of the relationships between measures of bloat and energy consumption.

In this chapter we systematically analyze and expose the effect of energy proportionality and bottleneck resources on the impact of bloat reduction, using multiple metrics for evaluating the impact of bloat – peak power, equi-performance power (Section 3.2), peak performance, and resulting energy efficiency measures. We first summarize experimental findings that reveal a wide variation in the impact of bloat, indicating that a software – only view is inadequate for assessing the energy-efficiency benefits from bloat reduction. Based on this motivation, the remaining part of this chapter develops an abstract model to enable a generalized exploration of the effect of bloat on power-performance, from a whole system perspective. The analysis shows that such a whole system perspective is required for an informed assessment of the impact of bloat and its reduction. The nature and extent of impact is dependent on whether a bloat-impacted resource is a bottleneck resource as well as on the energy proportionality of not just the bloat-impacted resource but other key resources as well.

---

[2]Most of the content in this chapter is republished with permission of ACM, from [Does lean imply green?: A study of the power performance implications of Java runtime bloat. Suparna Bhattacharya, Karthick Rajamani, K. Gopinath, and Manish Gupta. In Proceedings of the 12th ACM SIGMETRICS/PERFORMANCE joint international conference on Measurement and Modeling of Computer Systems (SIGMETRICS '12). ACM, New York, NY, USA, 259–270. 2012.]

[3]Energy proportional [51] hardware consumes power in proportion to actual resource utilization.

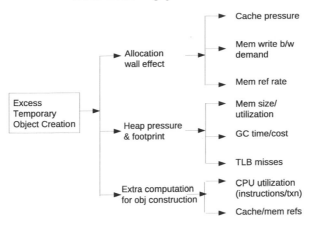

FIGURE 3.1: Effects of Java temporary object creation bloat.

3.1 The Effects of Java Runtime Bloat on System Resources

In order to perform a systematic investigation of the impact of bloat on power-performance, let us first focus on a detailed study of a single type of bloat that is very common in Java applications, is relatively easy to vary, and exhibits a diversity of effects impacting multiple system resources: the presence of *excess temporary objects*.

The Java programming model encourages the creation of many short-lived objects – these are allocated on the heap[4] and may be passed across framework layers, only to be used briefly and eventually garbage collected. For example, temporary objects are generated repeatedly during data transformations required for framework based reuse. This can lead to the problem of object churn, i.e., a high volume of temporary objects [345] spreading out the memory access footprint of the application (because the memory is not reused until after the next garbage collection cycle). Excessive allocation of these temporary objects, which we refer to henceforth as temporary objects bloat, is a common type of Java runtime bloat; it causes excess memory usage and has deep and pervasive effects on the performance of a system.

Figure 3.1 depicts some of the potential effects of temporary object bloat in Java. We discuss these briefly.

3.1.1 Allocation wall effect

Researchers have previously observed [435] that a large volume of temporary object allocations causes high memory write bandwidth consumption

---

[4]State-of-the-art escape analysis techniques in modern production JVMs can only optimize a small fraction of these allocations [345].

which limits scalability of certain classes of Java programs on multi-core platforms. The term "allocation wall" has been used to describe this phenomenon, as this slowdown is due to allocation overheads and not garbage collection cost.

3.1.2   Heap pressure effect

Creating many temporary objects can increase garbage collection overhead. Given a sufficiently sized JVM heap (typical in well-tuned JVM configurations) and a generational garbage collector, the percentage time spent in GC[5] becomes small enough that it could be ignored in comparison to other effects. Even so, de-bloating software to reduce temporary object allocation can reduce its footprint by enabling lower heap sizes to be used for the same application. Heap size reductions are beneficial from a power-performance perspective, when the memory footprint savings is significant enough to (i) enable inactive ranks of memory to be switched to a low power mode, (ii) enable use of a smaller memory configuration of the system for cost and power savings, or (iii) enable allocation of spared memory to another virtual partition for improved workload consolidation with same system configuration/power envelope.

3.1.3   Object construction computation overhead

If the construction of a temporary object involves a costly initialization sequence for setting up its fields, object creation can also have a significant instruction pathlength impact. Object construction computation overhead manifests as an increased CPU utilization due to increased instructions per transaction. It may or may not have a significant effect on the CPI[6] depending on the nature of instructions added.

3.1.4   Influence of system configuration

A single category of bloat may have multiple effects as described above, each of which, in turn, impacts physical resource utilization in different ways. A given effect may have a low impact in one situation but can become crucial under a different set of system/workload characteristics.

For example, the "allocation wall" effect [435] is observable in the presence of bottlenecks caused by constraints in memory system resources like cache, memory bandwidth, and latency. In this situation, it leads to increased (effective) CPU utilization through its impact on the CPI. This increased utilization impacts power consumption. However, if the system is configured with a processor DVFS[7] algorithm that is memory slack aware [412], it could exploit the available slack to run the processor at a lower speed; hence, the change in CPI

---

[5]Garbage collection.
[6]Cycles per instruction.
[7]Dynamic voltage and frequency scaling.

may have a relatively lower impact on the effective (scaled) CPU utilization and power consumption.

In other situations with relatively large on-chip caches and powerful memory subsystems, where memory performance is not a constraint, the allocation wall can effectively become irrelevant. Reducing bloat can still be beneficial in this case, due to another effect. Lower temporary object allocation rates enable a Java application to be configured to use a smaller heap size without an increase in GC time. If this heap size reduction is sufficient to enable the working set of the application to be entirely contained in a (huge) on-chip cache, off-chip memory accesses may be avoided altogether, potentially leading to significant power savings.

3.2   Insights from an Experimental Study

Researchers have empirically studied the impact of bloat reduction on power and performance in real systems by conducting a series of controlled experiments under varied system characteristics and bottleneck scenarios[68]. We summarize this study here.

The experiments introduced variation in temporary object bloat in a power-performance benchmark by increasing or decreasing the degree of object reuse, a well-known coding technique for reducing object creation in Java (Section 5.2.1). Two variations of reuse were applied: object memory reuse (which saves allocation costs) and object content reuse (which also saves object construction computation overhead). The impact was measured across multiple platforms with differing energy proportionality and memory system characteristics (Section 3.2.1). Specific experiments were then designed to vary software/system characteristics within a single platform to examine the impact of the availability of key resource(s) on the benefit of bloat reduction (Section 3.2.2). In addition to the energy-efficiency score reported by the benchmark,[8] the study also used comparisons of peak performance, power at peak performance, and power at *equi-performance* points.

> An *equi-performance power* comparison between two alternatives compares power measurements taken at the same performance level (application throughput) for the two alternatives. It also serves as an energy-efficiency measure at constant performance, useful in assessing the relative efficiency impact of bloat reduction when there is also a change in performance as a result of bloat (reduction).

---

[8]SPECpower_ssj2008, a server side Java ("ssj") energy-efficiency benchmark and the first commercial workload benchmark from SPEC [368] requiring power consumption and energy efficiency reports across the full range of system loads.

Table 3.1 summarizes the specific effects of temporary object bloat (described in the previous section) in terms of the variants and configuration constraints under which they could be manifested.

TABLE 3.1:   Effects of reducing temporary object bloat observable by comparing debloated code variants under specific configuration constraints

De-bloating Variant	No constraint	Memory constraint	Low heap constraint
Object memory reuse	Effect not perceptible	Alleviates Allocation wall	Allievates Heap pressure
Object content reuse	Saves Computation overhead (field initialization)	Allievates Allocation wall + computation overhead	Allievates Heap pressure + computation overhead

3.2.1   Multi-platform experiments and results

The results showed a surprisingly wide variation in power-performance impact of bloat across these platforms. For example, on the platform with the least powerful memory system and no power management, the lower bloat variant exhibited 1.59X the energy efficiency of the baseline and the higher bloat variant had just 0.61X the energy efficiency of the baseline at peak performance. The relatively poorer off-chip memory system of this system was responsible for a pronounced "allocation wall" effect (Section 3.1.1) which combined with the poor energy-proportionality characteristic (no active power management enabled) resulted in the biggest energy efficiency impact for bloat (reduction) among the three configurations. On the two larger server systems with powerful memory subsystems, the impact was comparatively less dramatic. The lower bloat variant (which uses object memory reuse) showed 1.11X and 1.06X the performance of the baseline on these two platforms, respectively, while the higher bloat variant (obtained by turning off object content reuse) showed 0.85X and 0.76X the performance of the baseline.

Further, the results showed how the variation becomes even more nuanced with the introduction of sophisticated levels of power management (i.e., increasingly steeper energy proportionality characteristics). Comparing equi-performance power numbers indicates the synergistic impact of bloat reduction and power management techniques such as dynamic voltage and frequency scaling (DVFS). As bloat is lowered from the higher bloat variant to the baseline variant and then further to the lower bloat variant, for a given performance more slack is introduced. The DVFS algorithm then works on the slack to translate that to increased power reduction and energy efficiency. Additionally, one can see how the effectiveness of the DVFS capability magnifies the impact of bloat reduction by comparing *the larger equi-performance power variation* (0.75X with lower bloat variant and 1.83X with the higher bloat variant) for the platform which has a bigger take down in power with load reduction (cubic). This is especially interesting as the same platform was earlier found to exhibit the least energy efficiency gains with the lower bloat variant at peak performance.

3.2.2 Single platform experiment variations: Cache pressure and power management

To gain a deeper understanding of these effects, the study next examined (a) the impact of cache pressure by varying it on a single platform to better isolate its impact from other characteristics and (b) the impact of energy proportionality by then enabling and disabling DVFS on the same platform.

Cache resources of a microprocessor often have a key impact on a work-load's performance. Runtime bloat can reduce the effectiveness of the caches leading to lower performance. Alternatively, the impact of bloat reduction can vary with the shortage/availability of cache capacity of the microprocessor. The experiments varied cache pressure by increasing the degree of hardware multi-threading (SMT[9]) from 2-way (SMT2) to 4-way (SMT4) and further by reducing the cache size to half (SMT4-halfcache).

In Figure 3.2 we see the impact of bloat reduction (comparing the lower bloat variant to the baseline) as cache capacity per thread was progressively halved from SMT2 (2MB/thread) to SMT4 (1MB/thread) to SMT4-halfcache (512KB/thread). The SMT modes were changed through the OS facilities for setting the mode and SMT4-halfcache is realized by booting the system with reduced L3 cache sizes. Difference in energy proportionality characteristics were examined by doing the runs at `fixed frequency` as well as with DVFS.

The results showed that bloat reduction has the highest benefit when cache capacity is most constrained, for SMT4-halfcache and benefits decrease going to SMT4 and then SMT2, as cache capacity becomes more plentiful. There is actually an increase in peak power (negative power savings) with bloat reduction as the underutilized compute resources are utilized more, improving throughput. This power increase is higher when using DVFS for adaptive power management than with fixed frequency operation because of the super-linear power characteristics with DVFS. Consequently, the energy-efficiency improvement mirrors the performance improvement for `fixed frequency` while energy-efficiency improvements are lower than the performance improvements for DVFS. As earlier, savings in equi-performance power (and equi-performance energy efficiency increase) is higher than the improvement in energy efficiency at peak performance. The savings in equi-performance power is highest for SMT4-halfcache as bloat reduction is most effective when the cache pressure is maximum (more severe a performance bottleneck).

With the lowest cache-constraint, SMT2, the performance impact is the lowest and we observe a slight decrease in system power consumption (positive

---

[9]Symmetric multi-threading.

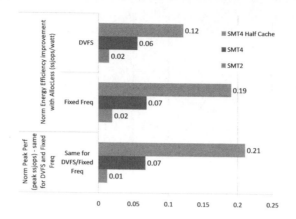

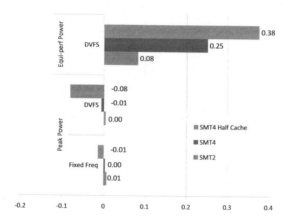

FIGURE 3.2: Cache capacity effect on improvement from bloat reduction[68]: SMT2 has most cache resource/thread, SMT4 has half SMT2's capacity/thread, SMT4-halfcache has half SMT4's capacity/thread. For energy efficiency and peak performance the numbers show the improvement (increase) for the lower bloat variant (*AllocLess*) relative to the baseline. For equi-performance power and peak power the quantities shown represent the saving (decrease) for the lower bloat variant relative to the baseline.

power savings), which is more pronounced at `fixed frequency`. Using the memory power statistics the study observed an 18% reduction in memory power consumption, which, however, translated to a lower overall system power savings because the memory component contributes to a small fraction of system power.

3.2.3 Key experimental observations

In summary, the experimental results confirm that reducing runtime software bloat in the form of excess temporary objects has significant energy efficiency impact, with up to 59% improvement at peak load seen on a memory performance constrained platform (1.59X energy efficiency for lower bloat variant over the baseline) and 45% equi-performance power savings on an energy proportional system (1.83X equi-performance power for the higher bloat variant compared to the baseline). The extent of impact was dependent on the bottleneck pressure with respect to the memory hierarchy resources and energy proportionality of the physical resources.

- Energy efficiency benefits at peak performance tend to be most pronounced when bloat affects a performance bottleneck and underutilized resources are less energy proportional, while equi-performance power savings are highest when impacted resources are super-linearly energy proportional.

- A higher degree of energy proportionality (e.g., power management in the form of processor DVFS) has a complementary impact with respect to addressing software bloat. On the one hand it can reduce the peak energy-efficiency degradation caused by bloat, if the energy proportionality is working on underutilized resources. On the other hand, it magnifies the equi-performance energy efficiency improvements brought about by reduction in bloat, if the energy proportionality is working on the resources that get freed up due to bloat reduction.

- Performance effects of bloat are most acute when memory hierarchy resources are already strained, e.g., bottlenecks due to low off-chip memory bandwidth, low allocated memory or cache pressure with smaller caches or higher degree of multi-threading.

- Power consumption at peak performance may increase or decrease with bloat reduction depending on the impact on resources underutilized in the presence of bloat.

Addressing runtime software bloat was also found to facilitate heap size reductions. The reduced bloat alternative performs better even with significantly reduced heap sizes.

3.3 Analyzing the Interplay of Bloat, Energy Proportionality, and System Bottlenecks

Could the experimental observations from this study be a manifestation of certain general implications that are valid for any category of bloat? To generalize these findings, we now describe the next part of this research, which constructs an abstract analytical model [67, 68] to better understand the power-performance implications of reducing bloat and when its effects are more pronounced. This is a highly simplified model relating the hardware resource pressure caused by bloat and its impact on system energy efficiency using a few fundamental operational laws [92, 196] of queueing theory.[10] Hence, it is generally applicable to any kind of resource bloat irrespective of its category.

Applications use a variety of hardware resources on any given system, e.g., processor cores, on-chip caches, off-chip memory, and disk storage. An imbalanced use of these resources can cause a performance bottleneck at one resource and underutilization of others. For example high cache miss rates caused by profligate use of objects due to bloat can cause underutilization of the processor cores. The power efficiency characteristics differ in nature and magnitude across resources types, e.g., CPU with a super-linear power versus load characteristic when using dynamic voltage and frequency scaling (DVFS) versus main memory with a linear power characteristic and high standby power. This difference leads to a very different impact on energy efficiency for bloat reduction depending on which resource(s) is impacted by bloat and which are underutilized.

When a resource becomes the performance bottleneck because of bloat, reduction of bloat can increase power consumption because of increase in throughput. The varied and sometimes non-linear power characteristics of resources with load can make the impact of bloat reduction on power difficult to assess when compared across different throughput levels. To enable comparison on an equal footing before and after bloat reduction, we also analyze comparisons of power at *equi-performance* levels with and without bloat.

3.3.1 Power efficiency impact quantified using a simple abstract model

Let $R_i, i = 1..N$, be various types of resources with service demands (time) D_i in the software without bloat. Let $b_i, i = 1..N$, be the overhead due to bloat

---

[10]The laws used express basic operational relationships, such as the utilization law/equality, hence widely applicable and general enough to hold without any assumptions about inter-arrival or service times; we do not use any additional assumptions from operational analysis beyond these laws, e.g., we do not rely on product form queuing network conditions and do not assume closed loop networks.

introduced in each of these resources in the bloated software, changing the service demands to $D_i(1 + b_i)$.

Using asymptotic bounds based on bottleneck analysis [196] to approximate achievable performance, peak throughput is given by $X = min_i(1/D_i)$ whereas peak throughput with bloat is $X_b = min_i(1/((1 + b_i)D_i))$.

Let $P_i(U_i)$ be the power consumed by resource R_i with utilization U_i (the exact relationship between utilization and power can be different for different resources). Utilization U_i of resource R_i is $D_i X$ [196], when running the non-bloated software and $D_i(1 + b_i)X_b$ with the bloated software.

Power without bloat, $P - \Sigma(P_i(U_i)) = \Sigma(P_i(D_i X))$
Power with bloat, $P_b = \Sigma(P_i(D_i(1 + b_i)X_b))$

Power efficiency (perf/watt metric) without bloat,

$$E = X/P = \frac{min_i(1/D_i)}{\Sigma(P_i(D_i X))}$$

Power efficiency with bloat,

$$E_b = X_b/P_b = \frac{min_i(1/((1 + b_i)D_i))}{\Sigma(P_i(D_i(1 + b_i)X_b))}$$

We are primarily interested in quantifying potential improvements from bloat reduction. Hence, we define the metrics of interest for relative throughput, peak power, power-efficiency, and equi-performance power of the non-bloated software *normalized with respect to that of the bloated software*.

Relative throughput with bloat reduction (> 1 is good),

$$\phi_x = (X/X_b) = \frac{min_i(1/D_i)}{min_i(1/((1 + b_i)D_i))} \tag{3.1}$$

Relative peak power with bloat reduction (< 1 is good),

$$\phi_p = \frac{P}{P_b} = \frac{\Sigma(P_i(D_i X))}{\Sigma(P_i(D_i(1 + b_i)X_b))} \tag{3.2}$$

Relative power-efficiency[11] with bloat reduction (> 1 is good),

$$\phi_e = \frac{X/P}{X_b/P_b} = (X/X_b)\frac{\Sigma(P_i(D_i(1 + b_i)X_b))}{\Sigma(P_i(D_i X))} = \frac{\phi_x}{\phi_p} \tag{3.3}$$

---

[11]Notice that in the case of SPECpower_ssj2008, the analysis extends to the energy-efficiency score across load levels (and not just the peak load), as the load levels are scaled by a fixed percentage with respect to achievable peak performance.

Relative equiperformance power with bloat reduction (i.e., comparing power consumed by original and bloated software at the same throughput X_b) (< 1 is good),

$$\phi_q = \frac{\Sigma(P_i(D_i X_b))}{\Sigma(P_i(D_i(1 + b_i)X_b))} \tag{3.4}$$

3.3.2   Effect of degrees of energy proportionality

An energy proportional hardware resource consumes power in proportion to actual resource utilization (a desirable property of server systems). In the presence of certain power management schemes (e.g., DVFS), this relationship may be super-linear; we characterize this with an exponent which we call the degree of energy proportionality of the resource.

Let us model the power consumed by each resource as $P_i(U_i) = a_i U_i^{\alpha_i} + c_i$, where α_i = degree of energy proportionality of resource R_i.

Let $P_{static} = \Sigma(c_i)$ be the static power of the hardware system, $P_{dyn} = \Sigma(a_i U_i^{\alpha_i})$ be the total dynamic (load dependent) power consumption. (α_i would be zero if resource R_i is non-energy proportional)

Let $L_i = a_i U_i^{\alpha_i}$ be the load-dependent power for resource R_i. Thus $P_{dyn} = \Sigma(L_i)$, $P = P_{static} + P_{dyn}$.

Relative peak power impact with bloat reduction:

$$\phi_p = \frac{\Sigma(L_i(D_i X_b)\phi_x^{\alpha_i}) + P_{static}}{\Sigma(L_i(D_i X_b)(1 + b_i)^{\alpha_i}) + P_{static}}$$

Defining f_i = fraction of load dependent power consumed by resource R_i (wrt total system power) when running bloated software, f_s = fraction of static power consumed with the bloated software ($f_s + \Sigma f_i = 1$), the above can be re-written as:

$$\phi_p = \Sigma(f_i(\frac{\phi_x}{1 + b_i})^{\alpha_i}) + f_s \tag{3.5}$$

As for the relative equiperformance power,

$$\phi_q = \Sigma(\frac{f_i}{(1 + b_i)^{\alpha_i}}) + f_s \tag{3.6}$$

3.3.3   System bottlenecks and bloat: A curious interaction

Consider the situation where bloat primarily affects the demand for a single resource. Let R_k be the bloated resource, then $b_k > 0$ and $b_i = 0, \forall i \neq k$. We show how the impact of bloat reduction depends on where the primary bottleneck is relative to the bloated resource. Table 3.2 summarizes the impact on performance, peak power, and equiperformance power.

TABLE 3.2: Effect of bloat reduction in different scenarios when bloat affects a single resource R_k

	Relative peak perf ϕ_x	Relative power at peak perf ϕ_p	Relative equi-perf power ϕ_q
Bloat at non-bottleneck resource	1 (same)	$\frac{f_k}{(1+b_k)^{\alpha_k}} + \Sigma_{i \neq k} f_i + f_s \leq 1$ (same if $\alpha_k = 0$, else decreases)	$\frac{f_k}{(1+b_k)^{\alpha_k}} + \Sigma_{i \neq k} f_i + f_s \leq 1$ (same if $\alpha_k = 0$, decreases otherwise)
Bloat at bottleneck resource	$1 + b_k$ (improves)	$f_k + \Sigma_{i \neq k}(f_i(1+b_k)^{\alpha_i}) + f_s \geq 1$ (same if $\alpha_i = 0, \forall i \neq k$, else increases)	same as above
Bloat reduction shifts bottleneck	$1 + b_{eff}$ (improves, but less)	$f_k(\frac{1+b_{eff}}{1+b_k})^{\alpha_k} + \Sigma_{i \neq k}(f_i(1 + b_{eff})^{\alpha_i}) + f_s$ (can increase or decrease or stay same)	same as above

3.3.3.1 Bloat at non-bottleneck resource

When bloat does not affect the bottleneck resource, $\phi_x = 1$, i.e., there is *no change in performance with bloat reduction*. Substituting in equation 3.5, we obtain:

$$\phi_p = \frac{f_k}{(1 + b_k)^{\alpha_k}} + \Sigma_{i \neq k} f_i + f_s \leq 1$$

Peak power *decreases* with bloat reduction, showing a higher improvement when the bloated resource has a steeper (larger α_k) power-to-load characteristic and consumes a higher fraction of system power (larger f_k).

Since there is no change in performance $\phi_e = 1/\phi_p$, i.e., the relative power-efficiency improves with bloat reduction. And $\phi_q = \phi_p$, i.e., the relative equiperformance power is the same as the relative power at peak performance.

3.3.3.2 Bloat at bottleneck resource

If bloat affects the bottleneck resource, i.e., $k = argmin_i(1/D_i)$, then $\phi_x = 1 + b_k > 1$, i.e., *throughput improves with bloat reduction*, maximum improvement being $1 + b_k$ when bloat is eliminated. Substituting in equation 3.5

$$\phi_p = f_k + \Sigma_{i \neq k}(f_i(1 + b_k)^{\alpha_i}) + f_s \geq 1$$

Peak power *increases* or remains the same depending on the power characteristics of the *non-bloated resources*, in contrast with the previous case. Reducing bloat allows more productive use of the bottleneck resource, improving peak throughput. If this increase in throughput increases the usage of resources that were underutilized earlier because of the bloat-affected bottleneck, and this increases their power consumption, then the power consumed by the application at peak throughput can increase.

$$\phi_e = (1 + b_k)/\phi_p <= 1 + b_k$$

Relative power efficiency improvement is less than or equal to the throughput gain. A steeper energy proportionality characteristic of the other (nonbloated) resources lowers the efficiency improvement from bloat reduction, especially if their power consumption is significant compared to the power consumption of the bloat-impacted bottleneck resource. The highest improvement from bloat reduction occurs in the case when the non-bottleneck resources are non-energy proportional (when $\alpha_i = 0, \forall i \neq k$).

3.3.3.3   Bloat reduction shifts bottleneck

If reducing bloat causes the bottleneck to shift from R_k to R_l, then, $1 < \phi_x < 1 + b_k$, i.e., throughput improves with bloat reduction (but to a lower extent than the previous case). Let us term $b_{eff} = \phi_x - 1$ as the effective bloat factor. Now the analysis for peak power and power efficiency is similar to the previous case, adjusting for b_{eff}.

Equiperformance power is impacted to the same extent in all the three cases above.

3.3.4   Summary

The extent to which bloat impacts power-performance may not always be obvious because bloat shifts the relative usage of different resources, some of which may have a non-linear power vs. load variation. Performance and equiperformance power are generally improved with bloat reduction. However, power at peak performance can increase with bloat reduction as a result of increased throughput following reduced pressure at a bottleneck resource. The degree of increase depends on the energy proportionality characteristics of the resources which are not bloated. Reducing bloat can also cause a shift in bottleneck from the bloated resource to another resource. In general, bloat can involve multiple resources, so we need to use equations 3.5 and 3.6 to assess the combined impact of these conditions.

3.3.5   Model predictions seen in experimental observations

Our model provides a generalized analysis for relating the power-performance impact of bloat reduction to resource bottlenecks and system energy proportionality. It highlights two important factors that determine the extent to which bloat impacts power-performance: (a) the relative energy proportionality of the system resources and (b) the extent to which the resource's usage is a bottleneck to performance.

TABLE 3.3: Trends from cross-platform experiments

System	Mem stress	Degree of energy proportionality	Peak Perf/ Eff. Score impact	Equi-perf power impact
X	High	Low	High	Low
Y	Low	Med	Med	Med
Z	Low	High	Low	High

Table 3.3 summarizes the trends seen in cross-platform experimental observations of the impact of bloat reduction (Section 3.2.1) through object memory reuse in the context of memory constraint (bottleneck strain) and energy proportionality differences between platforms. The higher the degree of energy proportionality (higher α_k), the higher the equiperformance power savings from bloat reduction in Table 3.2. This is supported by the increasing impact on equiperformance power in Table 3.3 going from System X to System Z. On the other hand, the higher the extent of bottleneck, the higher the performance improvement with bloat reduction in Table 3.2, consequently the higher the increase in efficiency. This is supported by the higher impact on energy-efficiency for System X over System Y and System Z.

1. **Peak power** can *increase or decrease* with reduction in bloat depending on whether it affects the bottleneck resource or other resources. In the situation where bloat affects the bottleneck, *the degree of impact can depend on the steepness of energy proportionality characteristic of the resources which are not bloated* and the fraction of system power consumed by them.

   In the cache pressure experiments (Figure 3.2), bloat-induced cache pressure causes underutilization of the compute resources for SMT4 and HC, hence throughput increases with bloat reduction. Consequently peak power actually rises (negative peak power savings) with reduction in bloat. This effect is more pronounced for the steeper power characteristics with DVFS than for fixed frequency. On the other hand, reducing bloat also reduces memory references and consequently memory power; the overall system power reduction due to this effect, however, is small as memory consumes a low fraction of system power on this system. In the SMT2 case, where the performance impact of bloat is lowest, this reduction translates into a slight reduction in overall system power.

2. **The energy efficiency** improvement from reducing bloat is likely to be most pronounced when bloat affects a bottleneck resource and the

non-bottleneck resources are not energy proportional (per Table 3.2). *Energy proportional hardware mitigates the effect of bloat on energy efficiency at peak performance* – Figure 3.2 confirms this, showing the gains in energy efficiency are greater when running the cores at fixed frequency than with DVFS. Similar behavior can be seen in Section 3.2.1 with System X getting a higher improvement in energy efficiency with bloat reduction (1.59) than System Y (1.1) or System Z (1.06).

3. While energy efficiency improvement at peak performance is higher with non-energy-proportional resources, *the improvement at equal performance can be significantly higher for energy proportional hardware.* This can be seen as significantly higher **equiperformance power savings** for DVFS in Figure 3.2 compared to the energy efficiency improvements at peak. The same figure also shows that having more energy proportional resources (DVFS) can yield significantly higher equiperformance energy savings (compared to fixed frequency).

3.4   Conclusions

In this chapter, we discussed a systematic empirical and analytical investigation of the impact of runtime resource bloat on energy efficiency. The findings show that lean does usually imply green, but to tell the shade, a whole system analysis is necessary.

In-depth experimental studies demonstrate how the benefits from reducing bloat could vary widely with hardware and software characteristics due to a curious interplay between bloat, energy proportionality, and system bottlenecks. These implications can be understood more completely using a simplified abstract model for relating the power-performance impact of bloat reduction to energy proportionality and resource bottlenecks for "what-if" analysis. For example, on the one hand, energy proportionality of a bloat-impacted resource can amplify benefits of reducing bloat. On the other hand, energy proportionality of the remaining resources can shrink those benefits when bloat affects a bottleneck.

The complexity of modern software and system layers make it impractical to compute the exact power-performance impact of run-time bloat (reduction) analytically. However, the analytical model described here is still useful in reasoning about implications of reducing bloat. It surfaces certain aspects of the complex behavior that determine the impact of bloat, by abstracting operational relationships at bottleneck zones. This creates a foundation for reasoning quantitatively about the impact of bloat reduction on system power, performance, and energy efficiency.

Part II

The Antidote: Resource Proportional Software Design

Chapter 4

Resource Proportional Software Design Principles to Reduce Propensity for Bloat

The last two chapters introduced the problem of software bloat and its impact on system power-performance. We noted how origin of this issue lies in the inherent tension between flexibility, productivity, and efficiency, but we also observed that the degree and its impact can vary significantly.

In this chapter we explore some core principles that could enable system and software designers to reduce propensity for bloat without sacrificing productivity or flexibility. The rest of the chapters in Part II describe practical strategies that programmers and tool developers can use to apply these principles when coding and optimizing software components and applications. Part

III of the book dives deeper into case studies and opportunities for architects and system software engineers to apply these principles when designing software stacks for emerging systems.

4.1   Insights from Energy Proportional Hardware Design

The concept of energy-proportional computing [51] has become popular as a principled design approach to achieve significant energy savings in server systems [389, 382, 49]. The benefits of energy proportional design are based on two key observations:

- computing servers in data centers must often be provisioned with enough hardware resources to support occasional bursts of high peak demand, but tend to require only a small fraction of the available computing capacity in typical situations.

- provisioning resources that are unused most of the time may cause energy wastage but the wastage could be reduced if the corresponding hardware components (servers, cooling equipment, power supplies, etc.) were designed to *consume power only to the extent they are actually utilized*, i.e., if the components were truly energy proportional.

Thus, designing components to be more energy proportional can enable large energy savings [51] without sacrificing the ability to meet peak demands.

4.2   Resource Proportional Design of Software Features

Applying analogous insights to the design of software components, we notice a parallel between the energy wastage due to hardware overprovisioning in a non-energy-proportional system as described above and overheads arising from the functional overprovisioning of software.

- Enterprise applications must support extremely demanding levels of variability and interoperability, but they actually exploit only a small fraction of this versatility in a typical deployment situation.

- Provisioning functionality that is unused most of the time causes resource wastage, but the wastage could be reduced if software features could be designed so that the runtime resource usage of an application is only a function of the exploited fraction of features.[1] Then efficiency would be achievable without losing flexibility.

With this view, we propose a notion called *resource proportionality* of software features to enable a principled design approach to reduce energy wastage due to software bloat. Resource proportional features allow software to be provisioned to support a very high level of versatility in terms of features (and non-functional concerns) but still incur minimum bloat by consuming computing resources in proportion to what would have been required to support only the features (concerns) actually exploited in a given deployment scenario.

There are three key steps in resource proportional software design:

1. Provision software components for peak functionality with a wide feature set, so it is extremely flexible.

2. Design each software component to use resources in proportion to actual features used.

3. Account for non-uniform distribution of feature utilization scenarios, e.g., prioritize feature configurations which are most commonly deployed.

The second step involves minimizing the runtime overhead expended on unused built-in generality. While some costs in provisioning for generality cannot be completely eliminated, could we draw on some insights from architecture research where similar problems have been studied in a more structured setting? For example, researchers with Stanford's ELM (Efficient Low-Power Microprocessor) project discovered that it was data and instruction supply overhead in general-purpose programmable embedded processors, not inefficiencies in the core logic, that accounted for the 50X energy-efficiency gap between these processors and hard-wired media ASICs [113].

---

[1]I.e., the remaining fraction does not induce a runtime overhead.

Insights from computer architecture research

The operational perspective developed in this chapter is inspired by related work in hardware or system architecture research on the trade-off between energy efficiency and flexibility.

Requirements-aware energy scaledown Energy scale-down studies [254, 113] have demonstrated how the power consumption of similar operations can vary significantly across different systems and device. For example, Mayo and Ranganathan [254] report results of an email reply benchmark which consumes 16W on a laptop, 1.44W on a handheld, and 0.473W on a cellphone. Implementing certain requirements-aware energy scale-down optimizations (e.g., in the display component) enabled significant reductions (more than factor of 2 improvement) in their experience.

Optimizing data and instruction supply overheads in supporting generality In contrast with approaches based on scaling down requirements to save energy, the Stanford ELM project [113] explored the extent to which efficiency gains could be attained without sacrificing generality and programmability. On analyzing sources of the difference in energy/operation between hardwired media ASICs (approx. 5pJ/op) vs. programmable embedded processors (approx. 250pJ/op), Dally et al. found that data and instruction supply overheads alone account for 70% of this difference. By devising techniques to optimize instruction and data supply energy costs (e.g., using a deeper cache hierarchy with explicit control) in their design of the Stanford ELM microprocessor, they report a 23X improvement in energy efficiency, closing the gap with ASICs to within 3X.

In the case of software, structural interactions due to excess concerns are likely to include data transformations, parsing, bookkeeping, method lookups, indirection chains, condition checks, and other overheads incurred in accessing actual data and core logic implemented.

How do we identify and address similar inefficiencies in software stacks to help bridge the gap between large framework-based applications and custom-built programs providing the same functionality? As software is different from hardware in its characteristics, let us analyze what makes it non-resource proportional.

4.3   How Software Becomes Non-resource Proportional

Compare the code examples shown in Figure 4.1 and Figure 4.2. Figure 4.1 shows a simple Buffer class which supports basic set and get operations, and Figure 4.2 shows a richer implementation which provides a logged restorable buffer. The differences between the two implementations (i.e., the extra code required to support the additional capabilities, logging and restorability) are marked in gray (Figure 4.2).

```
public class Buffer {
      int buf = 0;
      public Buffer(int x) {
            buf = x;
      }
      int get() {
            int tmp3 = buf;
            return tmp3;
      }
      void set(int x) {
            buf = x;
      }
}
```

FIGURE 4.1: Simplified example of a basic Buffer class supporting operations to set and get a single integer value.

```
1:public class Buffer {
2:       int buf = 0;
3:       int back = 0;                    <----- potentialbloat
4:       public Buffer(int x) {
5:              buf = x;
6:       }
7:       void logit() {
8:              System.out.println("LOG: buf = " + buf);
9:              System.out.println("LOG: back = " + back);
10:      }
11:      int get() {
12:             int tmp3 = buf;
13:             logit();                   <----- potentialbloat
14:             return tmp3;
15:      }
16:      void set(int x) {
17:             int tmp = buf;      <---- potentialbloat
18:             back = tmp;         <---- potentialbloat
19:             buf = x;
20:             logit();                   <---- potentialbloat
21:      }
22:      void restore() {
23:             int tmp2 = back;
24:             logit();
25:             buf = tmp2;
26:      }
27: }
```

FIGURE 4.2: Example that expands the Buffer class implementation to support a Logged Restorable Buffer [242, 217]. The restore function allows the previous setting of the buffer to be restored and the logit function is used to log the buffer operations, get, set, and restore for tracing purposes. The lines in gray mark the code added to support this expanded functionality. Of these, the lines annotated as *potentialbloat* are sources of runtime overhead that make the code non-resource proportional (in scenarios where logging and restorability are not required).

For example, the code for the function `void restore()` and all code that applies to the variable `back` are statements added to support the restore feature which allows the contents of a buffer to be restored to its previous value when required.

Suppose the richer Buffer class shown in Figure 4.2 is included by a server application and invoked for every request, a scenario where the buffer does not need to be logged and also need not be restorable. In this use case, logging is an excess feature but the `logit()` method is called nevertheless, incurring a heavy runtime overhead. Restore is also an excess feature, but we notice that the restore() method by itself does not induce a runtime execution overhead in this situation as it would not be called. The `get()` functionality of the buffer is also unaffected by the optional feature restore. On the other hand, the `set()` method does incur the runtime overhead of saving the previous value. The statements `tmp = this.buf; back = tmp;` in this method, therefore, are a potential source of execution bloat in this usage scenario. Secondly, the extra memory allocation and initialization of the back field, i.e., the statement `int back = 0;` is another potential source of bloat due to the optional restore feature.

Thus, it is these particular sources of bloat (lines labelled with the arrow *potentialbloat* in Figure 4.2) which cause the buffer component to consume more resources than required when neither logging nor restore capabilities are utilized.

Using this example, we can develop our intuition about conditions leading to sources of bloat that makes a component non-resource proportional.

1. Interaction between necessary software features and less necessary features introduces extra code statements and object state (a very common example is data transformations to support wide interfaces)

2. The overheads of these extra statements get amplified when the effect of interactions is incurred repeatedly (e.g., in a loop) or involve costly operations. Further, when some features are used more heavily and others are used less heavily, the interactions between them are the ones that matter the most.

3. The impact is more acute when it affects bottleneck resources or components with super-linear power vs. utilization characteristics, as described previously in Chapter 3.

We can now combine these insights to build a systematic approach to reason about bloat and the different levels at which it could be mitigated.

As illustrated in Figure 4.3, the following are some of the factors[2] that jointly determine how runtime bloat arises from overprovisioned features and impacts power-performance.

---

[2]Designated by labelled arrows in the figure.

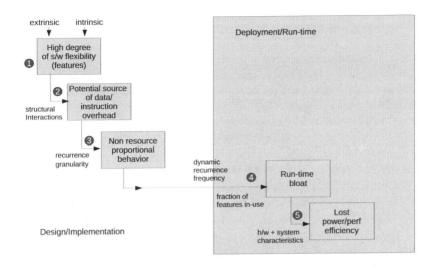

FIGURE 4.3: High level cause effect flow for guiding optimization.

Software design or code characteristics leading to non-resource proportional components

- Structural interactions between optional features and the rest of the code, which represent code statements or fields that are potential bloat contributors

- Structural recurrence granularity of these interactions (e.g., nesting depth of loops containing these potential contributors) compared to the rest of the code

Software deployment or runtime environment characteristics causing non-resource proportionality to manifest as bloat

- Features unexploited by the deployment scenario (excess features)

- Dynamic recurrence frequency of the relevant structural interactions corresponding to these unexploited features

- Hardware system characteristics (resource bottlenecks and energy proportionality)

4.4   Defining Resource Proportionality with Respect to Feature Utilization to Predict Bloat Propensity

In order to enable a principled approach to assess the propensity for bloat of a given software implementation, we now outline an approach for quantifying bloat propensity and develop the notion of resource proportionality that we introduced earlier in this chapter. This provides a foundation for performing a simple "what-if" analysis to compare alternative strategies for bloat reduction.

In general, the term **resource proportionality** reflects *the degree to which resources used by a system scale in proportion to function accomplished.*

In performance (scale up) and energy efficiency (scale down) studies, the function accomplished is typically interpreted in terms of load, i.e., volume of transactions (work) processed. In these cases, non-proportional behavior may arise due to the presence of bottlenecks and start-up overheads (which hurt scale up) or unreleased resources (which hurt scale-down).

In the context of this book, where we are more concerned with more efficient means to support software flexibility and leverage emerging systems, function accomplished is interpreted in terms of features exploited (utilized). Thus, the resources used by the system should be in-proportion with features that are actually utilized (rather than all the features that the system is configured to support). In this case, as illustrated in Figure 4.3, non-resource proportional behavior can arise when the structural interactions due to an unexploited feature occur at a recurrence granularity that is significant compared to the rest of the code.

However, at the outset, it is not clear how this intuitive notion of resource proportionality in terms of feature utilization should be quantified. Features may not only be of different sizes but are also qualitatively different. Thus, unlike load based measures, the number of features exploited cannot be used as a uniform scale against which resource proportionality can be measured quantitatively. In the following subsection, we describe how we address this issue to arrive at suitable measures for feature utilization, resource amplification proportionality, and propensity for bloat.

Insights from Power-aware VLSI design

Power-awareness metric Bharadwaj et al. [58] introduced a power-awareness metric for evaluating and constructing power-aware VLSI designs, where power-awareness ϕ is a measure of how close a system is to the most power efficient system that could be designed specifically for a given operating point, summarized across all operating points in the desired range of scenarios supported, weighted by their likelihoods. According to their definition, "a system H is perfectly power aware iff its energy dissipation in scenario s_i is no greater than that of a dedicated system H_{s_i} constructed to execute scenario s_i as efficiently as possible."

$$\phi = \frac{\Sigma_{scenarios} E(H_{perfect}, s_i) d_i}{\Sigma_{scenarios} E(H, s_i) d_i}$$

where d_i is the likelihood (distribution) of scenario s_i

This metric ensures that a higher weight is given to most common (typical) scenarios. Thus it is possible that the most energy proportional system may not necessarily be the most power-aware alternative.

Our characterization of resource proportionality in software features and the resource bloat propensity factor are modeled using similar principles. However, basing measures of bloat on the notion of a dedicated software system H_{s_i} constructed to execute scenario s_i as efficiently as possible is impractical (e.g., consider Blum's speedup theorem). Further, we are mainly interested in systematic approaches for avoiding overheads that arise in connection with overprovisioning of features as opposed to algorithmic efficiency issues. Hence, we chose to constrain the baselines for desired scenarios to a well-defined set of specialized variants that implement the exact subset of features required to execute each scenario s_i.

Feature utilization scale: To construct a scale of feature utilization, we first compute the resource usage of a specialized version of software that is configured only with the utilized features. We can now assess how resource proportional the fully configured version of the software is by plotting against this scale the actual resource usage of the fully configured version as its utilized features are increased.

Consider a set of possible component designs:

- capable of addressing one or more feature spaces in F ($F_{1..n}$)

- consuming resources R

- under a set of deployment scenarios, S ($S_{1..m}$) with inputs I (a vector) computing outputs O under conditions C

Let
R^{F_1} = resource expended by a component specialized for a feature space F_1
R^{F_2} = resource expended by a component specialized for a feature space F_2
. . .

$R^{F_{1..n}}$ = resource expended by a component generalized across a feature space covering $F_1, F_2, ...F_n$

For each deployment scenario S_j (characterized by a set of inputs and operating conditions):
Let $F_j = F(S_j)$ be the feature space required for scenario S_j

4.4.1   Effect of using a generalized component in this scenario

The resource amplification (proportionality) due to generalization is the resource usage R^{actual} of the generalized component in this deployment scenario compared to that of a component design specialized just for the feature space required:
$$R^{actual(S_j)} = R^{F_{1..n}}(S_j)$$
$$R^{specialized(S_j)} = R^{F_j}(S_j)$$

Resource overhead $R^{overhead(S_j)} = R^{actual(S_j)} - R^{specialized(S_j)}$
Resource amplification (proportionality)

$$\alpha(S_j) = \frac{R^{actual(S_j)}}{R^{specialized(S_j)}} \tag{4.1}$$

If the resource amplification proportionality factor of the generalized component $\alpha <= k$ across the chosen input space or chosen set of scenarios, we refer to this as a k-RPD component.

We define the bloat propensity factor b as the relative increase in resource consumption attributed to the overhead introduced by additional features present in the generalized component, as compared to the resources needed to support only the features required.

Bloat propensity

$$b(S_j) = \frac{R^{actual(S_j)} - R^{specialized(S_j)}}{R^{specialized(S_j)}} = \alpha(S_j) - 1 \tag{4.2}$$

The system impact, e.g., power-performance effects of this resource amplifiation across all resources (say $R_1 \ldots R_q$) for a given deployment scenario S_j, could be studied using the models for bottleneck analysis and power characteristics described in Chapter 3. For example, the equiperformance power amplification in this scenario is $P_j^{actual}/P_j^{specialized}$. Notice that the power-perf models are typically not linear; hence power amplification is not the same resource amplification.

4.4.2   Weighted RP accounting for scenario distribution

If we plot R^{actual} vs. $R^{specialized}$ across scenarios $S_{1..m}$ we obtain a resource proportionality curve. We can also plot P^{actual} vs. $P^{specialized}$ to study the corresponding power amplification. Since all scenarios are not equally common (e.g., scenarios involving exception handling, slow paths, or uncommon configuration settings may occur infrequently), a weighted resource proportionality analysis could be used to account for relative distribution (or importance) of scenarios.

Let w_j be the weight (e.g., probability) assigned to scenario S_j, where $\Sigma w_j = 1$.

Average resource (amplification) proportionality

$$\alpha = \frac{\Sigma w_j R^{actual(S_j)}}{\Sigma w_j R^{specialized(S_j)}}$$

Average power amplification $\eta = \frac{\Sigma w_j P_j^{actual}}{\Sigma w_j P_j^{specialized}}$ reflects the net increase in energy consumption at equiperformance.

4.4.3 Effect of adding features not required for a scenario

Consider the effect of extending a component originally specialized for scenario S_j to also support a particular feature space F_v that does not overlap with F_j. (New features are typically added to enable the component to address some other scenarios, e.g., say S_v.)

The resource amplification (proportionality) and bloat propensity of F_v in scenario S_j can be computed by comparing the resource usage of the extended component which is designed to support both F_v and F_j, to that of a component specialized just for F_j, the feature space required for the scenario S_j.

Feature Resource amplification $\alpha_v(S_j) = \frac{R^{F_j v}(S_j)}{R^{specialized(S_j)}}$

The bloat propensity $b_v(S_j)$ of a feature space F_v is the relative increase in resource consumption attributed to the overhead induced by that feature space over that needed to support only the features required in scenario S_j.

Feature Bloat propensity $b_v(S_j) = \frac{R^{F_j v}(S_j) - R^{specialized(S_j)}}{R^{specialized(S_j)}} = \alpha_v(S_j) - 1$

If the unbloated version of the software which exploits features F_j consumes r amount of resources, then the bloated version that is overprovisioned with additional features F_v consumes $r(1 + b_v)$ amount of resources.

Notice that sometimes the addition of a feature, such as caching, prefetching (or readahead), pre-computation, etc., could reduce consumption of one resource (e.g., disk IO, CPU), potentially at the cost of slightly increased consumption of other resource (e.g., memory). The same feature if used in a scenario where caching or prefetching is not beneficial could only induce overheads. This can lead to interesting interaction effects to account for.

Further, different deployment scenarios may not be fully independent, e.g., when they operate on common external data files. Hence a feature such as encryption which is added for a specific new deployment scenario S_v may also affect the requirements for earlier deployment scenarios which now need to interpret encrypted data in common external files. This case can be modelled as a change in the feature space $F_j = F(S_j)$ for these deployment scenarios or as the addition of new deployment scenarios that require reading encrypted files. A change in an underlying technology (e.g., NVM replacing hard disks or SSD), as discussed in Chapter 8, can also render features such as caching or marshalling redundant, thus manifesting bloat propensity.

Various such nuances arise when considering non-functional requirements in RPD and are discussed in Chapter 7.

4.4.4   Bloat relative to actual resource consumed by a component

The percentage bloat β incurred due to additional features present in the generalized component with respect to the actual resource consumption observed is the fraction of resource consumption attributed to the overhead:

Percentage bloat

$$\beta(S_j) = \frac{R^{overhead(S_j)}}{R^{actual(S_j)}} \tag{4.3}$$

4.4.5   Computing bloat propensity when $R^{specialized}$ is not directly available

One difficulty with adopting the above measures in practice is the requirement for a specialized version of software that is configured only with the utilized features in order to compute $R^{specialized}$. This is impractical for applications which are not created using a product line or feature-oriented programming approach (such as the use of pre-processor directives or externally maintained information that can be used to unconfigure features at compile time). However, it is possible to obtain an estimate of bloat propensity by *profiling the resources consumed by structural interactions* induced by additional features.

Let $r(s)$ be the resource usage directly attributed to a statement s during program execution and let $r^{cum}(\psi)$ be the cumulative resource usage collectively attributed to a set of statements ψ, i.e., to all statements $s \in \psi$ and the methods called by these statements.

$$r^{cum}(\psi) = \sum_{\{s|s \in \psi \text{ OR } ancestor(s) \in \psi\}} r(s)$$

If we can locate the structural interaction statements which are potential bloat contributors due to a set of optional features F (e.g., the lines labeled as *potentialbloat* in Figure 4.2; such statements can be detected using the techniques discussed later in Chapter 5), then the resource bloat (overhead) contributed by these features (when they are not needed) can be computed as follows:

$$R^{overhead} \gtrapprox r^{cum}(\{s \mid potentialbloat(s)\})$$

$$\beta \gtrapprox \frac{r^{cum}(\{s \mid potentialbloat(s)\})}{R^{actual}} \tag{4.4}$$

$$b \gtrapprox \frac{r^{cum}(\{s \mid potentialbloat(s)\})}{R^{actual} - r^{cum}(\{s \mid potentialbloat(s)\})} \tag{4.5}$$

where *potentialbloat(s)* is true when a statement s directly contributes to bloat.

Thus, we can lower the bloat propensity factor of a given software implementation in two ways:

- by reducing *the number of potential bloat contributing statements* (as described in Section 5.1)

- by reducing *their cumulative resource utilization $r^{cum}(s)$ overhead* (e.g., by optimizing recurring costs induced by s as in Section 5.2).

4.4.6 Trade-off between feature exploitation and provisioning overhead

As we have seen, the actual resource usage of a generalized component design varies across deployment scenarios $S_{1..m}$ as:

$$R^{actual(S_j)} = R^{overhead(S_j)} + R^{specialized(S_j)}$$

i.e., $R^{overhead}$ determines the extent to which resource usage deviates from being exactly proportional ($k = 1$) with respect to feature utilization (deviation from a 1-RPD design).

Now, consider a change in the deployment scenario from S_j to S'_j, after which features F_v are exploited in addition to feature space F_j.

$$R^{actual(S'_j)} = R^{exploit}(F_v) + R^{actual(S_j)}$$
$$= R^{exploit}(F_v) + R^{overhead(S_j)} + R^{specialized(S_j)}$$

where $R^{exploit}(F_v)$ is the additional resource cost incurred when features F_v are exploited in addition to feature space F_j.

When a feature is not exploited, its contribution to resource consumption is determined by its provisioning overhead; when it is exploited, then its contribution to resource consumption is determined by its exploitation cost in addition to its provisioning overhead. With *dynamically reconfigurable features*, for example, the provisioning overhead may be lower ($R^{overhead}_{dyn} < R^{overhead}$), reducing the bloat propensity factor due to unused features. However, the exploitation cost may include a feature activation overhead in addition to base in-use resource demand ($R_{dyn}^{exploit} > R^{exploit}$).

$$R_{dyn}^{actual}(F_{jv}) = R_{dyn}^{exploit}(F_v) + R_{dyn}^{overhead}(F_v) + R^{specialized}(F_j)$$

4.4.7   Resource proportionality characteristics

4.4.7.1   Scenarios with montonically ordered feature spaces

For illustration, let us first consider a very simplistic setting with a sequence of scenarios $S_{1..m}$ that utilize monotonically increasing feature spaces $F_1 \subset F_2 \dots F_{m-1} \subset F_m$ of a software component implementation generalized across feature spaces $F_{1..m}$. Let $R^{full} = R^{F_m}$ be the resource usage incurred in the deployment scenario S_m where all features supported by the generalized component are utilized (i.e., at 100% feature utilization). Let us define the normalized feature utilization in this setting as $\mu(S_j) = \frac{R^{specialized(S_j)}}{R^{full}}$.

Figure 4.4 illustrates examples of different types of resource proportionality characteristics which may be exhibited by different types of generalized software implementations in this setting. On the x-axis we plot feature utilization μ for an increasing subset of features and on the y-axis we plot the corresponding resource usage R^{actual} normalized wrt to R^{full}.

Non-proportional depicts an implementation whose features are completely non-resource proportional, i.e., the resource usage cannot be scaled down even when only a small fraction of its features are exploited. *A* depicts a typical implementation with some amount of bloat, where some overhead is incurred due to unexploited features, but resource usage can still be scaled down to an extent when less features are used, e.g., according to a sub-linear resource proportionality characteristic. *B* depicts an implementation with a steep resource proportionality characteristic, e.g., using dynamically reconfigurable features, where the resource consumption increases sharply as more features are used because of the feature activation overhead. Such an implementation incurs the lowest bloat but is resource efficient only as long as the fraction of features exploited is small.

4.4.7.2   General scenarios (unordered feature spaces)

In general scenarios, there is no monotonic ordering inclusion across feature spaces exploited in different scenarios, e.g., some scenarios may use disjoint feature sets, some may use a few overlapping features, features variations may manifest at different granularities, and all features provisioned in a component may not be exercised together in any scenario. Thus instead of forcing a linear feature utilization scale, we can plot $R^{specialized}$ vs. R^{actual} for each scenario as shown in Figure 4.5. The relative gap between the two indicates the bloat propensity of the actual implementation.

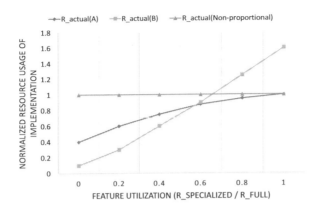

FIGURE 4.4: Different types of resource proportionality characteristics. The curves plot the normalized actual resource usage of a fully generalized implementation as the feature utilization μ increases. The resource usage (y-axis) is normalized wrt to the resource usage when all the features are utilized.

4.5   Resource Proportional Optimization Control Points for Bloat Mitigation

Sources of non-resource proportionality can be contained in many different ways by applying suitable interventions at any of the several potential control points marked by the numbered circles in Figure 4.3 and elaborated in Table 4.1. The first involves careful design and can benefit from good programming models and abstractions. The initial effort invested is significant in return for future productivity scaling through reuse. The second can affect development productivity but the effort can be reduced with the aid of tools. The third is very suitable for automated transformations but may require expensive analysis. The fourth can affect deployment productivity as it requires some way to determine which features are likely to be actually used at runtime.

For a given deployment, the largest gains from de-bloating are expected for those structural interactions which are due to unexploited features, have a high dynamic recurrence frequency, and either affect a steeply energy proportional

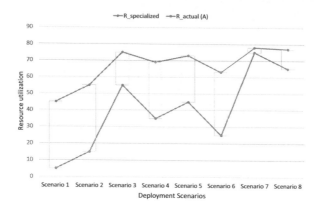

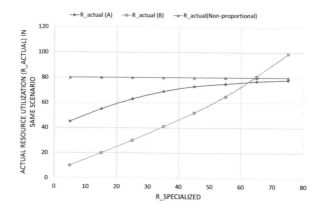

FIGURE 4.5: Resource proportionality characteristics depicting the differences in resource utilization of a specialized vs. generalized implementation for different scenarios (a) in no particular order and (b) ordered by $R^{specialized}$.

power-hungry resource or affect a bottleneck resource in a system where non-bottleneck resources are power-hungry and non-energy proportional. A systematic de-bloating strategy would focus on these opportunities first, via automated optimization and then using bloat diagnosis tools to guide manual code changes if still worthwhile. As detailed information about unexploited features or concerns is difficult to obtain automatically, an alternative strategy would be to first identify expensive inter-method structural interactions which have

TABLE 4.1: Different levels of intervention to contain the impact of non-resource proportionality

	RP optimization	Applicable techniques	RP model
1	Expand variability with reuse (Step 1 in RPD, Section 4.2)	Flexible programming models, abstractions, parameterization	Scales up feature space in Section 4.4
2	Reduce structural interactions from optional features and optimize incidental concerns	Use bloat diagnostic tools to aid manual optimization (Section 5.1)	Reduces number of statements s in Equation 4.4 (Section 4.4.5)
3	Amortize recurring overheads	Automated compiler/runtime optimizations, e.g., object reuse, memoization and layer folding [228], partial evaluation (Section 5.2)	Reduces r_{cum} in Equation 4.4 (Section 4.4.5)
4	Enable high overhead features on-demand	Runtime feature adaptation, lazy evaluation (also see Section 5.3)	Trade-offs explored in Section 4.4.6
5	Tune system power-performance	Power manage non-bloated resources (if they are power hungry) and reduce bottleneck strain for bloated resources	Power-performance model in Section 3.3.3

significant power-performance impact and then manually determine if methods that induce those interactions correspond to unexploited features.

In the next two chapters we describe RPD strategies and tools that could be used when writing software for new components or when refactoring existing code to improve its resource proportionality.

4.6 Conclusions

The underlying principle of RPD is that if software features were resource proportional, energy wastage due to bloat need not be an inevitable consequence of overprovisioning flexibility even on emerging systems where technology evolution may render some features as irrelevant. Our insights show how the propensity for bloat of a software implementation and its power performance effects depends on characteristics of the program structure and the underlying hardware system, such as the recurrence granularity of structural interactions between features, system bottlenecks, and the energy proportionality of hardware resources. By controlling these characteristics, software features could, in principle, be made more resource proportional. However, we note that it may not be straightforward to achieve this in practice for existing software implementations. We propose quantitative measures to assess the propensity for bloat of an implementation as a foundation for performing a simple "what-if" comparison of the trade-offs and implications of alternate strategies for bloat reduction. We highlighted multiple control points that could potentially be used to devise systematic strategies for containing non-resource proportionality to achieve runtime efficiency software without necessarily sacrificing flexibility or productivity.

Chapter 5

Resource Proportional Design Strategies I: What Component and Tool Developers Can Do

Software engineering principles for designing flexible reusable components focus on modularity and abstraction as a means of expanding variability in order to support a wide feature set. These principles continue to be useful as a first step in RPD (provisioning for maximum flexibility), but do not automatically translate into RP designs.

For example, the separation of concerns[405] design principle aims for high cohesion within a module (comprising interdependent features) and low coupling between modules (comprising relatively independent capabilities which may be combined as needed). Modularization is typically expressed using programming constructs such as methods, classes, frameworks, and microservices, which offer various levels of de-coupling between modules.

Ironically, as illustrated by the examples in Chapter 2, the indirections and wider interfaces introduced to achieve this de-coupling increase the overheads of reuse, leading to resource use amplification. Further, the cohesion introduced to avoid costs for combining finer grained concerns to support variability increases the possibility of inadvertently pulling in non-essential functionality when reused in a different context. Information hiding and encapsulation allows independence between modules but can encourage bloated designs.

In this chapter we discuss some practical RPD strategies that could be employed to tackle this dilemma when developing a new component or adding new features to a component. We also propose techniques that could be employed to create tools that would aid RPD analysis to support these strategies.

5.1 Strategy 1: Minimize Interactions between Independently Usable Features without Sacrificing Efficient Reuse

The first strategy is to minimize code and data interactions between features that need not always be used together without adding indirections. This could be (a) embedded as coding practices and design patterns, (b) implemented as a formal discipline using a software engineering model and tools such as feature oriented programming, or (c) aided by new RPD analysis tools that automatically highlight candidate code and data interaction sources introduced by the component features being developed.

5.1.1 Development practice: Abstracting a minimal core of base features – Lessons from the Linux kernel

One way to approach this is to code every fine-grained feature in a way that avoids structural interactions due to any optional feature on other features which can be used independently of the optional feature.

When there is a hierarchy of features such as a base feature and series of extensions which build on previous extensions, the structural interactions due to the extensions can be avoided by making the base feature and the extensions independently accessible. This ensures that code or data corresponding to an extension can only consume resources during execution if the extension is explicitly invoked. Indeed this kind of a minimalist and incremental coding approach tends to be used in Linux kernel development where efficiency, flexibility, and maintainability of the code are all critical considerations.

For example, in the "embedded anchor" design pattern[82], structures for a common feature such as linked lists, wait queues, and base inodes are directly embedded in caller-specific structures, where methods for the common features operate on chains of the embedded structures (e.g., list head) without any awareness of the enclosing structure, which remains accessible to the caller using a containerof() macro. The cost of allocating a given enclosed structure is only incurred explicitly by callers who maintain the additional fields in the structure, and thus extensions do not impose a structural interaction overhead on the base feature. The "no mid-layer" design pattern[83] avoids introducing intermediate layers that implement common functionality such as the page cache (readahead and write-behind) for filesystems and IO scheduling (elevator algorithms) for block devices. This is because middle layers could impose an overhead or needless complexity in situations where caching or IO scheduling is not relevant. Instead these capabilities are provided as helper libraries which are explicitly leveraged as appropriate by filesystem or block driver-specific logic.

As a result, it has been an easier lift to evolve existing Linux filesystems such as ext4 and XFS to utilize emerging technologies, e.g., support for direct access to persistent memory for files (DAX) [111, 214]. For files configured with DAX, filesystem operation function pointers are set up to DAX-specific variations which replace the use of common page cache and block IO submission logic with memory based accesses on reads or writes and direct memory mapping to persistent memory address ranges corresponding to desired filesystem blocks on memory mapped reads or writes. Notice, however, that optimizing for persistent memory usage patterns requires more than merely replacing the logic above, e.g., addressing overheads such as the cost of memory mapped file I/O which weren't as prominent with slower technologies[107], but could benefit from similar caching idioms, e.g., map-ahead and a mapping cache. More discussion on the nuances of resource proportional software design as system technology evolves is covered in Part III, Chapter 8 and 9.

In an object oriented programming language, inheritance from a base class or mix-in style interfaces [84] can be used to separately derive extensions from a base feature without introducing structural interaction overhead for code paths where the extension is not used.

However, a lot of careful thinking is often required in order to systematically build features in this way. In particular, deciding what the minimal core (base feature) should be is not always immediately obvious, e.g., a linear hierarchy among features may not be easily apparent. In such situations a transformation in perspective may be required to logically restructure code into a well engineered sequence of feature increments.

For example, consider the design of locks for kernel data structures. Fine-grained locking reduces structural interaction impact on concurrency constraints where coarse-grained locking can lead to needless contention between independent data structures or independent features when deployed in SMP environments with a large number of processing cores/threads. On the other hand, in low contention environments, when several data structures are traversed (even if independent) in a code path, the lock acquisition instructions (and cache line impact) for each fine-grained lock can add up to a needless structural interaction, as the locks are rarely contended. The read-copy-update (RCU) [257, 259] locking primitive in the Linux kernel optimizes these common spin lock acquisition overheads in read-paths by deferring the cost to updaters. RCU uses a clever mechanism that relies on tracking natural transition boundaries such as context switches or system call returns on every CPU to determine when existing readers are done accessing copies of old values and updates can be safely marked complete. Choosing this novel near-lock-free-read approach as a base feature minimizes both forms of structural interaction overheads in the most common (read-heavy) cases, although deferred updates can introduce other kinds of subtle interactions in certain scenarios[342, 313].

The evolution of crash dump support in Linux reflects another example of the dilemma involved in architecting an independently useful minimal self-contained base feature that allows enough flexibility for different variants to be supported without imposing structural interaction overheads. Different variants may need to selectively preserve different portions of the crash data (e.g., stack trace, kernel-only memory pages, application memory state, subsystem-specific analysis log state, etc.) on different dump targets (e.g., different disk media or over a network to a different system, sometimes even across a firewall). Early implementations of crash dump experimented with either restricting the solution to specialized target types or introducing plugins to allow multiple dump targets and filter mechanisms. However adding these plugins incurs complexity when the system is already in a vulnerable state after a crash and also duplicates driver-specific or data saving infrastructure just for the crash dump path. Instead of limiting flexibility or introducing such structural interaction complexity, the Linux kernel adopted a different approach[163]: the minimal base feature for crash dump builds on a feature called kexec to carefully soft boot a new kernel while simply preserving the memory and register state of the previous kernel in-place and making it visible to the new kernel's user space as a core file (/proc/vmcore). This allows reuse of existing space tools such as gdb/crash to analyze the memory right in-place even before saving the dump or perform the actual dump to disk/network using standard user space utilities, e.g., including file copy to disk/network using *cp*, *scp* commands. No structural interactions are incurred by the base implementation to support such variety. Notice how this transformed (resource proportional design) perspective also enables greater flexibility than the earlier implementations.

What would it take to standardize such an approach consistently? The development culture in the Linux community encourages contributors to submit every feature enhancement to the Linux kernel as a series of small increments (a patchset), where each increment is a self-contained minimal code change (called a "patch") that is independently useful and does exactly one thing. This is hard to achieve in practice as the developer attention, deep insight, and expert judgment that must be devoted to ensure an optimal sequence of increments may be viable only in specialized areas such as core system software kernel development. However, the underlying principle has a lot of similarity to the discipline of feature oriented software development, described next.

5.1.2 A formal discipline for labeling feature interactions due to optional features: Insights from FOP (feature oriented programming)

The state of the art in software engineering research offers a rich variety of techniques for identifying, locating, analyzing, annotating, and separating software concerns and features. For example, feature oriented programming (FOP)[316] allows programs to be composed of a set of features.

Feature oriented software development (FOSD) models [316, 34], software product line engineering [310], and aspect oriented programming [216] provide disciplined mechanisms to construct software with explicitly built-in concern assignment information.[1] A specialized variant can be generated for any combination of features allowed by the feature model rules of the program.

A software product line is a family of programs where each program is defined by a unique composition of features.

---

[1] *Concern or feature **assignments*** associate concerns and their properties, i.e., <u>concern intent</u>, with the source code components, methods, or even statements where they are implemented, i.e., <u>concern extent</u>.

Feature interactions in FOSD

Feature oriented programming introduced the notion of *feature interactions* to factor the adaptations that are required to compose two or more features. The feature interaction problem was first explored by telecommunications researchers with a primary focus on runtime manifestation of composition effects that are not apparent when two features are used in isolation. Prehofer showed how feature interactions could be elegantly represented as code extensions (or derivatives of interacting features) which are statically separable from the base program and the increments needed for each feature individually [316]. These extensions or derivatives are called structural feature interactions.

Figure 5.1 shows the FOP version of the example we introduced earlier in Figure 4.2 of a Logged Restorable Buffer from [242, 217]. The component has 3 features: BUFFER (the basic buffer functionality), LOG (logs the buffer operations, get and set, for tracing purposes), and RESTORE (allows the previous version of the buffer to be restored). The program statements have been annotated with colors (white background: BUFFER, dark shade: RESTORE, light shade: LOG) corresponding to the features they were introduced to support, to label concern extents. BUFFER is the base feature (a buffer constructor, set and get methods), while LOG and RESTORE are (composable) derived features that extend the base buffer to support a logged buffer and a restorable buffer, respectively. The feature model for the program expresses these relationships.

For such software, we can identify bloat by using the concern (feature) assignment information and the rules that specify optional features, provided the granularity of features is defined at a sufficiently fine and detailed level. In the above example, the statements labeled as interaction(BUFFER, RESTORE) or interaction(LOG, BUFFER) are the structural interactions due to the optional extensions RESTORE and LOG that could induce execution bloat in programs where only the basic BUFFER feature is needed.

Notice that because these features are built systematically as program increments, FOP tools can be used to automatically generate a specialized subprogram that implements a desired precise subset of features, e.g., a basic buffer (Figure 5.2) or a restorable buffer (Figure 5.3). These specialized variants do not incur structural interaction overheads of any features that are not explicitly selected.

```
1:public class Buffer {
2:       int buf = 0;
3:       int back = 0;
4:       public Buffer(int x) {
5:               buf = x;
6:       }
7:       void logit() {
8:               System.out.println("LOG: buf = " + buf);
9:               System.out.println("LOG: back = " + back);
10:      }
11:      int get() {
12:              int tmp3 = buf;
13:              logit();                    <----- interaction(LOG, BUFFER)
14:              return tmp3;
15:      }
16:      void set(int x) {
17:              int tmp = buf;      <---- interaction(BUFFER, RESTORE)
18:              back = tmp;         <---- interaction(BUFFER, RESTORE)
19:              buf = x;
20:              logit();                    <---- interaction(LOG, BUFFER)
21:      }
22:      void restore() {
23:              int tmp2 = back;  <---- interaction(BUFFER, RESTORE)
24:              logit();                    <---- interaction(LOG, RESTORE)
25:              buf = tmp2;         <---- interaction (BUFFER, RESTORE)
26:      }
27: }
```

FIGURE 5.1: Example: Logged Restorable Buffer from [242, 217].

```
public class Buffer {
      int buf = 0;
      public Buffer(int x) {
              buf = x;
      }
      int get() {
              int tmp3 = buf;
              return tmp3;
      }
      void set(int x) {
              buf = x;
      }
}
```

FIGURE 5.2: Variant that specializes just the Basic buffer from Figure 5.1.

Pros and Cons of using the feature oriented programming approach for resource proportional software:

The key benefit of FOP is the preservation of explicit concern/feature annotations and properties, which allows automated RPD analysis:

```
public class Buffer {
     int buf = 0;
       int back = 0;
     public Buffer(int x) {
            buf = x;
     }
   int get() {
            int tmp3 = buf;
            return tmp3;
   }
   void set(int x) {
              int tmp = buf;      <---- interaction(BUFFER, RESTORE)

              back = tmp;         <---- interaction(BUFFER, RESTORE)
              buf = x;
   }
     void restore() {

             int tmp2 = back; <---- interaction(BUFFER, RESTORE)

             buf = tmp2;         <---- interaction (BUFFER, RESTORE)
   }
}
```

FIGURE 5.3: Variant that specializes a Restorable Buffer from Figure 5.1.

- feature to code mapping annotations help precisely identify the presence of structural interactions between features

- feature models capture optional features and dependencies, which helps determine which of the structural interactions are between features that may not always get used together, and hence need to be minimized

- Typically, it is helpful to have further information to determine which/when features are being exploited. Some automation of this analysis may be possible, e.g., when a non-interacting code extent of a feature is executed, it signifies that the feature is being explicitly invoked.

A practical challenge with adopting FOP is the development effort needed to provide relevant granularity for guiding resource proportional design:

- RPD would need FOP with a very detailed resolution of features/concerns for bloat analysis (including fine-grained diffused concerns) compared to what is typical for product lines.

- High resolution FOP requires too much upfront effort for programmers, hence scaling development is difficult in practice (note that an MDD[2] level FOP approach may be more practical for other FOSD use cases, but the resulting granularity may not be fine enough for bloat analysis).

- Full FOP may also be an overkill for RPD, since not all features matter for bloat and not all structural interactions are relevant (e.g., interactions between features which are almost always enabled together)

---

[2]Model driven development.

or expensive (e.g., some interactions may not occur frequently enough to matter, such as code that is only involved in the configuration/initialization path)

5.1.3   RPD analysis tool: Aid detection of structural interactions using Concern Augmented Program Analysis (CAPA)

Since a majority of software components are not created in this feature oriented fashion, there has been substantial research on semi-automatic techniques to aid concern identification and analysis [335], addressing closely related problems such as feature location [123, 343], aspect mining [212, 253, 46, 350], code search, concept assignment [69, 73], and change impact analysis [35, 153]. For instance, a concern location scheme may follow a query based approach (e.g., a directed search from given seed or pattern [343]) while an aspect mining algorithm may employ a generative approach [253] (e.g., discover structural patterns indicative of cross-cutting concerns).[3]

In practice, the nature of concern information that is easily recoverable may be incomplete or too coarse grained for bloat detection, although possibly good enough for program understanding and maintenance.

For example, suppose the `Buffer` class in Figure 5.1 is not constructed using FOP. Thus feature annotations for the statements (including the interactions) are not supplied to start with and only *coarse-grained concern information may be available about some classes or methods*, such as the `Buffer` class and the `restore` method. As the `Buffer` class can be reused in scenarios where the buffer need not be restorable, support for restore is an optional concern. Under such usage scenarios, `Buffer` is an overgeneral component (class) and restore is an excess concern. However, we noticed in Section 4.3 that:

- the `restore()` method by itself does not induce a runtime execution overhead in this situation as it would not be called.

- the `get()` functionality of the buffer is also unaffected by the excess concern restore.

- the `set()` method does incur the runtime overhead of saving the previous value. The statements `tmp= this.buf; back = tmp;` in this method, therefore, are a potential source of execution bloat in this usage scenario.

- the extra memory allocation and initialization of the back field, i.e., the statement `int back = 0;` is another potential source of bloat due to the optional restore concern.

---

[3]Section 5.1.3 contains significant portions republished with permission of ACM, from [Combining concern input with program analysis for bloat detection. Suparna Bhattacharya, Kanchi Gopinath, and Mangala Gowri Nanda. In Proceedings of the 2013 ACM SIGPLAN international conference on Object oriented programming systems languages and applications (OOPSLA '13). ACM, New York, NY, USA, 745–764. 2013]; permission conveyed through Copyright Clearance Center, Inc.

These potential sources of bloat make this implementation of `Buffer` non-resource proportional. Both of these sources correspond to labeled structural interactions in the FOP annotated version of the component.

An RPD tool developer could combine this coarse grained concern[4] information with program analysis ("Concern Augmented Program Analysis" or CAPA) to systematically automate the analysis of a software component for such candidate structural interactions that are potential sources of non-resource proportionality [65]. This would help component developers modify their design and implementation to avoid introducing these interactions.

The basic intuition behind such a tool is as follows: some statements corresponding to each feature can represent structurally intertwined code between multiple features. Such a statement is a candidate non-resource proportional source that contributes to potential execution bloat when:

- it corresponds to an optional feature, but

- occurs in a method that belongs to an essential feature.

Let us study the key components of a tool that could be used to identify these statements as detailed in [65]. The tool first builds a static analysis-based representation of a component by performing an automatic decomposition according to the following simplified feature refinement model based purely on structural dependencies (e.g., data and control dependence). The analysis expects that:

1. Each method in the component is introduced by a potential feature. (The information about which feature introduced the method is not used in this static analysis stage but is part of separately supplied concern information.)

2. Statements inside each method can represent structurally intertwined code between multiple features (and constitute a resolution of this feature interaction, i.e., the extra code needed to make the features work correctly together).

3. Each potential feature-specific extension either uses an additional input field or method or affects an additional output field.

4. Potential interactions between feature-specific extensions in different methods can occur when a statement in one method uses a field and a statement in another method defines (updates) the field.

The closer a programmer's development approach corresponds to these expectations, the better the chances that the statements identified by the tool

---

[4]An example of a coarse-grained concern mapping could be "the `Buffer` class was introduced to provide the BUFFER feature, and the `restore()` method was introduced to add the RESTORE feature," without finer-grained information about which statements in the class specifically had to be modified to extend it for restorability.

would correspond to candidate feature interactions. Using this, the tool may now perform the following steps:

1. Reverse engineer each potential feature extension based on program dependence assuming that every method could contribute a potential feature (Section 5.1.3.1: Computing microslices).

2. Build a graph that connects the uses and the definitions in the microslices (Section 5.1.3.2: Computing the microslice interaction graph (MSIG)).

3. Enrich the graph with externally supplied concern information (Section 5.1.3.3: Creating a concern augmented microslice interaction Graph).

   - Apply concern analysis information to group methods in terms of actual concern properties (optional or mandatory).

   - Overlay the MSIG with information about the optional or mandatory nature of each feature to generate a Concern Augmented MSIG (the CAMSIG).

4. Traverse the CAMSIG, applying heuristic rules in order to determine which feature extensions (microslices) may be in excess and hence potential sources of bloat (Section 5.1.3.4).

5.1.3.1 Computing microslices

The first step in the analysis is called *microslicing* as it breaks up each method into many fine (micro) slices which represent the smallest incremental units within the method that could possibly be assigned to different features or unique combinations of features. Each microslice can contain a non-contiguous set of statements, which are computed as follows:

Starting with every input field variable in a method, we compute a forward intra-procedural slice[5] and starting with every output field variable we compute a backward intra-procedural slice.[6] These slices are then partitioned into microslices such that the input and output combination for each slice (or rather, $\mu slice$) is unique.

For the advanced reader: A formal definition of microslices and the forward and backward slicing steps involved in computing a microslice are described in detail in [65]. For example, the slicing criteria are obtained using escape analysis, where input fields correspond to escape-in and output fields to escape-out or formalout nodes for a Java method.

---

[5]Set of statements affected by the input field.
[6]Set of statements affecting that output field.

A microslice is uniquely characterized either by the list of statements it contains (program line numbers), or by the enclosing method name, the subset of input fields affecting those statements and the subset of output fields affected by those statements. In the examples that follow (including Figure 5.4), we describe microslices using the following notation which captures both characterizations:

$$\langle$$

method name,

$in = \{$input fields and methods affecting the microslice$\}$,

$out = \{$output fields affected by the microslice$\}$,

$\mu slice = \{$list of statements (line numbers)$\}$

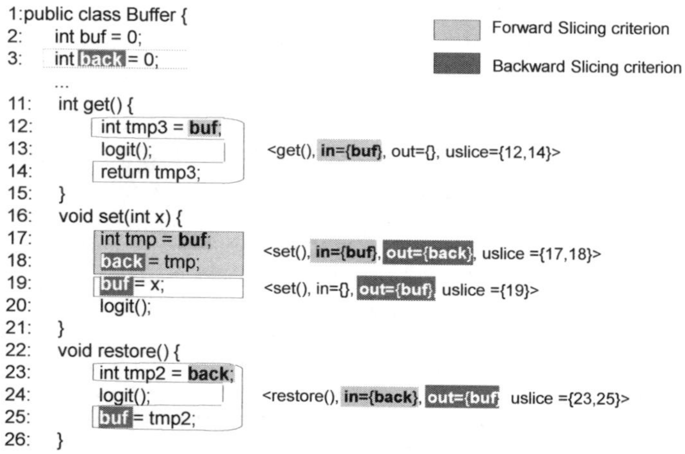

$$\rangle$$

```
1:public class Buffer {
2:    int buf = 0;
3:    int back = 0;
      ...
11:   int get() {
12:       int tmp3 = buf;
13:       logit();            <get(), in={buf}, out={}, uslice={12,14}>
14:       return tmp3;
15:   }
16:   void set(int x) {
17:       int tmp = buf;
18:       back = tmp;          <set(), in={buf}, out={back}, uslice ={17,18}>
19:       buf = x;             <set(), in={}, out={buf} uslice ={19}>
20:       logit();
21:   }
22:   void restore() {
23:       int tmp2 = back;
24:       logit();            <restore(), in={back}, out={buf} uslice ={23,25}>
25:       buf = tmp2;
26:   }
```

▨ Forward Slicing criterion

▨ Backward Slicing criterion

FIGURE 5.4: Micro-slicing to identify potential feature interactions: A few sample microslices computed for Figure 5.1.

EXAMPLE 1.

In Figure 5.1, the `restore()` method contains the microslices:
$\langle restore(), in = \{back\}, out = \{buf\}, \mu slice = \{23, 25\}\rangle$
$\langle restore(), in = \{logit()\}, out = \{\}, \mu slice = \{24\}\rangle$

Similarly, in Figure 5.1, the set() method contains the microslices:
$\langle set(), in = \{buf\}, out = \{back\}, \mu slice = \{17, 18\}\rangle$
$\langle set(), in = \{\}, out = \{buf\}, \mu slice = \{19\}\rangle$
$\langle set(), in = \{logit()\}, out = \{\}, \mu slice = \{20\}\rangle$

Suppose line 19 in the program is replaced by `this.buf = tmp + x;`. Now line 17 affects two heap variables, `buf` and `back`, unlike line 18 which only affects back. Thus lines 17 and 18 would no longer belong to the same microslice. Hence, the microslices in this case would be:

$\langle set(), in = \{buf\}, out = \{buf, back\}, \mu slice = \{17\}\rangle$
$\langle set(), in = \{buf\}, out = \{back\}, \mu slice = \{18\}\rangle$
$\langle set(), in = \{buf\}, out = \{buf\}, \mu slice = \{19\}\rangle$
$\langle set(), in = \{logit()\}, out = \{\}, \mu slice = \{20\}\rangle$

Thus for each unique combination of input and output data, we compute a unique microslice, and each statement belongs to exactly one microslice. Complete details of computing microslices can be found in[65].

5.1.3.2 Computing the microslice interaction graph

Potential dependencies between features in different methods can occur when

- the code for a feature in one method uses a field and

- the code for another feature in a second method defines (updates) the field.

This helps determine the related features.

For each field or object that is shared between methods, we find the methods where the field is used as an input and identify the microslices that are in a forward slice of that input criterion. Likewise, we find the methods where the same field or object is an output and identify the microslices that are in a backward slice affecting that output criterion. Now we create a directed edge from each microslice node in the second set to each microslice node in the first set. This creates a directed graph where a (target node) microslice is a descendant of the microslices it might require (e.g., parent nodes generate state that is stored in fields which might later be used by statements in the child nodes). A leaf node has no outgoing edges. We term this the Microslice Interaction Graph (MSIG).

EXAMPLE 2.

In Figure 5.5, for the microslices represented by:

$\langle set(), in = \{buf\}, out = \{back\}, \mu slice = \{17, 18\}\rangle$
$\langle restore(), in = \{back\}, out = \{buf\}, \mu slice = \{23, 25\}\rangle$

The $\mu slice = \{23, 25\}$ has an incoming edge from the $\mu slice = \{17, 18\}$ labeled "back" and an outgoing edge labeled "buf." The complete MSIG for the example in Figure 5.1 can be found in Figure 5.5 and details of generating the MSIG can be found in [65].

5.1.3.3 Computing the Concern Augmented microslice interaction graph

Having computed the MSIG, we now need to augment this graph with Concern related information to generate what we term the Concern Augmented MSIG (CAMSIG) graph.

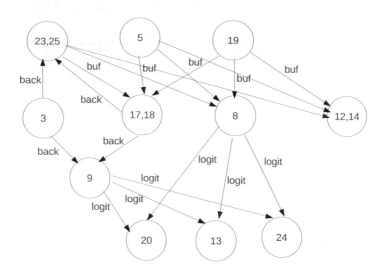

FIGURE 5.5: Microslice Interaction Graph for Figure 5.1. The nodes in the graph represent microslices as a list of line numbers corresponding to the statements comprising the microslice.

This step requires some form of coarse-grained concern information to be available. Programmers or users of the tool should supply concern information to the extent they can, e.g., they may specify a coarse-grained grouping of methods or classes that represent related concerns or features which are likely to be optional in some deployment scenarios. By default, existing modularity indicators such as component package names, classes, and methods are used as-is for initial potential concern grouping.

Let us assume that we are given a concern-based abstraction that partitions the concerns of a component into two groups – mandatory (always required) and optional. The exact mechanism used to realize this grouping may range from a purely manual assignment by an expert to automated classification rules based on an analysis of representative client programs. We also have information about which methods are introduced by these concerns, for example using a concern analysis tool. We perform a Concern Augmented program analysis using this information and the statically computed MSIG to identify candidate excess statements that cause execution bloat when none of the specified optional concerns are required. We apply some heuristic rules. For example, a microslice in the CAMSIG, that is inside a method which corresponds to a mandatory concern, but is used solely by nodes along CAMSIG paths that eventually contain an optional node, is considered to be potential bloat.

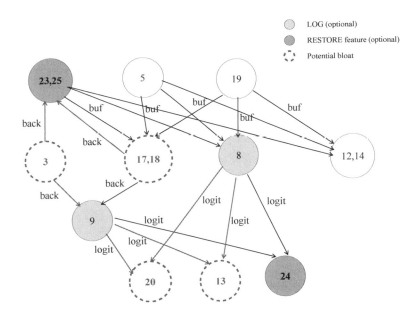

FIGURE 5.6: Concern Augmented Microslice Interaction Graph corresponding to Figure 5.5. The shaded nodes represent microslices in methods belonging to optional features (LOG: light gray and RESTORE: dark gray). The arrows represent potential USE interactions (where the destination node USEs the source node's feature).

EXAMPLE 3.

Consider the $\mu slice = \{17, 18\}$ in Figure 5.6. If we go forward along all outgoing edges, then we always reach an optional node. Thus, this microslice is marked as potential bloat, for the given optional concerns.

The complete Concern Augmented MSIG for the example in Figure 5.1 can be found in Figure 5.6. More details on generating the Concern Augmented MSIG are described in [65].

Note that disentangling execution bloat is usually more involved than merely moving around such bloat contributing statements or creating specialized versions of a function. This tool therefore takes a view similar to most existing techniques for bloat analysis which are intended as an aid for humans; it helps component developers estimate candidate bloat statements and does not attempt to perform automated de-bloating.

5.1.3.4   Putting it together: The CAPA tool

First, an automatic decomposition of each method is performed using a combination of intra-procedural forward and backward slicing to partition the statements into candidate feature-specific statements (microslices) as we illustrated Section 5.1.3.1 above.

A cross-method Microslice Interaction Graph (MSIG) is then built to connect associated statements (interacting microslices) in other methods for each of these candidate feature statements as described in the example in Section 5.1.3.2. The cross-method search space can be restricted through *may use feature* hints from the programmer for better accuracy and speed, especially during static analysis.

After this, the tool uses the programmer-specified concern group information to perform a concern augmented program analysis to annotate potential bloat contributor statements (as illustrated in the example in Section 5.1.3.3).

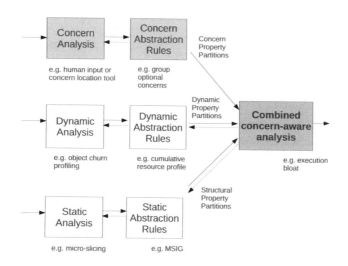

FIGURE 5.7: Concern Augmented Program Analysis.

Figure 5.7 shows an example of a concern augmented program analysis that combines static and dynamic program analysis with concern partition analysis. Here, each different type of analysis (static, dynamic, concern) can implement multiple partitioning functions and analysis functions.

For example, the microslicing analysis can be viewed as a partitioning function that partitions the statements of a program into microslices and an analysis function that computes the structural interactions (i.e., the *may use* and *may be used by* relations) associated with each partition.

The concern analysis separately implements partitioning in terms of concern assignment information (e.g., it groups methods introduced by the same concern). We can then define concern abstraction rules which group concerns such that each group has concerns that must be available together but different groups may be mutually optional. As this kind of information may not be available for all methods, a conservative approximation is applied to a default partition which has all the unassigned statements.

Instead of introducing new programming constructs, existing concern specification constructs and feature modeling tool could be used by the programmer to supply concern information. This information as well as context about which optional features are used in a given deployment scenario should be made available as runtime configuration state for dynamic analysis and optimization. For example, any time the component is invoked as part of a program, a pointer to this runtime configuration state could be maintained as part of the invocation context. For example, the Buffer component in Figure 5.1 may be invoked in two ways by the same program, one where only the basic BUFFER feature is required and another where the RESTORE feature is also required. The invocation context in each case would reflect the corresponding feature configuration state.

In the last block shown in the figure the results of the different types of analyses are combined in a task-specific manner (expressible as a datalog formula over analysis functions and partitioning predicates, for example Figure 5.8) to arrive at an output partitioning, an output analysis function, and its results (aggregated) for each partition. For example, the statements in the microslice partitions associated with structural interactions across methods corresponding to mutually optional groups of concerns (or across a mandatory and an optional concern) are relevant for execution bloat. Thus, a combination of static analysis and concern analysis can be used to find a list of these statements.

$$
\begin{aligned}
optional(s) : &-inmethod(s, m), inconcern(m, c), isoptional(c) \\
mandatory(s) : &-inmethod(s, m), inconcern(m, c), ismandatory(c) \\
potentialbloat(s) : &-mandatory(s), usedby(s, null), mustuse(s, s'), \\
&\quad optional(s') \\
potentialbloat(s) : &-mandatory(s), usedby(s, s'), optional(s')
\end{aligned}
\tag{5.1}
$$

FIGURE 5.8: Rules expressed as a Datalog formula: Each rule in the formula states that the left hand side relation holds for the tuple inside the braces if all the relations on the right hand side hold. Here $s =$ statement, $m =$ method, $c =$ concern.

The output of this analysis information could further be combined with the results of measuring dynamic resource usage profiles (i.e., the output of the dynamic analysis phase in Fig 5.7) to compute an estimate of the execution overhead due to bloat and further narrow down the list to the top bloat contributing statements. In the final step, these annotations are verified and corrected by the component developer. As part of the verification process, the tool could be used to generate a reduced feature subprogram that excludes these annotated statements and test that subprogram.

Availability of concern information in CAPA thus enables annotations that directly focus on unnecessary or incidental execution overhead due to software (feature/concern) bloat. This connects the user level view software bloat in terms of feature bloat to deeper forms of runtime inefficiencies highlighted by previous research on bloat.

5.1.3.5   Example: Big endian to little endian conversion

```
1:        public string applyConvertors(String s, boolean isBigE) {
2-10:       ...
11:           if (isBigE) {
12:               tmp1 = str2int(s);
13:               tmp2 = big2LittleEndian(tmp1);
14:               result = int2str(tmp2);
15:           }
16-30:      ...
30:       }
31:
```

FIGURE 5.9: Big endian to little endian transformation.

Typical causes of bloat such as misuse of reuse, designing without context, and just-in-case programming result in excess code for supporting more capabilities and conditions (concerns) than strictly required. This manifests at runtime as extra checks, bloated transformations, and the creation of heavyweight data structures for simple tasks.

In Figure 5.9 the code to canonicalize data from big endian to little endian (line 11-15, including the cost of parsing and formatting from string to integer and back, intermediate objects generated and checking isBigE) implements a software concern that may be unnecessary in deployment situations when all systems or data sources involved are uniformly big endian. Note that if the method big2LittleEndian() is labeled as an optional concern, the analysis also deduces that not only line 12, but also lines 11, 13, and 14 are potential sources of bloat associated with this optional concern.

5.1.4   Interactions due to hidden features

It may not always be possible to attribute sources of disproportionate resource consumption to specific structural interactions due to explicitly identified features. For example, using a more complex data structure or algorithm that addresses a wide range of conditions can effectively be interpreted as a structurally diffused interaction that cannot necessarily be isolated to particular statements or fields. Further, not all features or concerns are likely to be explicitly stated or apparent from class or method names. This is especially true for non-functional concerns, internal attributes of the architecture, design and implementation constructs, and implicit assumptions about the runtime system, deployment environment, and underlying hardware characteristics. Both diffused interactions due to known features and the interactions arising due to such hidden features need to be optimized when developing a component using RPD.

5.2 Strategy 2: Reduce Recurring Overheads due to Incidental Sources of Bloat

The second strategy can be applied to mitigate the RPD impact of structural interactions that are hard to avoid or even detect, by focusing instead on optimization practices that reduce recurring processing overheads such as intermediate objects constructed in a loop due to the incidental sources of bloat. For example the resource cost of using over general data structures and transformations (e.g., Decimal parsing for a Date object, string copies to pre-fill XML logs for conditional reporting, or using a Java TreeMap where an Array would have sufficed) is more prominent if incurred on every request or data element processed or within heavily nested loops. Adopting performance optimizations such as object reuse, caching, or memoization (as coding practices, compiler transformations, and runtime optimizations) can help reduce such repetitive expenses and thus also improve a component's resource proportionality.

5.2.1 Object reuse and result caching (amortize data construction overheads)

Managed languages such as Java relieve programmers from the responsibility of deciding when to free objects or managing the scope and lifetime of data structures on the assumption that the abandoned objects would be efficiently garbage collected by the underlying runtime. The availability of this "hidden" feature promotes a philosophy of "create and go forth" where, by design, the programming model encourages the creation of many short-lived objects – these are allocated on the heap[7] and may be passed across framework layers, only to be used briefly and eventually garbage collected. For example, temporary objects are generated repeatedly during data transformations required for framework based reuse, thus amplifying structural interaction overheads which introduce excess transformations. As illustrated by Jack Shirazi in his book *Java Performance Tuning* [351], creating too many temporary objects results in higher garbage collection overhead, object construction costs, and higher memory system stress resulting in an increase in processing time and

---

[7]State of the art escape analysis techniques in modern production JVMs can only optimize a small fraction of these allocations [345].

memory consumption.[8] At the end of the chapter on object creation Shirazi recommends a list of performance improvement strategies of which we reproduce a few here:

- Reduce the number of temporary objects being used, especially in loops.

- Avoid creating temporary objects within frequently called methods.

- Reuse objects where possible.

- Empty collection objects before reusing them. (Do not shrink them unless they are very large.)

Implementing these strategies manually gets fairly complicated when objects are generated inside a function that may be deeply nested several framework layers away from closest enclosing loops where they could be reused. Sometimes reuse of objects may be safe only under some conditions or some code paths and not others. Even state of the art escape analysis algorithms in runtime optimizers have been shown to succeed in leveraging only 10% of opportunities for saving temporary object allocations[345]. Hence, a variety of tools have been created by researchers to help developers identify opportunities for reuse using static and dynamic analysis [414, 285, 422, 91].

Notice that higher degrees of reuse offer greater benefits but are also more challenging to automate safely. For example, reusing object contents (and memoizing method results) saves computation costs of object construction (and data transformations)[285] resulting in much higher potential savings than object memory reuse, but the latter is typically safer to apply across a wide set of conditions[66]. Xu also identifies a third form of reuse, referred to as shape reusability[414] which can save some part of the construction cost of compound objects. It is also possible to automate code transformations for reusing certain types of objects, e.g., containers and strings across loops[66] as shown in Figure 5.10.

Most such tools tend to involve costly program analysis and are currently more suited for offline use rather than in runtime optimizers. However, this is not a problem when developing or enhancing components with new features, where static analysis based tools such as [66] are especially helpful. Dynamic analysis based approaches can also be used if there is an executable program or set of test cases that invoke the features of components in various ways. For RPD development, in particular, both types of tools could be enhanced to narrow the analysis state space exploration and automated transformations to focus on code paths impacted by candidate structural interactions whose impact needs to be minimized.

---

[8]Paragraph adapted by permission from Springer Nature: [Bhattacharya S., Nanda M.G., Gopinath K., Gupta M. (2011) Reuse, Recycle to De-bloat Software. In: Mezini M. (eds) ECOOP 2011 – Object-Oriented Programming. ECOOP 2011. Lecture Notes in Computer Science, vol 6813. Springer, Berlin, Heidelberg]

Kth level loop reuse preserves objects across inner loop iterations, by **cycling** through sequence of reuse slots
- Allocate each slot on first use
- **Reset** index at **K-1 loop header**

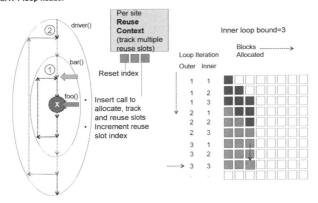

FIGURE 5.10: Code transformation example for object reuse across nested loops. A (thread-local) reuse context area and allocation site-specific reuse methods are generated by the transformation. At each listed allocation site to convert for reuse, the call to **new** is replaced with a reuse method call which performs allocation tracking and reuse. For objects to be reused across a level k loop, at the statement preceding a listed allocation site's level $k - 1$ loop header, code is inserted to reset the reuse slot index for the allocation site and specify the maximum reuse slots.

5.2.2   Adaptive selection and replacement of collection data structures

Another form recurring cost is incurred in maintaining data structures for object collections. The problem of selecting the most efficient collection data structure for a given use case can involve a complex set of trade-offs between execution time complexity and space usage to support the most common operations anticipated, and has received a lot of research attention [137, 421, 415, 112]. For example a LinkedList and an Array address different ends of the spectrum in terms of variability and random insert speed vs. compactness and lookup speed. A Java TreeMap or a HashMap can provide a very high amount of flexibility for key value collections, but both incur high per-entry (data element) overhead, especially in situations where that flexibility is not needed. This extra flexibility (e.g., fast random insertion) can be viewed as a hidden feature which induces a structural interaction overhead that is hard to optimize. The ideal solution would be to design collection data structures that are more resource proportional. See Section 5.4.2 for a discussion on exploring potential approaches to achieve that.

However, when using existing (non-resource proportional) collection data structures during component development, static analysis tools can help detect inefficient collection usage patterns to guide the developer in making the right choices. For example, a tool could highlight underutilized collections which incur the overhead of a full collection allocation to store just a small number of elements or overpopulated collections which contain a large proportion of entries which are never looked up[421]. Further, dynamic analysis tools such as Chameleon[137] and Coco[415] provide offline collection data structure replacement recommendations or adaptively select the most suitable implementation for a given collection and even switch between different collection structures online.

5.3   Strategy 3: Activate or Deactivate High Overhead Features On-demand

For high overhead features which are rarely used, in situations where strategy 1 is very difficult to apply and strategy 2 is not effective, the third strategy is to introduce a means to conditionally switch from a specialized variant of the program without the feature to one where the feature is enabled. The activation and de-activation of the feature typically incurs a cost, which can range from something as simple as turning a configuration flag on(off) to activating aspects in an AOP program to a more involved approach offering reconfigurability between two significantly different variants (in FOP parlance) of the component optimized with and without the high cost feature, respectively.

5.3.1   Insights from AOP (aspect oriented programming)

Aspect oriented programming[216] offers an interesting approach to (statically or dynamically) introduce cross-cutting concerns such as logging, transaction support, security verification, metrics collection, fault tolerance, etc., that typically are intertwined with other features resulting in a large number of structural interactions distributed very widely across a component or program. Aspects are coded and maintained separately from the core component and woven in on-demand as advice that would be applied (before, after, or around) to selected join points in a program (e.g., method calls, field reads/writes, exceptions) that match a pointcut specification.

For example, a sample pointcut specification in Spring Framework is described in Chapter 6 of the Spring Java/J2EE Application Framework Reference documentation[328]:

```
@Pointcut("execution(* transfer(..))")// the pointcut expression
private void anyOldTransfer() {}// the pointcut
```

This specification matches the execution of any method whose name ends with "transfer" and refers to it as the pointcut "anyOldTransfer." Subsequently, one can associate advice with this pointcut, e.g., one can specify a function which is run either before, after, or around method executions that match the specification and thus extend their capabilities.

Desired aspects can be woven in statically at compile time, at load time, or dynamically at runtime (e.g., the Spring Framework supports weaving in aspects at runtime [328]). This provides a mechanism to turn off structural interactions imposed by a costly feature when it is not required.

Maintaining consistency between the core program and the aspects as the component evolves is a challenge with AOP and other dynamic interception approaches to extend a program with additional features. Some recent FOSD tools support a virtual (vs. physical) separation of concerns via disciplined use of standard static code switches using preprocessor directives or runtime configuration parameters, with mechanisms to programmatically manage variants with feature models[209].

5.3.2 Practical considerations (80-20% rule vs. pure RPD)

Enabling runtime activation and de-activation can be costly both in terms of runtime weaving overheads, development effort, and maintainability. Hence this strategy needs to be applied judiciously in RPD, typically only for uncommon features with a high overhead on frequently used features that cannot be mitigated using other strategies. Or consider a component has a few specific variants (combination of features) that are expected to be used very heavily, where variants optimized to support just the common combinations of features consume significantly lower resources than the fully configured set of features. This clearly means that the features in the component are not implemented in a resource proportional fashion. However, introducing the ability to switch between the optimized variants vs. full-configuration on-demand would allow a restricted form of near resource proportional behavior to be exhibited across most expected deployment scenarios. Notice that adaptive replacement of collection data structures online[415] as described in the previous section are specialized examples of conditionally switching between variants of a collection component.

To relieve component developers of the burden of making RPD optimization decisions manually, there is a need to design RPD tools that could be used as part of the development process to apply these strategies systematically, possibly even embedded in IDEs[9] or build tools.

---

[9]Integrated Development Environment.

5.4   Strategy 4: Design Programming Constructs and Runtimes with Resource Proportionality Awareness

Instead of attempting to construct software in a way that avoids all structural interactions that are potential bloat contributors, it might be simpler to focus only on avoiding those interactions that are likely to be expensive in terms of resource (and power) consumption. However, these may not be known or easy to determine at the time of software construction as the resource consumption characteristics depend on runtime parameters.

What if we could construct software in a way that exposes potential sources of bloat and co-design the underlying (hardware and runtime) system in a way that enables it to detect, measure, and possibly even manage the runtime impact of bloat to ensure resource proportionality?

This leads us to our fourth strategy, which could be framed as a 2 step challenge:

- Design programming constructs and tools to expose bloat by making the overheads of flexibility explicit, preferably in a way that enables bloat to be moved out of line if desired

- Design systems to monitor and mitigate bloat automatically for optimizing resource efficiency, preferably using mechanisms that can adapt to changes in deployment scenarios

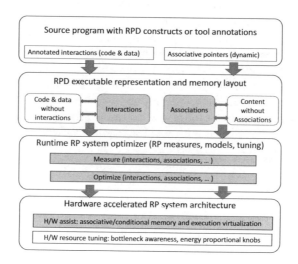

FIGURE 5.11: RP aware runtime stack: programming constructs, annotations passed to the runtime, measurement, and optimization.

Figure 5.11 illustrates how such an RP-aware stack may be built. It shows two examples of RP annotation constructs (annotating code interactions and associative data pointers) and how they may be translated into an executable representation and memory layout of the program, which is accessible to the underlying runtime optimizer. The RP runtime optimizer can use this information to measure the resource amplification impact. These measures could then be used as basis for performing RP-aware deployment and runtime optimizations. In the future, we anticipate that some of these optimizations could leverage new hardware mechanisms designed for RP systems.

5.4.1 Annotating code with line of sight into sources of overheads

If components are developed from scratch using feature oriented programming with a fine-grained feature mapping, then virtual concern separation tools can be used to automatically generate annotations marking structural interactions that could be potential bloat contributors based on dependencies and inter-feature relationships described in the feature model. The annotations can be used by the underlying runtime to profile the highest contributors to non-resource proportionality and also report the interacting features. If any of the features are known to be optional for a given deployment, then the runtime optimizer could even potentially be directed to skip those instructions and fields. A study of the library specialization problem[24] compares various mechanisms available to programmers to express fine-grained features in a manner that allows automatic generation of a low overhead version of a component with a reduced subset of features.

As discussed previously in Section 5.1.2, a fine-grained FOP development approach may be too effort intensive to adopt as a scalable development practice. In these cases, the tooling described in Section 5.1.3 could instead be extended to annotate approximate potential interactions. Note that this analysis is imprecise and hence, while the runtime profile estimates based on the approximate annotations can guide RPD, they cannot be directly used for automatic runtime optimization (such as skipping bloat contributing statements and fields) unless the annotations are verified and corrected by the component developer.

5.4.2 Alternate data and program representations

The above annotations are useful for making the overheads of flexibility explicit, which enables RP profiling and helps guide RPD decisions during component construction. However, the ideal of automatic system optimizations that move bloat out of line is more challenging to achieve. Just as virtual memory paging mechanisms provide a large virtual address space but only utilize physical memory for data that is actually required, is it possible to devise system architectural abstractions that would support fine-grained resource proportional virtualization of code and data bloat at runtime? In other words, what would it take to allow a large potential feature space to be provisioned (including diffused interactions and hidden features) while using physical resources (CPU, memory, network) only for features and data structure flexibility that is actually required?

5.4.2.1   Separating code bloat

When exact annotations of structural interactions due to optional features are available, then these statements and fields could be placed in a different code segment or instruction area and object heap by the compiler. This keeps the potential excess code out of line when it isn't required in a way that allows the runtime to introduce fixup jump statements to fault in the optional interactions if needed using a mechanism similar to that used for exception code segments (for example, the implementation of copy_from_user in the Linux kernel uses this approach.)[10] The faults could be triggered based on runtime configuration state which signifies which optional features are used in a given deployment context.

Another computation model to avoid the cost of needless structural interactions at runtime is to make the interpreter and JIT perform conditional execution of statements based on feature enablement settings in the current execution context. The physical realization of this approach can be accelerated through hardware acceleration support, e.g., mechanism to tag instructions with a set of execution conditions and filter a subset of instructions to execute based on those conditions.

5.4.2.2   Separating data structure bloat

Addressing schemes that enable layouts to clearly separate out actual data and bloat into different memory regions would be attractive as they would enable these areas to be managed differently.

A key challenge that this problem presents, especially for the kind of data structure bloat patterns described in Chapter 2, is that the distinction between actual data (e.g., content stored in collections) and bloat (data structure overheads) is a matter of logical interpretation (unlike holes of unused data in memory), not something that the operating system or underlying architecture can determine and optimize for transparently.

Some common patterns of memory data bloat in real world industrial Java applications described in the literature [264] include:

---

[10] copy_from_user is a Linux kernel function which is used to copy data from a user space buffer into kernel space pages, e.g., during a write() system call. If the user space address is invalid, an exception gets raised, but instead of causing a fault in the kernel, the exception handler is set up to transfer control to a code segment that causes the function to return an error. The statements in this code segment are logically part of the copy_from_user function and hence can directly issue return -EFAULT, but are placed out of line in a separate memory area as they are never invoked under normal conditions.

1. The use of fine-grained modeling resulting in highly delegated data structure designs, where multiple objects are used to represent a single entity, e.g., anecdote 1 in [264] illustrates an example of a single request structure modelled using 34 objects. This incurs a heavy representation cost due to high per object overheads (JVM object headers) and pointer costs of several levels of delegation.

2. Choosing the wrong collection data structure or oversizing collections because of inability to foresee the most suitable option for the actual usage context. For example, using a Java `Treemap <Double, Double>` with 80% memory bloat in a situation where parallel double arrays could have sufficed because all entries are loaded upfront[266] is a typical case of overestimating the level of flexibility demanded by the real usage scenario and a lack of visibility into costs incurred (five pointers per entry plus the overhead of boxing the primitive type double into a Double object). Creating a large number of small or empty collections results in high overheads because of oversized pre-allocation defaults (e.g., 16 entries for an ArrayList) and per-collection infrastructure costs.

3. Data duplication and transformations across frameworks, layers, or products, e.g., Java and non-Java code, databases and object caching frameworks, SOAP/XML, and Java objects [264]. Duplication of data amongst multiple caches and pools separated by insulation abstractions at different layers results in space wastage and complex configuration choices.

4. Inefficient implementations of dynamic types, inclusion of structural information (such as field names) with data leading to bloated representations, and interpretation costs. Anecdote 15 in [264] incurs significant dynamic dispatch overheads even for types that are never expected to change across requests.

Many of these cases and other similar patterns are symptoms of designing without context and optimizing for the uncommon case without awareness of costs incurred. Given that inefficient space usage is closely connected with data structure layout constraints, it is worth exploring whether there are alternate addressing schemes which have a lower propensity for bloat.

A closer look at some of these space inefficient structures indicates one kind of systemic constraint which encourages memory bloat: the emphasis on spatial locality for grouping related data and the pointer cost of indirections for correlation of data which cannot be implicitly related arithmetically by spatial location.

- Arrays and hash maps need to be oversized because extensibility is constrained by need for contiguity of entries or spatial redistribution.

- Object headers are spatially co-located with objects. This can contribute to high per object costs for small objects.

- Delegation costs are heavy due to a combination of pointer and object header overheads (pointer indirections being the price paid for lack of assurance of spatial contiguity).

- Pointer based indexing and linkage overheads contribute to per-entry overheads in collections.

- Need for contiguous space for entries in nested collections much of which is wasted given sparse occupancy of the effective multilevel index space.

- Association of structural information with data through spatial inclusion results in bloated representations.

Memory hierarchy and virtual memory models used in most systems today employ addressing schemes where memory content is accessed by location in physical/virtual address spaces, i.e., data is typically referenced (uniquely) by where it is placed. While virtual addressing abstracts away the actual underlying physical location/address, it only does so in large units of contiguous memory locations, i.e., pages. Thus spatial contiguity cannot be quite abstracted away to the extent required for reducing data structure bloat using these kinds of addressing schemes.

An ideal RPD based solution to the problem of data structure bloat should retain all the advantages of flexibility offered by modeling principles adopted by Java developers while automatically obliviating the source of bloat.

5.4.2.3   New programming construct to represent associative pointers

Associations are needed between keys and values, between objects and their types (header), between contained substructures or related objects (parent and children in a tree/graph, siblings in a list, ordered entries in a sorted array). RPD principles also apply when many of the above associations are maintained just in case, and do not get utilized in practice. Several associations follow very similar rules and hence can be easily compressed (e.g., the type of objects in a collection are usually the same, and could benefit from a bulk representation per collection).

Programming constructs with a special representation for associative pointers would help make these associations explicit to the runtime system allowing separation of actual data content from associative structures and object type headers, and hence enable tools to profile overheads (including data structure health metrics) at runtime.

Repeated association creation in a loop could be optimized using logic similar to the object reuse technique described in Section 5.2.1. If an association is not used in a given deployment scenario or a property of the association (e.g., order) is not explicitly used, then feature code to physically realize the association or the desired property (e.g., sorting) could be skipped or deferred.

Incidentally, the ability to efficiently abstract and recall associations is a key strength of human memory, including the tendency to forget associations that are no longer accessed.

The physical realization of the associations (and their compression) could also be delegated to the runtime system and in the future, it may even utilize new forms of hardware level association virtualization support, such as content addressable virtual memory references[64, 103] and fine-grained copy-on-write to reduce physical data copies across framework layers. Here are some desirable characteristics of an addressing scheme for implementing bloat resilient or RPD friendly structures:

1. Freedom from spatial locality constraints, i.e., location independent addressability at fine granularity (object or field level)

2. Efficient representation of associations between related data, including sparse multimaps (1:1, 1:many, many:1)

3. Compatibility support for co-existence with existing addressing schemes, including the ability to take advantage of spatial locality where suitable

4. Efficient support for ordered relations

5. Simple to manage

6. Low overhead implementation and compressability

5.4.2.4   Research topic: Content addressable data layout for associative structures

An existing alternate addressing scheme which inherently supports some of these desirable characteristics if exploited to its full potential is the model of content addressability, which may be implemented using associative memory structures such as hardware TCAMs.

Content addressable memory (CAM/TCAM) is an associative memory array equipped with a fast dedicated parallel search circuitry implemented in hardware. This enables memory to be accessed by specifying a search key on the memory content instead of specifying the address of a memory location [293]. In Ternary CAMs (TCAMs) entries in the array are stored so that any bit position can optionally be set to "X," a don't care (wildcard) bit instead of a 0 or a 1, to allow compact and flexible data representation schemes. A hardware TCAM can consume an order of magnitude more energy per search than reads in traditional memory.[11] Researchers have proposed a virtual TCAM hierarchy[64] that could allow a more resource proportional approach to leverage physical CAM/TCAM resources.

---

[11]With resistive memory based TCAMs, the energy costs are expected to be lower [172, 324].

The TCAM abstraction is a powerful primitive for representing associative data structures and applications because it has the ability to *simultaneously search through a large number of subspaces of a higher dimensional space in one shot*. For example, each subspace can be compactly represented as one (or a few) TCAM entries using the don't care bits to cover ranges that constitute it. A search key can also contain don't care bits and thus be used to refer to a subspace without requiring the consituent entries to be stored contiguously.

EXAMPLE:

Consider the example (described by Mitchell and Sevitsky [264]) of a Java application which uses a TreeMap to represent a map from *double* to *double*, containing over a million entries. The application does not use the sorted property of the TreeMap until the map is fully populated, yet this structure incurs a high asymptotic memory bloat factor of 80%.

The above computations assume a 32-bit JVM with 12 byte object headers and 8 byte alignment (exact overheads vary with the JVM and the architecture, e.g., 32 bit vs. 64 bit). A detailed examination of the collection representation for $TreeMap < Double, Double >$ under these assumptions indicates [265]:

- a fixed overhead of 48 bytes for the collection header

- a per entry overhead of 40 bytes, comprising five pointers of 4 bytes each for the key, value, left, right, and parent links, 1 byte for a boolean flag, a 12 byte object header, and padding to align the entry at a 8 byte boundary

- object header overhead of 12 bytes each for the *Double* key and *Double* value object mapped by each entry, plus additional 4 bytes of padding to align each to an 8 byte boundary, resulting in an overhead of 16 bytes for each key and each value. The *double* primitive contained in these objects occupy 8 bytes each. Adding the 12 byte header increases the space consumption to 20 bytes each and the 8 byte alignment padding further increases the total space used to 24 bytes each.

According to this analysis, the actual content size for each double key-value pair is 16 bytes (8 bytes each) and the corresponding representation overhead for the mapping, the headers, and alignment padding is 72 bytes $(40 + 16 * 2)$. Thus, the bloat factor of this representation is 82%, ignoring the fixed cost of 48 bytes for the collection header when the number of entries n mapped is very large.

A more suitable alternative would have been to represent the data with parallel *double* arrays, which have less than 2% memory bloat [264]. This, however, assumes that the developer has advance knowledge of the usage pattern and an awareness of the costs, which may not always be the case in framework driven application components.

Using a CAM based map representation can afford more flexibility while saving space. This representation also provides a schema that separates actual data content from overhead [63].

Let us consider a scheme where the *double* is directly used as a key in the CAM entries (instead of a hashcode), and the value is inlined in the associative RAM.

Instead of allocating dedicated CAM space for every map (which would be inefficient for lots of small collections), the map id (collection identifier) is also included in the content key space. The entire collection can be retrieved (unordered) using a ternary search on the map id with key bits set to don't care. Further it is possible to associate a bulk type for all objects in the collection. This will require an additional type bit to be included in the content word, so that a single ternary entry with the map-id, with key bits set to don't care, can be associated with the default type for all objects in the collection. For large collections, the map id could be made implicit (by splitting CAM space), or the map id could be stored only in each entry loaded in a physical TCAM and compressed when paged out.

Ordering is supported by implementing an intermediate content ordered memory area which is used to cache sorted groups (ranges) of CAM words as needed. Such caches may in turn be CAM indexed by map id and key ranges. We call these content ordered memory (COM) caches.

The net space savings may be calculated as follows and is summarized in Table 5.1 (where s = size of the content ordered cache).

Space Overheads Computation:

Consider a CAM mapping schema of the form: $\langle Mapid, double \rangle [double]$, where:

n = number of elements in the collection

D' = number of bytes consumed by the actual content (i.e., the double keys and values stored in the map)

f = fraction of keys cached in the content ordered memory (COM) cache $(0 < f < 1)$

J' = number of bytes consumed by the cumulative overhead incurred when using the CAM based representation. If the MapId is included in every CAM entry and takes up 4 bytes, the overhead is $4n$ + size of an additional CAM entry for bulk typing + size of the sorting cache (COM)

S' = $1 + \frac{J'}{D'}$ = resource amplification (scaling) factor in memory consumption due to the overheads (as defined in [266])

With this CAM based representation:[12]

$D' = 16n$

$J' = 12 + 4n + 8fn$

$S' = 1.25 + \frac{f}{2} + \frac{0.75}{n}$

With the original data representation,[13] the cumulative overhead J and scaling formula S were:

$D = 16n$

$J = 48 + 40n + 32n$

$S = 5.5 + \frac{3}{n}$

Relative improvement achieved using the CAM representation:

$\frac{J'}{J} = \frac{4+8f}{72}$ (for large n)

$\frac{1}{18} < \frac{J'}{J} < \frac{1}{6}$

$2.6 < \frac{S}{S'} < 4.4$

Thus, the CAM based representation lowers the overheads to less than one-sixth that of the original representation and improves the space usage by 2.6 to 4.4 times (i.e., it requires about one fourth of the space occupied by the original representation).

TABLE 5.1:   Example: Space overheads vs. savings of using a CAM based representation of a Java TreeMap (from *double* to *double*) with 1 million entries [63]. Each line in the table corresponds to a specific optimization scheme addressing one aspect of the data representation cost in the original layout.

Scheme	Overhead	Saving
Key-Val Map (Ordered)	4MB + 8s	20MB
Bulk Type	0.13MB	24MB

Table 5.1 illustrates different categories of savings achieved for a TreeMap with 1 million entries, excluding alignment padding. For example using CAM based key-value mapping saves the pointer costs in each entry (5 pointers of 4 bytes each), while adding the overhead of a mapid (4 bytes) embedded in each CAM entry. Bulk typing saves the per entry object header costs (12 bytes each of object header overhead for key and value, respectively, or 24 bytes per entry), while adding an additional type bit to each CAM entry (or 1M bits in

---

[12]The size of a CAM entry is 12 bytes, 4 bytes for the mapid and 8 bytes for the key (actual content); one extra CAM entry is needed to represent the bulk type causing a fixed overhead of 12 bytes; the size of the sorting cache is f * 8n, as 8 bytes are needed for each value and f is the fraction of values cached.

[13]Each entry maps a double 8 bytes key to a double 8 bytes value, i.e., $D = 8 * 2 = 16$ bytes of content and $J =$ fixed collection overhead of 48 bytes + n * per entry overhead of 40 bytes + n * 2 * object overhead of 16 bytes per key and per value $= 48 + 40n + 2 * 16n$.

total, i.e., 131 KB) [63]. Ordering using a content ordered cache adds overhead that is proportional to the number of cache entries, s multiplied by the space consumed by each cache entry which is an 8 bytes double value. In summary, with just these three optimizations:

Net savings $= 20 + 24 - 4 - 0.13 - 8s >= 31MB$

Actual data $= 16MB$, Overhead $<= 13MB$ (between $4 - 13MB$)

i.e., percentage bloat between $20\% - 45\%$

We note a significant space-time trade-off in the chosen size of the content ordered cache. What makes this interesting is that it is an optimization decision that can be taken at the system or runtime level (instead of application level). If we choose s such that bloat is 30%, then the overhead is 7MB, less than one-sixth of that with the original layout.

Further, if a particular property (e.g., ordering) is not used, then the corresponding overhead (e.g., COM cache) can be easily turned off by the runtime. Alternatively if greater flexibility is required, e.g., to allow some objects of a different type to be used as values, then additional type entries for the corresponding keys may be added.

One caveat in all of the above comparisons is that a TCAM is more costly and power hungry than regular memory, and hence a virtual TCAM hierarchy [64] is required. Also, some ternary values (e.g., don't care bits used for bulk typing) require more than 1 physical bit of regular memory when stored in any binary storage device such as DRAM or flash or hard disk.

5.5 Summary

Figure 5.12 summarizes the sequence of RPD strategies, techniques, and tools component developers could systematically use to construct a new RP component or add RP features to an existing component.

When components are being developed from scratch, it is most effective to introduce resource proportionality bottom up especially when developing low level base functionality. This can be achieved by following a minimalist incremental development practice that preferably avoids or at least explicitly exposes feature interactions that could cause bloat (Strategy 1). The use of disciplined programming models such as feature oriented programming (FOP) at this stage is very beneficial. However, consistently maintaining fine-grained feature labels could also get tedious to sustain and scale up as a development approach, which would hurt productivity. Instead, the processes can be aided by building an RPD analysis tool that combines coarse-grained concern (feature) information with fine-grained static analysis to automatically highlight candidate feature interactions that could be potential bloat contributors. Interactions that cannot be avoided are exposed as inputs for Strategy 2 as well as an annotation source for Strategy 4.

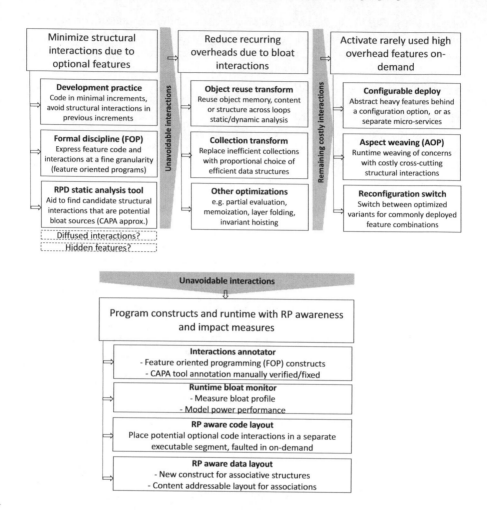

FIGURE 5.12: RPD workflow for component developers (applicable during component construction or new feature addition).

Strategy 2 focuses on mitigating the impact of these potential bloat interactions through well-known optimization practices or code transformations (e.g., object reuse, memoization), amortizing their recurring overheads when executed repeatedly in loops or when used in collection data structures (affecting many elements or instances). Such optimizations can be guided by static/dynamic analysis tools and some may be applied online at runtime (e.g., online replacement of collection data structures). Remaining interaction overheads that cannot be reduced using these techniques are exposed as inputs for Strategy 3, along with the list of the contributing optional features. Notice that measuring the overhead of specific interactions requires some form of dynamic analysis or profiling (subsequently generalized as the first step in Strategy 4) using a representative test program.

Strategy 3 can be viewed as a fallback option that attempts to isolate features that are rarely exploited in practice but impose a heavy overhead even when not in use (i.e., if they induce heavy structural interactions which are hard to avoid or mitigate). Wrapping these interactions within a runtime (feature) configuration option check is probably the simplest example. Aspect oriented programming can be a more effective approach as it completely separates the aspect code from the main component and only weaves it in at runtime when needed. Or if the interaction is highly localized (rather than cross-cutting), the feature could even be moved to a separately dynamically loadable component or microservice. Note that turning the feature on (or off) may be expensive because of the extra indirections for reconfigurability (see graph in Chapter 4); thus this approach is suitable only for features that aren't used most of the time or rarely switched on a given deployment scenario.

Strategy 4 is a step towards a recommended long term direction: creating programming constructs and runtime system infrastructure with resource proportionality awareness built-in. One approach is to generate code annotations that provide a runtime system ability to distinguish, measure, and manage potential bloat sources differently from core function. The annotations may be provided through FOP constructs or a list of potential bloat interactions identified by the CAPA tool. As the CAPA tool annotations are approximate, manual verification and corrections may need to be applied by the component developer. To reduce the burden on developers, this step could be performed iteratively. An initial approximate measure of the potential bloat profile is obtained using the unverified list of annotations. From this initial profile, only those annotations with significant power performance impact are selected for verification and retained thereafter.

The runtime system can now monitor the bloat profile by measuring various performance metrics and report the impact of these annotated statements whenever the component is deployed by an application. Further, power-performance modeling approaches discussed in Chapter 3 can be used to study/predict impact under different hardware resource bottleneck conditions and system power management strategies.

Based on these measures, system level optimizations can be developed to mitigate the impact. For example, a RP executable code layout maintains *bloat contributing code* out-of-line in a separate segment (while still leaving it intertwined in the source code for logical coherence). Such code can be faulted in on-demand as an exception when the caller of the component actually needs the optional feature that introduced the corresponding feature interaction.

Separating data structure bloat gets challenging when representation choices involve both hidden (unarticulated) features and diffused interactions. A special programming construct for associative pointers offers sufficient expressiveness to cover a wide range of scenarios while exposing more opportunity for runtime optimization and monitoring. For example, content addressable hardware acceleration mechanisms for fine-grained paging, compaction, and copy-on-write could be used to represent and manage the associative structures.

Chapter 6

Resource Proportional Design Strategies II: Refactoring Existing Software for Improved Resource Proportionality

In this chapter we describe some practical RP optimization strategies that could be applied to an existing application built using components that weren't originally designed to be resource proportional. Chapters 7 and 10 cover broad RPD strategies for optimization of deeply layered stacks including RP architecture implications.

Here are a few reasons to improve RP of existing software:

1. To increase performance, reduce resource utilization, or save power in typical deployment scenarios (where rich feature sets and just-in-case abstractions are likely to be underutilized)

2. To future proof software with respect to technology changes (e.g., the transition from disk to flash to non-volatile memory (NVM) described

121

in Chapter 8) or deployment environment changes (reorganized stack layers, homogeneous vs. heterogeneous systems) affecting underlying assumptions which might make some features unnecessary (e.g., caching, queuing, marshalling, serialization, validation, error recovery)

3. To reduce the surface area of bugs and security leaks by restricting incidental interactions beyond features that are explicitly in-use

Given that fine-grained interactions matter the most for RPD, it can be very challenging and effort intensive to refactor large complex unannotated applications which are likely to contain numerous such interactions. A top down impact driven approach is likely to be more effective in this case; hence we now adopt a workflow sequence that almost reverses the order we recommended in Chapter 5 for constructing RP components from the ground up.

6.1   Strategy 1: Whole System Impact Analysis to Identify Candidate Resources and Indicators of Bloat that Are Likely to Matter the Most

The first step in an impact driven RP optimization strategy requires some form of an overall "what if" impact analysis. The focus of the analysis depends on the key objective.

For example, if the goal is to achieve power-performance improvement then a study of application's utilization of system resources, bottlenecks, and power consumption (per Chapter 3) helps determine the type of RP optimizations that would be most effective.

If the intent is to future proof for changes in storage technology or deployment environment conditions and layers, then analyzing implementation specific concerns such as caching or validation logic (which could potentially involve hidden features or diffused interactions) and their relative cost under different system parameters helps determine which areas in the code deserve the most attention. Several examples of these trade-offs are worked out in Chapters 8 and 9.

If the main interest is to reduce the surface area of code exposed, analyzing the percentage of unutilized code (program instructions or methods executed) provides a measure of code size bloat.

6.1.1   Resource utilization, bottleneck analysis and power, performance models

For RP improvements to translate into improved system power-performance, standard wisdom and principles of performance analysis, tuning, and optimization continue to apply. Literature on the subject of performance engineering, energy optimization, performance understanding, diagnosis, and profiling methodologies is fairly vast and outside the primary focus of this book. A number of tools have been proposed to aid the process in various ways, such as characterizing application behavior by profiling performance along different dimensions, highlighting regions of code or data that have high resource consumption, identifying candidate hotspots, detecting a variety of performance antipatterns, analyzing signs of disproportionate activity, recommending potential solutions, code transformations, and prioritizing optimization opportunities.

As a basic step, it is helpful to collect system resource utilization metrics and identify the most prominent bottleneck resources impacting the performance (and power/cost) limits of the application.

- Processor (CPU) utilization

- Memory size (Java heap and non-heap)

- Memory bandwidth (reads and writes)

- IO bandwidth (reads and writes)

- IOPs (read and write requests)

- Network bandwidth (bytes sent and received)

- Network operations (requests sent and received)

If there is a single dominant bottleneck, focusing on RP improvements that de-bloat that resource first is likely to be most effective. If the application has a balanced utilization across multiple resources, then the bottleneck could shift on de-bloating; hence it is advisable to focus on all the resources close to the bottleneck. In addition, RP improvements should also be targeted at the most power-hungry and most costly resources. The *what-if analysis* model discussed in Chapter 3 helps study trade-offs, especially when used in conjunction with tools such as Mantis[232], SPEED[168], path frequency estimators[89] that derive performance models, or resource consumption models for existing programs (using static/dynamic/statistical analysis) [232, 168, 89]. For example, Mantis[232] builds a machine learning performance model along with a program slice to compute the value of the features of the model (program variables that determine performance) for given inputs. A combination of these models can be used to assess how the resource utilization reduced through de-bloating and the corresponding power-performance savings varies with program input conditions and system characteristics.

6.1.2   Measure indicators of bloat

For the resources which matter, an assessment of potential excesses or redundancies may be obtained using runtime state analysis (tools) to measure bell weathers of non-resource proportional execution characteristics. For example:

- **Data structure health** measures derived from heap snapshots [263] to assess memory inefficiencies in data structures, e.g., a high overhead compared to the actual content stored is a sign of bloat that matters when memory space utilization is important

- **Temporary object generation** measures collected with simple tools (GC volume of temporary objects generated by different methods) and deeper analysis with advanced tools (e.g., object reuse and lifetime analysis [288, 416]) if the volume is significant, as it matters in situations where memory write bandwidth utilization could become a potential bottleneck

- **Copy profiling** derived using (heavyweight) abstract dynamic slicing tools[417, 419] (not intended for production settings) to identify overheads arising from a high proportion of object copies or copy chains, which could be addressed to improve RP in both computation and object creation costs

- **Low utility data structures** derived using (heavyweight) abstract dynamic slicing tools[418, 419] (not intended for production settings) to locate objects with disproportionately expensive construction costs compared to usage (utility), which could be refactored for improving RP in computation

- **Cacheable data and memoization** metrics [414] to detect opportunities for saving recurring computation costs through reuse of object content or memoization [119] (result caching) of method calls, which can indirectly improve RP in computation by mitigating repeated overheads

- **Computation health** measures derived using rule based or statistical pattern mining of stack traces, profiles, source code, or execution logs (e.g. using frequent sequence mining, NLP[1], or topic models) to classify computation expenses attributable to different types of operations or concerns, such as proportion of computation spent on ancillary concerns. These might include serialization/de-serialization, transactions, and caching. Computation expenses may be classified using tools such as WAIT[30], StackMine[176], and Anti-pattern mining [300, 399].

---

[1] Natural language processing.

- **Code coverage and utilization** profiles in terms of proportion of total methods and instructions executed in each component to detect unutilized code. For example, according to [200] about 40-80% of methods in library classes are typically used

- **IO amplification**, e.g., vs. in-memory data content, including representation overhead, duplicate bytes due to copies, and transformations persisted or transmitted

- **Protocol health**, e.g., bytes expended in protocol overhead vs. data content exchanged

The output from some of these tools indicates refactoring opportunities which could be further prioritized using performance optimization advisors such as Speedoo[102].

6.2 Strategy 2: Replacing Entire Components or Features with a Different Implementation

Interpreting detailed results from the above tools and incorporating refactoring suggestions in a way that best fits the semantic intent requires considerable effort and domain expertise given the complexity of the code base. This is especially burdensome for those who aren't the primary developers or maintainers of the original application code and components. A simpler starting point in such cases is to replace or re-design components which have a non-resource proportional propensity, e.g., expensive message protocols or expensive serialization-deserialization logic.

6.2.1 Example: Serialization-Deserialization

For example, in Apache Spark, studies have shown significant improvement from replacing the Java serializer with the faster and more compact Kyro serializer (with pre-registered classes). Even further gains have been demonstrated using a JVM based special purpose Skyway[282] method for moving data between heaps on different nodes. In general, Google protocol buffers and Apache Thrift provide greater effciency than XML or JSON based serialization. The Simple Binary Encoding (SBE)[381] codec has been specially designed for high frequency trading environments which require order of magnitude lower latency. SBE has been reported to have 25X higher throughput and predictable low latency compared to protocol buffers (e.g., 25ns vs. 1000ns for a typical market message) using a combination of specific restrictions that help it adhere to low latency design principles such as native-type mapping, copy-free, allocation-free, and word-aligned streaming access [318].

The choice of a specific serializer tends to involve giving up some (typically unnecessary) flexibility for efficiency. When the anticipated efficiency gains are substantial, it may be worth adopting protocol customization approaches to build a RP serialization/deserialization component. Such a component could enable features to be dynamically selected/specialized using strategies in Chapter 5 to preserve maximal scope for flexibility without an excessive tax. Google flatbuffers [317] is an example of a serialization approach that preserves flexibility and efficiency, and provides RP access to specific fields without having to deserialize an entire large object.

6.2.2   Example: Collection replacement

Another example of low level component selection to improve RP of existing software is the replacement of collection data structures with more efficient choices recommended by static/dynamic/runtime tools (or even automatically applied by tools such as Coco[415]), as discussed in Chapter 5.

6.2.3   Example: Trimming unused code (bloatware mitigation)

A different form of component replacement is automatic generation of a customized JRE which excludes code (methods or instructions) that would never be executed by applications running in a given deployment environment. JRed[200] uses static analysis to generate customized application bytecode that has been shown to reduce the code size of Java application code by 44% and JRE core rt.jar by 82% on average to reduce the attack surface for security vulnerabilities by about 50%. The unused methods also provide cues about unutilized features and could potentially also be used to further de-bloat the component of structural interactions they impose on other methods (as discussed in Section 6.4.1). Instead of removing the extra code entirely as JRed does, a resource proportional transformation would place the unused methods in a separate protected section where it can only be activated explicitly through a trusted mechanism.

6.3   Strategy 3: Reduce Recurring Overheads Due to Incidental Sources of Bloat

6.3.1   Object reuse and memoization

Section 5.2.1 described several techniques to amortize recurring computations and object generation such as object reuse, memoization, and caching. Such a strategy is particularly suitable when dealing with a large existing code base as these optimizations help mitigate the impact of bloat without requiring feature annotations or awareness of structural interactions. Further many of these techniques can be applied automatically[66, 288] or semi-automatically, wherever object profiling or static and dynamic analysis tools such as Cachetor[285] a MemoizeIt[119] indicate an opportunity.

As illustrated by the experiments in Chapter 3 object memory reuse can have a significant power-performance impact when the rate of temporary objects generation creates a memory write bandwidth bottleneck or heap pressure when memory size is a constraint. Reusing object content or object shape reuse and memoization also saves computation (Section 5.2.1).

6.4   Strategy 4: Refactor Code to Minimize Structural Interactions

The CAPA techniques described in Section 5.1.3 aid the detection of candidate structural interactions in existing software. However, it is advisable to employ this strategy selectively when working with pre-existing code as accuracy may be more limited in this case. The first challenge is that features in the components may not be implemented according to the microslicing model assumed. The second challenge is how to perform the concern partitioning step that identifies methods belonging to optional concerns in the absence of any concern annotations (even coarse-grained ones) or prior information about which features are optional. Both steps require extra effort and domain knowledge.[2]

Once the structural interactions are identified, the refactoring approach used to minimize these interactions could include a combination of optimization practices previously described in Chapter 5 such as (a) code specialization, (b) design patterns (embedded anchor, no-mid-layer, FOP style minimal increments), (c) copy on write structures, and (d) runtime concern separation mechanisms (e.g., control-plane splitting, configuration variables, feature gating[101], aspect weaving).

As the burden of manual verification and refactoring can still be fairly significant, we use some of the analysis aids discussed earlier to guide CAPA attention.

6.4.1   Using optional feature indicators

- Unutilized methods detected by tools such as JRed[200] could be treated as seeds for optional features

---

[2]Especially as the original developers of the application and framework components are typically not involved at this stage.

- Low utility data structures, expensive copy chains, and methods with high copy profiles as detected by abstract dynamic slicing tools described in [419] are likely to contain interactions with features which are optional in the context of usage. Common examples are a Date parser that uses decimal formatting methods or an entire Gregorian calendar generated just for computing the current time in milliseconds.

6.4.2   Using concern analysis tools

Semi-automatic techniques that aid concern identification[335] (such as *feature location* [123, 343], *aspect mining* [212, 253, 46, 350], code search, *concept assignment* [69, 73], and *change impact analysis* [35, 153]) may be helpful in finding methods which are invoked incidentally but likely to be optional. These approaches may exploit additional artifacts beyond program source and executable traces, often combining traditional program analysis with information retrieval, natural language processing, formal concept analysis, and some form of human input [434, 331, 132, 350, 430].

For example, generative approach tools such as FINT (Fan-In Tool) can identify candidate cross-cutting concerns (aspects) such as Undo and Persistence in JHotDraw or Lifecycle in Tomcat by analyzing metrics such as high fan-in methods [253]. Statistical topic models such as Latent Dirichlet Allocation (LDA) can also be used to discover potential source code concerns. The $FLAT^3$ (Feature Location and Textual Tracing Tool) [343] can be used to locate code that implements a feature such as "file saving" using a combination of two approaches, (a) a text document index using Apache Lucene to search for classes, methods, and fields containing keywords "file saving" and (b) a dynamic trace of the methods that are executed when the file save operation is performed.

6.5   Summary

The implementation style present in much existing framework based software reflects decisions made by programmers who have grown up with the oft mis-interpreted[3] adage that premature optimization is the root of all evil [222]. However, as the reader may have surmised from the difficulties examined in

---

[3]From the "The Fallacy of Premature Optimization" [189] "Software engineers use the Pareto Principle (also known as the 80/20 rule) to delay concern about software performance, mistakenly believing that performance problems will be easy to solve at the end of the software development cycle. This belief ignores the fact that the 20 percent of the code that takes 80 percent of the execution time is probably spread throughout the source code and is not easy to surgically modify."

this chapter, code needs to be constructed to be RP optimizable even if we want to postpone optimization decisions (late binding is fine but loss of context is not). In other words, RP awareness must be built from ground up; otherwise it can be very hard and messy to fix later.

When we must refactor existing software originally built by other developers, it is more practical to improve resource proportionality top down based on an assessment of impact. Figure 6.1 summarizes the sequence of strategies, techniques, and tools described in this chapter to aid refactoring of an existing component or application to improve its RP.

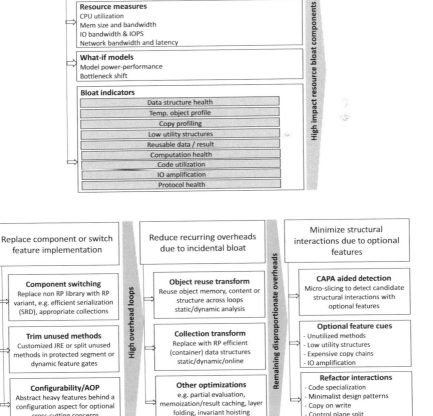

FIGURE 6.1: RP refactoring workflow for existing software.

We start with a whole system impact analysis and bloat profiling tools to identify which resources and opportunities to focus on. Once high impact resources and bloat indicators affecting those resources are determined (e.g., data structure health for memory usage, computation health for CPU usage, protocol health for network usage, code utilization for security exposure), the first strategy is to find efficient RP alternatives to replace high resource bloat (non-RP) components, such as a serialization-deserialization library. If a single alternative does not offer sufficient variability, a configurable mechanism to switch between an ensemble of alternative components may also help effective RP. One such application of this approach is covered later in 12.2.2 where we examine how to design RP allocation and caching for a radically non-uniform memory system.

Replacing entire components may achieve a few big wins, but a lot of opportunities are likely to remain throughout the code base. Bloat indicators such as temporary object profile and reusable data metrics guide us to the second strategy, which focuses on high overhead loops (e.g., loops which generate a high volume of temporary objects, leading to memory bandwidth bottlenecks or repeated computation) where recurring costs can be amortized through reuse and memoization, often using automated or semi-automated code transformation.

Any remaining disproportionate overheads seen at this point using dynamic analyses such as copy profiling, low utility data structures, and IO amplification would guide the third strategy, which is to detect and refactor structural interactions from optional features that potentially lead to these overheads. The detection, verification, and refactoring steps may require significant manual effort and domain expertise, but can be aided by the focused use of CAPA techniques described in Section 5.1.3 together with optional feature method detection cues based on ranked profiling results from the above analyses and the use of state-of-the-art concern analysis tools.

Chapter 7

Implications of a Resource Proportional Design

Up to a point, it is better to let the snags [bugs? bloat?] *be there than to spend such time in design that there are none (how many decades would this course take?)*
A.M. Turing, Proposals for ACE (1945)

7.1 Introduction

In previous chapters, we have discussed software bloat in framework based languages as well as bloat due to technological changes. Summarizing, a system is non-resource proportional when a runtime overhead is induced by features provisioned but not utilized. It is also non-resource proportional due to request load variations (for example, the number of requests, transactions, or users) if resource consumption does not decrease when load is reduced or entirely eliminated (due to, for example, static overheads).

Intuitively, a RP design expends resources proportional to the activity in the system, the proportionality factor being technology dependent. If the system has adaptive properties or self-reorganizing properties, some components of the system track the changes in the system and a RPD for that part would then expend resources proportional to the changes. Similarly, if there are stochastic aspects, a good design would expend resources proportional to the effort needed to handle the variations along with the noise in the system. A RPD systems design, therefore, can be a multi-level design with a RPD for the parts that are not changing ("structural") and a RPD for parts that track changes and that handle stochasticity or noise ("information theoretic") in the system. Relevant features of the system can be at each of these levels; this can be in addition to the control and data plane levels.

In a systems context, features necessary in a deployment may or may not be utilized as some of them may be

- duplicated across layers, e.g.,

  - double logging or journaling at filesystem, device, and application level

  - security at multiple levels in a network stack (L2, L3, or application levels), or security mechanisms at OS, hypervisor, VM levels, or at JVM, OS, hypervisor, VM levels

  - resource virtualization at OS, hypervisor, and runtime levels

  These are usually treated as separate features but may remain unutilized, underutilized, or redundant in one or more of the layers.

- conditionally utilized, say, on a failure (e.g., due to recovery, or for availability and consistency) and hence necessary, but incur a cost if the failure condition does not arise.

Hence, with layered systems, non-RP aspects are possible (see also Section 10.1.1.1). In general, many of the software patterns discussed as design elements for reusable software in [149], while very useful in constructing large scale software, need examination in terms of RP behavior.

If layering is not present as a structural part of the design, we have the issue of how to activate the subsystems ("brokering") (Section 7.1.1) and its RP aspect. We also need a simple model of when a systems design is RP (Section 7.1.2) and also mechanisms by which a non-RP design can be steered towards a RP design (Section 7.1.3). The first part concerns the design of the system at a high level, the second one captures the core of what RP is and what issues are relevant, the third concerns the dynamics of a RP design. We discuss them in turn.

7.1.1 Resource proportionality and high level design: Internal vs. external brokering

If a design has to be resource proportional, then a complete design may be a union of multiple designs for the various configuration spaces brokered internally at runtime, i.e., all the various designs exist independently but are bundled into a unit (possibly with some sharing just as in shared libraries) and each "optimal" design activated in sequence as demanded by the current context.[1] Or it could be brokered externally, i.e., various designs exist separately without any relation to each other and any one of them activated based on current context.

If internally brokered, we may have a bloated system due to its inherent structure and hence not likely to be resource proportional. A good example is the bloat often present in libraries.

If however the system is externally brokered, we need an external manager or scripting mechanism to select the next component for execution. Cooperative middleware is needed; historically, CORBA and its descendants (such as COM, ORB) are the manifestations[181]. Due to the generality or unsuitability of these generic services, many solutions have instead been proposed with similar features but tailored to a specific domain; a good discussion for the control systems domain is [164]. The most recent incarnation seems to be the microservices architecture[31] or the related serverless models.[2] Microservices are better regarded as services developed and maintained by "micro" teams; hence a polyglot culture (use of multiple languages) is often present. However, newer models or patterns may become necessary[332]:

- "circuit breakers": prevent a network or service failure from cascading to other services by calling a fallback method if failures cross a threshold

- "sagas": instead of using ACID semantics that can be unworkable in a distributed context, sagas maintain data consistency across services using carefully written compensating transactions on failures of transactions and an eventually consistent model[151].

However, there is no insight into all of the design space available with externally brokered models if a microservices-like model is used across organizational boundaries. Hence a competitive market place model (with models incorporating brokers, auctions, and principles such as "survival of the fittest") may be necessary. At a broader level, examples of such market mechanisms are

---

[1]These correspond to patterns such as facade or proxy design patterns[149].

[2]Microservices tend to be smaller than serverless functions with lightweight protocols such as REST, JSON, AMQP (messaging protocol), and SOAP. The modules are part of the binary that gets instantiated; it keeps running on the compute platform even if it is in an idle state. There is often a registry where all services are registered (SOA). A distinction often made is that, in serverless computing, instantiation of the computation happens only when triggered (payment only on actuals) unlike in microservices where the modules are always running.

discussed in [120] for handling vulnerabilities in software, in [301] for selecting cloud systems, and in [323] for an insurance model for archival storage.

The serverless model introduces newer issues such as the "serverless trilemma"[48] at a theoretical level. For example, there is a possibility of double billing when a function f_c is realized as a composition of functions f_a and f_b as each of them can be independently scheduled and costed while observing the dependencies.[3] Similarly, asynchronous composition does not satisfy the substitution principle as the (immediate) return value of the asynchronous function cannot be used for the actual function value. Furthermore, the composition has to be orchestrated through external brokering without the mediation of client that initiated the composition; otherwise, the substitution principle is again violated.

7.1.2   A simple model for systems resource proportionality

For the purposes of a simple quantitative discussion in this section on what constitutes a RPD, we use a simpler notation than the fully developed notation in Chapter 4 assuming that features (at whatever level) are available to predict the system using these features.

With respect to feature variations, consider a system with feature space F and inputs X (a vector) computing an output Y and consuming resources R1, R2,.... Note that some of the input space can be treated as part of the feature space discretized so that a design can be specific to that input space.

Simplifying by assuming only one resource R and specializing the design D1 and D2 for features $f1$ and $f2$, let resources consumed be $r10$ and $r02$ under designs D1 and D2 for an input i. If we do not have a specialized design for features $f1$ and $f2$ but have a composite design for both $f1$ and $f2$, let the resource consumed be $r12$.

If across the chosen input space, the ratio $r12/r10$ and $r12/r02$ is $<= k$, we have a k-RPD for that input space. Otherwise, we call it a non-k-RPD or even a bloated design. Note that one can extend this to more than 2 features.

Often in practice, there is a fast path for, say, $f1$ and a slow path for the other. If feature $f1$, say, is encountered very often in practice, the ratio $r12/r02$ can be high while ensuring that $r12/r10$ is close to 1 in a good design.

If a and b are the proportions of the only two features encountered, $a+b = 1$, then if $r12 <= k(a*r10 + b*r02)$, we have k-RPD. If $k <= 1$, then there are mutual benefits arising from the two features together. If $k > 1$, then the two features interact negatively. Note that it is possible that, when features are exercised in some proportion in a given run, there are positive or negative effects (for example, caching effects); the effective $r12$ then depends on a and b. If we have to handle a continuum of choices for a and b, integration over a or b with the corresponding value for $r12$ can give us a composite $r12$.

---

[3]This is possible, for example, if a DAG is being computed and state is not kept to avoid double billing. Keeping state changes the computation; information leak is a possibility too.

The above is a bit simplistic. When we add features, we may also get some benefits which are not taken into account in the above model. Note that for a feature like security or correctness, due to theoretical reasons, there is no perfect or complete solution. Hence, any new security feature, for example, has to be empirically evaluated by the cost incurred per unit vs. the economic loss (e.g., recall, replacement or repair costs per unit) incurred due to a breach. With a new feature, let the difference in the incremental benefit and cost be *c*; normalizing it to the cost of the resource, add this to *r12* to get a revised value for *r12*.

When there are more than two features, in principle, one can explore the above analysis by adding one feature at a time. Note that instead of linear proportionality factors, we can have cubic or other polynomials relating, for example, power or performance with utilization. With such non-linear factors, a simple resource proportional design with respect to one parameter (e.g., performance) may not be the most resource efficient with respect to other aspects (e.g., power) after a certain level of utilization of a critical resource as it can increase, for example, power usage. An important part of the design is then to ensure that such a high cost region is avoided by introducing delays or bloat, or, more generally, investigating non-work conserving solutions. With many non-uniform proportionality segments, it is better if trade-offs favor the most common utilization scenarios, and the utilization curve of a workload with respect to time is also taken into account.

7.1.2.1   A simple regression based model

A simple analysis is possible if the system metrics of interest are amenable to regression based modeling, more specifically multivariate linear regression; non-linear aspects can be in principle handled by new variables that capture higher-order terms, or with the use of piecewise linear segments[4]. To simplify the discussion, we will assume that we are modeling the control plane; including the data plane will make for a much more complex analysis.

If a program is not RP due to a new feature in one of its dimensions (say, for energy consumed), it is likely due to some interactions with other features. For example, assuming regression is feasible, the performance or resource usage can be in one of the following forms (following [249][5]): if p_i are the parameters that correspond to the features, coefficients $x_i = [x_0, ..., x_n]$ have to be determined so that the predicted value of the performance or resource usage $\psi_i(x_i, p_i, n)$ is given by the right hand side (RHS) in the equation below. Depending on the level of modeling and simplicity needed, the RHS can be

---

[4]We discuss simple cases only here; for a more complete discussion, see [142, 261, 311].

[5]This approach uses Ordinary Least Squares (OLS) algorithm to model the function between the vector of parameters and the metric of interest.

one of the three simple forms below with either only linear terms, only linear and interaction terms, or with quadratic terms also:

(a) $\psi_i(x_i, p_i, n) = x_0 + x_1 * p_1 + ... + x_n * p_n$

(b) $\psi_i(x_i, p_i, n) = x_0 + x_1 * p_1 + ... + x_n * p_n + x_{l1} * p_1 * p_2 + ...$

(c) $\psi_i(x_i, p_i, n) = x_0 + x_1 * p_1 + ... + x_n * p_n + x_{l1} * p_1 * p_2 + ... + x_{m1} * p_1 * p_1 + ...$

Here, terms like x_{l1} represent the interaction terms between two parameters, whereas terms like x_{m1} represent quadratic behavior. (b) and (c) are not RP if the coefficients $x_{l1}, ...$ and $x_{m1}, ...$ are non-zero. With respect to bloat, consider adding one more feature for case (b):

(b') $\psi_i(x_i, p_i, n+1) = x_0 + x_1 * p_1 + ... + x_n * p_n + x_{n+1} * p_{n+1} + x_l * p_1 * p_2 + ... + p_{n+1}(y_1 * p_1 + y_2 * p_2 + ...)$

The y_i here represent the interaction terms of the new feature with the old ones. If now we set $p_{n+1} = 0$, it is still the same as the previous equation, hence there is no bloat. A very simple model that would display bloat could be instead:

(b") $\psi_i(x_i, p_i, n+1) = x_0 + x_1 * p_1 + ... + x_n * p_n + x_{n+1} * p_{n+1} + x_l * p_1 * p_2 + ... + p_{n+1}(y_1 * p_1 + y_2 * p_2 + ...) + z_0 + z_1 * p_1 + ...z_n * p_n$

The z_i represent the additional terms needed to model bloat in a simple way, keeping the rest of the terms the same. With $p_{n+1} = 0$, we have a simple model for bloat: we can now check if a feature is k-RPD by checking if $\psi_i(x_i, p_i, n+1)/\psi_i(x_i, p_i, n) <= k$.

7.1.3 Steering a system towards RP

If a system is not k-RPD, there are multiple ways of attempting to get the system to be back to k-RPD. For example, one can try a systems approach using bottleneck analysis to identify parameters that can be changed. Any sufficiently complex system (or sometimes even a simple system) usually has enough knobs or design options that are available for this purpose.[6] Using aggregate models such as Roofline[407] or Gables models[185], some parameter changes can be explored, such as bandwidth or operational intensity (flops/bytes) but these are open-loop models (i.e., without any strategy, in our context, to correct the system towards resource proportionality).

Or, we can use a control-theoretic approach to reconfigure the system so that it is again k-RPD[398, 100, 311]. For example, if adding checkpointing in a distributed system as a feature makes the program slower (due to costs, for example, of networking traffic added due to consistent snapshots or clock synchronization) by more than a k factor, either we can use a different configuration of the system[230], or use some knobs such as packet aggregation or

---

[6]See, for example, the many design options available for disaggregated memory interfaces[304].

even user-level networking to attempt to make available the slack necessary to bring the system back to the k-RPD level again.

In general, if a complex program is to be k-RPD over a large input space, a control loop may be essential. Note that control theory and machine learning (ML) typically have complementary roles: ML in general models the input and output relationships in a complex system, with control theory addressing the dynamics of the system. They can work together if control parameters can be learned. Using control theory, "optimal" performance-sensitive configurations can be attempted to be discovered given some constraints including some strict ones (such as avoiding out-of-memory errors)[398].

For controlling and realizing k-RPD behavior, additional constraints, for example in the 2 variable case, are set by $r10$ and $r02$ in the context of controlling $r12$ as $r12 <= k * r10$ and $r12 <= k * r02$; a similar set of constraints are needed for the general multivariate case. The strict constraints cannot be handled in ML contexts as they can only provide limited formal guarantees. While users can be expected to provide constraints on performance metrics (such as k, QoS), developers can be required to provide information to the system on which configurations need dynamic adjustment and which performance metrics are affected by a configuration. The setting and adjusting the configuration can then be handled by synthesizing controllers[398, 311, 100]. An important issue is how vigorously the controller should react to the current error (here, deviation from k-RPD). In one approach[398], measures of system stability are computed as the inverse of the average coefficient of variation across sampled configurations during model building phase. Instead of trying to reach the goal (here, the specific values in the system needed for satisfying the k-RPD property) using the control loop directly, a virtual goal is posited. This virtual goal is chosen to be not close to the actual goal if the system has a computed high instability measure (to avoid accidental overshoots) but otherwise is kept close to the desired goal for better resource utilization.

Note that we do not take into account data inputs as we are modeling the control plane only. If we need to model the dependence of the metric on the data inputs, a simple approach would be to make p_i encode the inputs but this might be possible in some cases only. More generally, one can use the approach of deep learning to map the inputs and configuration parameters to outputs to obtain a multi-layered representation but this requires training data. We give an outline of such an analysis here, assuming that control parameters can be distinguished from the data plane ones.

Let the program with n and $n + 1$ features be represented as F_n and F_{n+1}, and assume that both are differentiable with respect to the inputs. Construct DNNs (deep neural networks) D_n and D_{n+1}, respectively, for the two functions. Divide the output (metric of interest such as power) into some number of classes; this requires care and this is related to the k in k-RPD. To study bloat, assign a value to the $(n + 1)$th parameter (often zero or "default") so that the function F_{n+1} even with the additional parameter is still equivalent to F_n. Next, find the boundary where the two differ significantly by the following procedure[303]. Let $F_n(inputs)[class]$ be the class probability that F_n predicts the output to be in *class* on *inputs*; similarly for F_{n+1}. The optimization procedure, starting from some input vector (initialized from a seed vector), varies the input to maximize the class difference of the output between F_n and F_{n+1} and also at the same time maximize the neuron coverage (i.e., go above activation threshold) by varying the $(n+1)$th parameter and other parameters that heavily interact with it.[7] As we search, any updates to the inputs are checked for (system wide) feasibility before proceeding further. If we find inputs where the output classes differ (i.e., one output class is k-RPD and the other is not), we are done (i.e., system is not k-RPD). Otherwise, we can further refine the classes and repeat the procedure until we run out of the budget for searching. We cannot declare the system as k-RPD in all cases as we have not searched the full input space and we could have a false positive case; however, such an approximate search may suffice in most situations to be able to declare the system as tentatively k-RPD.

7.1.4   Difficulties in realizing k-RPD

Even though we have sketched a way of steering a system towards RPD, in practice it is not simple. With respect to functional requirements, performance variability is an important issue as it can have a complex performance envelope. Two main aspects of this problem are due to load variations and feature variations. For a RPD, resource estimation, allocation, and monitoring are necessary.

With respect to load variations, any complex system has many subsystems each with its own possibly multidimensional configuration space. For a truly RPD, each configuration has its own requirements with respect to resources (such as those to be orchestrated in serverless computing). Hence, as we move from one system configuration to another, the resources required may change abruptly in one or more dimensions.[8] In the extreme (and interesting case), the performance envelope may be extremely non-linear. If this situation obtains, then performance anomalies are difficult to debug; hence the need to research blackbox RPD[398, 338] vs. whitebox RPD approaches.

Assuming a certain style of coding, dynamic program analysis may be useful for detecting, for example, bloat in the blackbox case. In the case of programs (and not that well suited to whole systems), a simple approach could be to generate some traces,[9] find summaries of methods using randomly generated test cases, identify potential wasteful methods (repeated redundant operations specifically) using these summaries, and then attempt test generation that exposes redundant operations such as redundant traversal of loops[121]. However, for achieving RP behavior, we need a deeper and more

---

[7]Due to the differentiability of the functions assumed, the gradient ascent method can be used here and a Lagrangian used for optimization; for details see [303].

[8]A simple example would be the setting of the voltage in DVFS (dynamic voltage and frequency scaling).

[9]Note that this step itself is non-trivial and requires deep understanding of the program to get it right, for example, in the context of programs that employ GC[275, 117].

robust approach such as a control-theoretic approach so as to track the system dynamics and also so that some constraints (such as QoS) can be met[398] which we have just discussed. In an another yet similar approach, performance goals are set, hard goals listed (such as, for example, no OOM (out of memory) errors), an approximate dynamic model of the software system is automatically constructed, and controllers synthesized[141, 142] for managing its non-functional requirements. It is possible in such an approach to still provide formal guarantees of the system's dynamic behavior[141] by updating the model continuously to track changes in the execution environment and overcome any negative effects of the initial approximation chosen.

More specifically, consider the cases where bloat is possible if network queues are not managed properly: this can result, for example, in losses in throughput and latency increases. A good approach is the application of backpressure to reduce the size of queues. An admission control system, for example, prevents new entries if the system is already at its capacity; this is an example of a backpressure that is binary in nature. In a more general setting, the queuing system can have queued elements that are active entities (for example, a microservices pipeline with circuit breakers). In such settings it has been shown (specifically, in traffic studies) that both forward and backward sensing (of queue sizes) contributes to stability of the system but a system with only forward sensing is not stable[187].[10] Furthermore, from a latency perspective, dropping packets at the tail of the queue in approaches like RED[145](random early detection) to limit queue sizes has limited usefulness as buffer bloat[286] can still result. As the drop in RED is at the end of the queue, the system still has no information about how much delay the successful packets have endured to come to the head of the queue. The RED approach assumes an uncorrelated Poisson arrival process. But in a reliable transport protocol such as TCP that has a feedback loop with a self-clocking property, arrivals can well be correlated, and can result in a standing queue that cannot be dissipated. A good solution is therefore to use the packet-sojourn time (calculated over a window) through the queue and drop the packet (at the head of the queue!) if higher than a set fraction of the round-trip time RTT (say, 5%), and continue dropping till the situation clears.

A more tricky case can arise in the context of fault tolerance. Consider a system with two new features: a caching subsystem and fault tolerance to failures in the caching subsystem. With caching, throughput of the system can be higher significantly; admission control, if present, will kick in at a much higher level of input requests than without caching. If the system tries to function in spite of a failure in the caching subsystem (say, corruption of the data), the queues will build up significantly as the admission control will not kick in at lower levels of requests nor can anything be done about the queues

---

[10]Assuming active entities in the queues, each entity adjusts its processing based on the difference in the processing already computed and the relative difference in the processing power of the entity ahead. In bilateral control, in contrast, the same information of the entity behind is also used.

already present. The queues could become sufficiently large that the system cannot recover. Here k in resource proportionality can be arbitrarily large. The problem is that admission control cannot be allowed to kick in earlier in the absence of failure without reducing throughput of the system; otherwise, some peaks in traffic might have not been let in.[11]

In highly available systems, redundancy is needed for recovery. Duplication of resources is often common but this is not often resource proportional. However, error coding (e.g., RAID5, RAID6) can be much more effective as we do not need to duplicate the resources completely; we need only an additional fraction of the resources. If a failure occurs, lost data can be recovered using linear operations on available data. Another approach is the use of resilient distributed datasets (RDD)[429] where the processed data is encoded with provenance.[12] On failure, the provenance information is used to reconstruct the missing data; this is also likely to be RP.

In a more general setting, if non-linear operations on coded data have already been computed, surviving processed data cannot be combined to recover the lost but already processed data, due to the non-linearity. If the processed data is needed for purposes of prediction (or, more generally, in the context of approximate computing), then the coding and encoding functions needed can be learned if represented as a DNN and trained together with the DNN for the processing needed for prediction. By thus using learning to discover the codes based on the training data, we can have a more RP design[224].

7.1.4.1 Searching a large RP design space

Exploring a large search space efficiently is non-trivial. Sampling is a well known technique. A sampling approach such as Simultaneous Perturbation Stochastic Analysis[362] has been used to show that the large Hadoop configuration space can be searched effectively[230]. In this approach, instead of perturbing one parameter at a time and estimating the gradients for all the parameters (with a cost of O(number of parameters p)), multiple parameters are simultaneously perturbed (cost $O(1)$) while making sure that the distribution of the random perturbation vector satisfies certain conditions.[13] The $O(p)$ improvement can be realized unless the increased number of iterations needed for convergence in the SPSA outweighs the savings due to simultaneous perturbations.

Given that the search space is huge for a RP design, other statistical techniques may also be useful. In some systems, it has been observed that

---

[11] Note that the microservices circuit breaker pattern, unless carefully designed with at least two levels, may not be able to cope with this problem.

[12] Provenance is the information that can be used to recreate the data if lost.

[13] Such as smoothness of the function being optimized and perturbation vector being independent and symmetrically distributed about the zero with finite inverse moments (see [362] for details). Uniform and normal distributions of perturbations do not fit the bill but symmetric Bernoulli distributions that take both positive and negative values do.

input configuration space clusters correlate to performance space clusters[150]. Across all such correlated cluster spaces, one can pose an optimization problem for RPD as choosing a cluster subspace with the most linear structure. The statistical technique of Kernel Correlated Component Analysis (KCCA) is useful for discovering the correlations.

Since performance metrics can be non-linear in terms of inputs, a well known solution is the use of kernels in ML for handling non-linearity. But since they still use the same variables as the original formulation, we may still need newer structures or concepts to improve the effectiveness of the predictions. This is handled implicitly in the DNN model by the layers that sample as well as build newer values of interest in some intermediate layer as a function of some earlier inputs.

7.1.5   Summary

Given the above discussion, we next study the implications of a RPD in the following areas:

- RPD over time and with respect to specific styles of RPD designs

- RPD impact on security

- RPD impact on some other systemwide properties (such as real time and correctness).

7.2   RPD over Time

7.2.1   Impact on "optimal" RPDs

In general, a RPD that is close to optimal will unravel with time as different components may be or will be innovated at different rates. Hence, it may be difficult to make RPDs robust with time. If changes happen in some parts of the design at highly unequal, especially non-linear, rates with respect to time (for example, exponential rates of increased logic capability in terms of transistors, given by Moore's law or Dennard's scaling in the past), the imbalance can be structural.

Investigating bloat/resource proportionality of a complex heterogenous design requires careful exploration of the utilization (in general, design space) of each of the subcomponents (for example, dynamic voltage and frequency scaling (DVFS) in CPUs with power having a polynomial relationship with voltage and frequency) as discussed in Section 3.3. In smartphones, for example, power consumption depends, at a more detailed level, on CPU/GPU frequency and CPU/GPU utilization, screen brightness level, WiFi/3G/LTE

FSM (finite-state-machine) and signal strength, WiFi beacon and WiFi status, cellular paging and cellular status, and whether SOC (system-on-chip) is suspended[201].

As time progresses, and technology regimes change, the zones of non-RPD (or bloat) will correspondingly change, assuming there is no complete redesign of the system. The latter may also be necessary in specific contexts; for example, in the case that we discuss next.

7.2.1.1   Impact on "microservices"-based RPDs

As technology evolves, core CPU logic processing typically improves faster than peripheral or interfacing logic processing/software due to economies of scale.[14] Hence, a RPD that depends heavily on microservices[31][15] as a way of handling various features present may require possibly extensive marshaling and protocol/data translations to activate specific processing sequences. Over a period of time, such interfacing logic/software can steadily become more expensive compared to CPU processing and the earlier RPD design may no longer be tenable, and has to be rearchitected.

As a concrete example, we can take a look at the tail latencies for services. As we move from monolithic services to microservices, instead of the 100 ms (millisecond) to 300 ms tail latency (at the 99th percentile level, say) service level objectives (SLOs) (such as approximately 300 ms for web search), sub-millisecond tail latencies (such as approximately 100 microseconds for protocol routing) are required[365], as many microservices need to be invoked serially for an equivalent service. This requires a careful rethinking of where the overheads are: whether inline or dispatch-based RPC processing is better, whether block based or poll based RPC reception is better, or whether synchronous or asynchronous processing is better[365]; we also need to consider all the combinations. Since the specific methods are load-dependent, an adaptive system is necessary along with a learning and predictive component to make the system usable.

Another solution is to split the system into a control plane[16] and a data plane, with the control plane typically managing the state of the system

---

[14]If control logic is spread out as many pieces of firmware in a large networked system, they might not all be updated in unison; this can give rise to an unbalanced system. With SDN, such an issue is less likely. Host Bus Adapters (HBA) in large storage designs are often a problem for the same reason; their innovation cycle is not as quick as that of the main processing system. Similarly, progress in memory density is likely to be faster in a widely used memory technology than in a specialized memory technology and thus may make possible, with time, what was infeasible in the past with the non-specialized memory technology.

[15]For example, a Facebook news feed service query may have a serial pipeline of many microservices[365] such as a spam filter, then a protocol router, a distributed social graph data store, a user database, etc.

[16]The control plane can do, for example, traffic shifting or coordinated dynamic power control across the many microservices, subsystems, and devices. Depending on resource proportionality of each resource (Section 3.3), optimal configurations appropriate with respect to power consumption, for example, may be far from expected ones.

and the data plane being stateless. This can possibly reduce the marshaling/translation costs as the control data may be available across the system in a standardized manner in such designs. A similar situation may be obtained with the data plane.

As the system goes through different (macro)states, the control plane has to track and allocate the resources as needed. Using microservices with a control plane architecture, there may be a need to adjust or allocate new resources as needed to track any changes in the control state especially the use of not recently used features; in a "serverless" design[10], these are handled by the resource providers directly. If there are significant latencies, the RP aspect may be compromised. To reduce such possibilities, control theoretic (or even, stochastic optimization) frameworks are needed in general.

So as technology evolves, to avoid the cost of low performance messaging or marshaling, more functionality is likely to be encapsulated into "units" even if it makes future modification less easy. This can be seen more commonly in the case of hardware where general purpose devices such as microcontrollers, CPUs, or storage devices add increased functionality with time. Another approach is to study flows of data and associated computation and make special pathways so that unnecessary data movement is avoided[276]. This is, by definition, workload dependent and hence changes over time are to be expected, and so the RPD needs to be reworked over time. We next consider a related issue.

7.2.2   RPD in the context of rapid change

In the model of system development until recently, the programming structures are assumed to be given or developed from first principles, and then tested on some data (some synthetically generated and some obtained in the field). Data enters the model only as a correctness check (testing) and for evaluating performance. This model cannot track rapid changes in the field as the program may be optimized for a different usage model.

To investigate RP designs in this model, let us consider a client/server model for simplicity. For each request, it is important to track the visit of each component of the processing of the request to a processing center/delay center. If there is a bottleneck device, queueing occurs at this device but with lower utilization for other devices. Depending on the RPD of these devices, the overall RP nature of the whole design can be determined. Technological change usually attempts to reduce the impact of the bottleneck device; hence, the bottleneck device changes as technology evolves. Hence, while cost-sensitive or performance-sensitive embedded systems/devices are designed to be efficient/RP, they remain so for only a particular epoch, and need to be redesigned with time or discarded completely.

Typically there is a division of labor in a mature economy into a small "capital goods" sector that builds general systems for manufacturing goods (which in turn are used for manufacturing varying consumer goods) and a large "consumer goods" sector that builds effective devices for specific needs.[17] Due to the lack of generality in the latter, the rate of obsolescence of consumer devices is much higher. Such a split is not so pronounced in programmable devices or software artifacts due to the "ease" of modification through software. Even here, systems software (such as IDEs, compilers, OSes, large frameworks, or libraries) may be categorizable as capital goods.

With the recent explosion of data-driven software, programs are being generated automatically from data, in specific domains such as speech/visual recognition, speech synthesis, and machine translation (essentially where approximate computing can be useful), using AI/ML techniques such as DNNs and the like. Instead of the earlier traditional "software 1.0" model of development (programmer written code), there is now a move to the "software 2.0" model[208] where code is generated automatically from data sets using techniques like stochastic gradient descent and backpropagation [159]. This data-driven model is meaningful in recent times as data is being generated by consumer devices in large amounts and hence an appropriate design strategy would be in the context of data abundance rather than data scarcity. The benefits in the specific domains of software[208] are the use of computationally homogenous (macro) operations such as dot products, thresholding at zero, and convolution that can then be realized in silicon[202]. Another advantage is the predictable or constant runtime and memory use. It can also provide compositionality (consider the example of coded computation discussed in Section 7.1.4) as backpropagation can be done across the constituents. From a RP perspective, what we have also is a generated model that is sensitive to the accuracy required in specific domains; for example, in the number of bits needed to express a value.

The downsides are the high cost of training, lack of explainability, and attacks with adversarial data. Another problem is that as data input space changes, we may need to retrain the system and get an altogether new model. However, this can be automated in contrast to programmer written code. A more serious issue is that there may not be much plasticity or dynamicity in the system;[18] any eventuality has to be present in the data sets used to train the model, otherwise adversarial input can be a problem. The cost of training and retraining, collection of the right data sets, and the defenses against adversarial data are thus some of the negatives. The traditional hand crafted software however can still attempt the coordination of the system, without having access to large data sets, through *ab initio* control-theoretic models, market-like mechanisms (competition, auctions, dynamic pricing,...) along with optimization of specific flows.

---

[17]Consider the computer industry. Initially, there were only servers and terminals, then servers and PCs appeared. Now, there are servers, desktops, laptops, pads, mobiles, and so on.

[18]This may be alleviated by streaming data regularly into the model and rebuilding the DNN.

7.3   RPD and Security: Resource Usage as a Side Channel

A RP design should keep only those data that need to be kept secure instead of everything. At an architectural level, in one such secure computing model,[19] an enclave is used to protect secrets and the architecture makes a guarantee that even a malicious kernel cannot observe the secret. However, these guarantees are not valid if side channels are present. Just as in Spectre[186], the race condition between the injected, speculatively executed memory references and the latency of the branch resolution can be used to leak secrets in the same model[98].

Any RP design therefore needs to seriously consider security aspects due to the possibility of sidechannels. If some operation has an identifiable signature with respect to sidechannels due to an RP aspect, the design is leaking some information. This has been well known for some time, for example, as timing and power sidechannels in cryptography. For example, in the DES or AES encryption, if part of the secret (or its reversible transformations) is used for deciding some set of iterations or cache accesses, that part of the secret can be inferred by using timing or electromagnetic (EM) channels[41, 251]. Hence a simple solution for an RPD that is not already designed with security in mind is to add sufficient noise so that such sidechannels are neutralized; this however comes at a cost.

However, there could be strategies that are aligned positively with both security and resource proportional design; this only requires that the "secrets" are not in the proportionality set. For example, redundancy is controlled in deduplicated storage by keeping copies to the minimum level required to handle failures; here security and resource proportionality go together as only the required copies need to be kept secure.

Another approach is to recognize information hiding and integrity-based defenses as part of a continuum and move across this continuum based on the actual probing attacks that occur so as to be RP. In the context of a system that resists code reuse attacks,[20] the former has low overheads but weak in protection while the latter has high overheads but strong[59] in protection. Due to bloat, the attack surface of the code increases, especially due to possibility of increased number of code reuse gadgets. Hence debloating is an important part of secure design, especially of large software systems that have evolved over some time[183]. While many debloating strategies have been developed, it is not the case that security is enhanced after every instance of debloating. Studies have shown that in some cases the number of gadgets can increase[81]. Just like the impact of bloat on performance can be counterintuitive (as it depends on whether the bottleneck device is affected, Section 3.3),

---

[19]The SGX model from Intel.

[20]In such attacks, user's code itself is carefully selected to craft malicious payloads without needing it to be injected from outside; such code sequences are called gadgets.

some debloating strategies can actually increase the attack surface as they can introduce code sequences such as trampolines[21] that were not present before.

Furthermore, a useful principle for security against DoS (denial-of-service) attacks could be that all legitimate users suffer no additional resource usage but non-legitimate users suffer highly non-proportional resource demands; this is similar to all legitimate users going through a fast path while all non-legitimate users are forced to go through a costly path. While this is difficult to engineer in many cases (as who is legitimate is not clear in open systems), one way is to make sure that all legitimate users suffer a (monetary) cost proportional to the gain in system security while non-legitimate users suffer much higher costs.

We give a few examples to further illustrate the problem:

- A desirable feature is the ease of "observability" in running systems to enable debugging, especially with open source systems. As a rule, this is unexceptional but this feature may need to be restricted to only the developer or to systems under development to avoid security breaches. The distinction made between capital goods and consumer goods may also be useful here; consumer goods should typically be in a locked down posture with respect to security.

A good example of a breach is the *mmap* attack[99] on a mobile phone. In this attack, UI (user interface) states are inferred in real time through analysis from a user (non-privileged) background app. The UI state is then hijacked (i.e., intercepted) to steal sensitive user input such as login credentials. This requires an attack app running on the victim device in the background.

The attack uses a shared memory side channel to detect certain transitions that can be used to infer window events in the target app. Shared memory is often used by window managers to receive window changes or updates from running apps.[22] Such changes are a "surprisingly clean" side channel that can be used for weaponizing an attack.

Assuming the VM (virtual memory) mappings are RP (i.e., specific chain of mappings are directly and necessarily related to an operation that is to be predicted), a HMM (Hidden Markov Model) model can be used to detect specific activity such as login using input methods, network events, content provider events, and CPU utilization. UI phishing can then be used, only during "login" sessions predicted, to capture passwords using both activity detection/inference without significantly raising suspicion.

---

[21] Trampolines or indirect jump vectors hold code addresses; if an execution jumps to one, it jumps (or bounces out) next.

[22] Shared VM size changes are from both app process and window compositor process which correspond to allocations/deallocations of a specific graphic buffer (with a size of 920x4KB pages on Samsung S3). This is the off-screen buffer in the Android shared memory space that is updated when shared files or libraries are mapped into VM. This can be sampled through shared vm or one can use the number of minor page faults in /proc/pid/stat.

GPU-based attacks have now become practical[277] for the same reason as one of the leakage vectors is also the memory allocation API (along with other sidechannels such as timing of allocation operations and snooping of shared GPU hardware performance metrics).

- More generally, state-of-the-art advanced ML techniques can be used to extract a signal that can compromise security or privacy; this signal can be made weak only with a non-RP design. For example, it is possible to study NFS (a network filesystem) traces to predict applications running (for, e.g., to distinguish between Win 10 booting vs. Win 7 or Solaris or Linux[423, 308]). Similarly, ML methods can be used for workload characterization; program phases (if resource proportional) can be detected[308] and may result in a security exposure. In general, "fingerprinting" is used to first "size" a system to determine its broad characteristics ("features") before the actual highly specific attack. For example, given the various parameters in the TCP/IP system, knowing a few helps determine whether we are interacting with a Windows 10 system or a Linux 4.x one, etc.

- At a more abstract level, another type of attack is possible. If a good RPD is available for some function, it is likely to be made into a service accessible from multiple users and at multiple levels. In essence, a good design may use many modules that are RP; or, equivalently, a good design in operation has possibly many "repeatable" RP subunits visited, chosen precisely for their efficacy.[23] Given this, an attacker can develop a signature based on the repetition patterns of the basic RPD subunits. One way to counter this is to possibly drop efficiency as a primary goal and go for n-version programming; note that this is also fraught with difficulties as the coordination between two differently designed systems may not be easy (as it happened, for example, in the Space shuttle[236]).

- RCU (read-copy-update) mechanism is a RP way of some OS kernels to ensure scalable synchronization (readers do not have to take a lock as any new writer gets a new copy of shared resource) and allows for a linear scaleup in multi-core designs. Early versions of kernels with this feature could be attacked with DoS as it was known that deferred reclamation of memory is an integral part of the design[342]. Even in current kernels, it is possible[313] as OOM (out-of-memory) situations can be made to happen if there is no proper coordination, for example, between slab caches and deferred free memory of RCU resulting in non-RP use of memory with potentially serious consequences.

---

[23] As an example from a completely different domain, chloroplasts and mitochondria are two outstanding innovations that have been incorporated in toto in essentially all plants and animals, respectively, as they exploit subtle quantum phenomena in electron transport to achieve high efficiencies. From a security perspective, any successful attack, say on mitochondria, can potentially bring down all animal forms. As no other structure is as efficient energy wise, they are inescapable in cells, except possibly in some sulfur-based life forms.

One solution to such attacks is obfuscation by adding, for example, noise to a "naive" RPD to avoid leaks:

- Cracking DES[71] has been possible in the past using differential power analysis or analyzing electromagnetic radiation. Robust designs need to add "noise" to prevent any signal extraction corresponding to keys.

- Another example of a side channel attack is on the use of "naive" encryption of packets in a wide area encrypted NFS. Given that the packet lengths of encrypted packets for read and writes, and other NFS operations, can be captured and analyzed, they can be combined with workload characterization to predict the actual operations with some accuracy. This problem is acute for cloud security and suitable counter-measures (such as padding) are necessary. Similar situations are possible if analysis is possible on access patterns due to encrypted search queries over an encrypted database.

7.3.1   RP remediation/countermeasures

It is clear that there is a cost to making a system secure. Since there is no theoretically clear definition of a secure system that can be used in practice[174, 160], the remediation has to be proportional to the threat. Given the difficulty of modeling these formally, an easier solution is to estimate the loss incurred without the security features and that of the cost of implementing security itself. Using an insurance model, one can incorporate the level of security appropriate to the loss/premia to be paid. Such costs can be used as constraints in the design.

For a principled secure design, non-interference principles (or some sim-plified version thereof) and the related information flow analyses or access control policies/mechanisms need to be an integral part of the design (at hardware, firmware, OS, software levels), a difficult proposition currently. For example, if information flow analysis is used as a security feature, it requires that every computation/branch (at the language level) requires flow label computation. If the costs for an information flow based security are higher than, for example, the cost constraints from an insurance model, the design may need be revisited in favor of simpler models (such as access control mod-els). Historically, simpler models for security such as SELinux have appeared first in Linux kernels instead of the not so common information flow based models.

One can make a design resource proportional first and then add controlled "noise" (or "obfuscate") to ensure resource usage is not a side channel. This is a 2-phase solution and may suffer from inefficiency; the RP aspect may also be lost. Or, one can design the RPD solution in tandem with the security concerns; this may have additional design complexity. The typical situation is that neither resource proportionality nor security is a high level or first level concern and both therefore need to be incorporated in the subsequent design iteration.

One general approach to mask a side channel signal is to parallelize the application, a useful strategy given the current multicore environment. If the scheduling of the threads is not easily reverse-engineered, an analysis to detect a signal across multiple threads is not easy. Similarly, multiplexing multiple computations with virtual machines can be useful. But we need to ensure that scheduling of the VMs themselves should avoid leaking information through side channels.

To ensure that clean sidechannels are not possible, one approach is to decrease the resolution of some critical parameter in the system. For example, to reduce the attacks due to Spectre and Meltdown in a browser, one solution is to reduce the resolution of the clock[272] but the attacker still has other avenues such as counting instructions. Similarly, if pagefaults in a secure enclave are visible to the host kernel (for example, in SGX), and the sequence of pagefaults reveals some secret (for example, the query in a key-value store), one solution is to use huge pages instead of the regular 4KB pages to reduce the resolution of the areas of memory that could have been accessed.

Another technique is the equivalent of dynamic binary rewriting: for Spectre, the use of indirect call promotion that transforms indirect calls into conditional direct calls by learning targets at runtime and patching to perform *just-in-time* promotion without the overhead of binary translation[437]. Such an approach also gives us a way of comparing the cost of this dynamic rewriting with the cost of avoiding Spectre by other means such as at the microarchitectural level and thus study the RP aspects.

In each of the following approaches, the (proportional) cost aspect is an important part of the design:

- Obfuscation techniques, e.g., "oblivious RAM"[158] and its costs.

- Zero-knowledge protocols may have extra rounds that increase costs. Costs also can accrue due to multiplexing a computation across crowds[386], principled anonymization, multiparty computation (MPC), or privacy preserving computation[130, 250, 325]. More complex design frameworks such as differential privacy are needed in the general case; this model guarantees a participant in a study that the results of the study are independent of his/her participation (essentially, disclosing nothing about the participant but learning useful information about a population). This is achieved by adding controlled (or, "proportional") noise to the answers to newer queries based on the queries already seen.[24]

---

[24]The fundamental law of information recovery says that very accurate answers to many queries will eventually destroy privacy; this applies to all privacy-preserving data analyses[130].

- Game theoretic aspects of how much information to reveal to others in a protocol. How much of the data is to be kept raw vs. anonymized? For example, consider anonymization of data across multiple "coopeting" players (producers, distributors, transmitters) in highly regulated industries such as electric power systems.

7.4   RPD and Other Systemwide Concerns

7.4.1   Real time systems

Real time systems, especially hard real time systems, are not easy to design as latencies in code or in subsystems can have stochastic aspects. For this reason, simpler behaviors are targeted in good designs eschewing complex patterns (for example, preferring scratchpad memory instead of caches). The design problem can arise at many levels and even at as simple a level as measuring time. Time measurement is not simple due to interrupts and due to power saving approaches in power constrained environments (such as smartphones), and these may interfere with measuring real time or virtual time as needed. Working with wrong measurements can give fictitious metrics such as rate of progress in the system or bandwidth available and give rise to counterproductive behavior if feedback is present in the system based on such metrics. In addition, to prevent system suspension during time critical work, smartphone OSs use techniques such as wakelock for this purpose in selected parts of the program. With an event driven programming model, high levels of concurrency, many versions of time available,[25] and incorrect use of wakelocks, correctness bugs abound that can result in non-RP behavior[201] such as draining the batteries instead of sleeping.

To guarantee RT properties, worst case execution time (WCET) sequences have to be taken into account. Whether such code sequences or traces are likely to be RP or not cannot be easily predicted. If we plot the performance metric of interest against the WCET sequences, there may be correlation (for example, energy consumed might increase with longer execution sequences as static overheads keep accumulating), inverse correlation (for example, average power requirements can decrease with longer sequences), simple convex or concave shaped correlations, or even more complex shapes.

To reduce WCET on some of the traces, some aspect (such as utilization) of some non-bottleneck device in the system may have to be changed appropriately as we may not be able to do that for the bottleneck device. Furthermore,

---

[25]Linux, for example, has versions such as CLOCK REALTIME, CLOCK BOOTTIME, CLOCK MONOTONIC, and CLOCK MONOTONIC RAW.

if it is possible to reduce bloat in the bottleneck device, it may or may not decrease its utilization; this depends on whether the input to this bottleneck device remains the same or increases. In the first case, WCETs may also be reduced but the full effect depends on the resource proportionality of various devices in the system. If utilization curve is highly nonlinear, the bottleneck device itself can change (Section 3.3). Care needs to be taken to constrain system behavior so as to avoid traces that violate resource proportionality; this will also help in meeting real time concerns.

To bring the structural aspect of this problem to the fore, we discuss one class of problems in the computer systems area that highlights the difficulties in RP design that are time-related and which occur not infrequently in soft RT systems (but not encountered or very uncommon in hard RT systems by design). One example of a non-resource proportional design is scanning the whole disk or SSD to fix local corruption; it could be due to, for example, a RAID1 failure, a file system crash, or a flash translation layer (FTL) recovery. A more RPD could be logging (called often "dirty region" logging in the specific case of RAID1) or copy-on-write (COW); more generally, error proportional recovery. However, logging or COW itself could be problematic (high latencies) or non-resource proportional. For example, if a technology does not have an efficient realization of update-in-place property (e.g., flash, or "punching" in hierarchical storage management (HSM) that ensures only one copy across the hierarchy) and there are also persistent pointers, any update of a persistent data structure, say a tree, near the leaves requires the pointer in the interior to the leaves to be updated as well all the way to the root. Due to such "update storms"[337, 127], latencies and QoS parameters can be affected adversely; to handle such specification requirements, non-RP designs may result. In COW-based designs, reference counter management may increase the overhead and result in similar storms; carefully designed data structures[127] are necessary so that the source structure is updated only once on the first update but later updates only require incremental updates to the immediate source COW structures and do not require any update to the source structure itself. In certain designs also, we see similar behavior such as in multilevel VM emulation with traps[240], or recursive VM faults with upcalls to thread libraries[33].

A RP system that is also soft real time is therefore likely to be difficult to design, especially a cost effective one, as there will be many different WCETs with different feature sets; this variability may make RP design uneconomical or problematic. If RPD and WCETs conflict (wrt power, for example), we may need to make RPD requirements less stringent (example, increase k in the k-RPD model) based on the performance metric vs. WCET curve. In the case of soft real time systems, devising an optimization problem with constraints may be useful; specifically, using this curve for joint or weighted optimization of RPD aspects and WCET RT aspects with costs. The costs could be based on how frequently RT guarantees are not met.

Other possibilities in the design are:

- Control and data plane separation: this may result in cleaner encapsulation of metadata and hence more predictability. It is not clear if layering in each plane is useful with respect to predictability.

- If thumb rules such as 80:20% rules are applicable in terms of code path execution or code running time, it is worth exploring if RPD is affected only for some part of the design (say, the 80%) and the remaining part is designed with a more lax RP model.

7.4.2   Correctness

To design and implement a correct RPD, a combinatorial explosion of feature sets needs to be explicitly tracked and the right resources allocated, and properties then need to be checked for correctness properties ("invariants") and also with respect to varying resources. Frameworks such as serverless computing or microservices make the correctness problem for the developer a bit simpler as the resource management is now the responsibility of vendor providing the framework. For the same reason, separation kernels[340] that are designed for information flow security and correctness do not allow any stochastic components in the design.[26]

A substantial part of a large well designed system is the set of assertions, or hooks, for dynamically collecting information for debugging. The cost aspect, just like for security, cannot be quantified easily as there is no theoretical or practical solution to proving a reasonably complex program correct. Hence, based on the cost of fixing incorrect programs in the field, a proportional cost for feasible verification as well for dynamic instrumentation infrastructure may be appropriate.

Here also, some possibilities in the design are (similar to the case with real time systems discussed before with minor changes):

- Control and data plane separation: this may result in cleaner encapsulation of metadata and hence the feasibility of a simpler description. Same with layering in each plane.

- If thumb rules such as 80:20% are applicable in terms of code coverage during execution, it is worth exploring if RPD is effected only for some part of the design (say, the 80%) and the remaining part is designed with a lax RPD model. This can simplify design and ensure or simplify verifiability.

---

[26]For example, the microkernel seL4 can be used as a separation kernel when augmented by a static partition-based scheduler that implements a static round-robin scheduling between partitions with fixed execution time slices.

Similarly, correctness needs to be investigated in the context of RPD with respect to other systemwide properties such as liveness/deadlock-free properties in the presence of locks and transactional or consensus properties in distributed systems. Proving such properties have been considered in specification/proof frameworks such as TLA+[234] which uses conditional atomic actions to describe transitions from one state to the next. Interestingly, in the context of distributed, concurrent, fault-tolerant subsystems where non-determinism is high, developing rule-based modules that use transitions similar to atomic actions in TLA+ has been shown to be effective[373]. This opens up the possibility that both code and correctness proofs can go together using a rule-based model for such systems. Note that such a model is likely to be RP for the following reason. As it is difficult to enumerate all the control flows from one step to another in such complex problem domains, the rule-based approach merely uses conditions ("state") to determine which actions execute next (thus control flow is implicitly specified rather than explicitly). As this is independent of how that state was reached, the code and specification are correspondingly simpler from a development and proof perspective. If the orchestration of the system (such as firing of conditions, activating the action modules) is RP, the whole system is also likely to be RP if the code modules are.

If verification of certain desired properties in a system is beyond the state of art, bug finding is an alternative approach. To understand the issue of proportional expenditure of effort in finding, isolating, or tolerating bugs in systems, consider first logical bugs. While there are many specialized tools for each type of logical bug (for example, race detection[144]), post facto reports of bugs may also be useful given the immediate context (crash reports, stack traces, for example). It has been found that stack traces are not useful in about half the cases as the predicates that are important for the crash may not be available; typically, pre-crash behavior, especially in the form of the value of some predicates, is important. Without knowing the number of bugs or the predicates that are important, we need to do an extensive and costly tracing of the predicates. One solution is to consider statistical bug hunting (or cooperative bug isolation[239]); this is especially suitable for a widely distributed large piece of software such as a browser. Detailed instrumentation for all users and at all the crucial areas in such a program can be costly. However, one can choose to statistically instrument each downloaded browser to only a few areas in the code (even if there is a need to instrument the code all over uniformly or for all important areas), resulting in a feature set, say, $(0,0,...,1,...1,...)$ where the ones correspond to code locations; they thus not only reveal the code locations but may also reveal a few predicates on a crash. Now the question boils down to ensuring that enough code locations (coverage) are instrumented and in enough copies of the downloaded browsers so that statistical valid inference is possible with respect to code quality. By using a form of likelihood ratio testing where a predicate being true increases the chance of failure over a prior, selecting predicates properly and eliminating correlated bug predictor predicates, a statistical inferencing is indeed possible[239]. Properly done, the next issue is the cost of such an approach. If this is only a small extra cost (say, 1-5%), then such instrumentation is reasonable.

In contrast, an example of a similar but effective RP solution for tracing a large distributed system is possibly the Dapper system[355]. Instead of a black-box scheme that only has message records (and that needs large amount of data in order to gain sufficient accuracy as it depends on statistical inference), Dapper uses an annotation method. This is possible since all applications in the system use the same thread model, control flow, and RPC system; hence, it is needed only to instrument a small set of common libraries, which is transparent to developers. The traces form a tree with each node in the tree being a simple log of timestamped records that has the start and end times, any RPC timing data, and zero or more application-specific annotations. In a typical trace, there is a single such log for each RPC, and each tier in the infrastructure adds one more level to the trace tree; note that it can contain data from multiple hosts. Clock skew issues across nodes are reduced by exploiting the fact that a RPC begin and end are temporally related. Deamons collect the data from the logfiles and are stored in a cell in a BigTable. The overhead is now related directly to the number of traces that a process samples per unit time; adaptive sampling ensures that workloads with low traffic boost their sampling rate while those with high traffic lower it to keep overheads under control. Such a design is needed for the microservices architecture[18] too; in addition, deduplication of exceptions or errors may also be required (as one error in one subsystem or microservice can result in other errors elsewhere). Furthermore, random sampling of both the machines within a large system as well as execution time within a machine can be useful for a low overhead system[204].

7.5   Conclusions

This chapter has looked at some of the implications of RPD as well as how to test for k-RPD, how to achieve it, the difficulties in RP behavior over time and the impact of RPD on systemwide properties such as security, realtime, and correctness. Recent advances in AI such as deep learning have a role to play as they provide effective way of modeling the relationship between the inputs and outputs in a data-driven way. If the system is cleanly separated into a data plane and a control plane, the dynamics of the system can be further managed using a control-theoretic framework.

Part III

Responding to Emerging Technologies: Designing Resource Proportional Systems

Chapter 8

Resource Proportional Programming for Persistent Memory

8.1   Characteristics of Emerging Persistent Memory Technologies

Having discussed RPD at higher levels of the system, let us now consider RP issues with respect to memory performance when a newer type of memory (PM) becomes available. In the next chapter, we will additionally consider interconnect technology for large memory deployments.

The market success of flash memory technologies has inspired significant new research investment in both flash and persistent memory (PM) technologies. PM is used in this book to refer to a category of technologies that provide non-volatile data retention (persistence) at near memory access times. Multiple emergent PM technologies (also called Storage Class Memory[270]) include Phase Change Memory (PCM), Spin Torque Transfer (STT), and Resistive RAM (ReRAM). Although these technologies have continued to evolve over time, they consistently demonstrate a range of characteristics that can perturb system resource proportionality in several ways.

- Decrease persistence latency times to those that are closer to memory speed access

- Increase density by an order of magnitude compared to current memory technologies

- Introduce asymmetrical read vs. write access latency

- Require more media specific functionality in the memory access path than what DRAM requires

As storage latencies decrease, at some point storage may become as fast as memory. Battery backed dynamic random access memory (DRAM) is an obvious example of a technology that is as fast as memory and persistent. Although battery backed DRAM has been available for a long time, the combination of memory-like latency, higher density, and lower cost makes more recent persistent memory technologies more and more attractive for write intensive applications such as databases.

Multiple persistent memory technologies continue to emerge from various manufacturers with distinct characteristics. Still, latencies are clustered in ranges that differentiate persistent memory from hard disk and flash technologies as shown in Table 8.1.[1]

TABLE 8.1:   Storage and Memory Latencies

Technology	Read Latency (μS)	Write Latency (μS)
Hard Disk	1000 - 10000	1000 - 10000
Flash SSD	10 - 100	10 - 2000
Persistent Memory	.05 -.5	.1 - 1
DRAM	.05	.05

Many of these technologies require more sophisticated processing in the data access path than DRAM. For example, DRAM mainly requires ECC while newer technologies often require support for wear leveling in addition.

---

[1]Disk and SSD latencies assume 4K data transfer Lengths while PM and DRAM assume 64 Bytes. That said, 4K data transfer time is a relatively small part of hard disk and SSD latency.

Wear leveling is required when access to data causes media wear to occur at a rate that is unacceptable relative to device life expectancy. By artificially distributing data access across all of the physical media in such a device its life expectancy can be extended to acceptable levels (e.g., 5 years) for most workloads.

As new technologies traverse the access time continuum from 10 μS down to 100 nS, a boundary is crossed that changes the most advantageous approach to integrating the technology into computer systems. Technologies with latencies near or above 1 μS are integrated as DMA (Direct Memory Access) devices wherein data flows in and out of memory buffers under the control of a DMA engine. The software to use such devices has the characteristics of Input/Output (IO) which generally involves the following process.

(1) Allocate a DRAM buffer to store data that is about to be read or written.

(2) Create a control block describing the DRAM buffer and a peripheral device and/or a second memory address range involved in the DMA. The control block may include additional instructions for a peripheral device such as a complete read or write command.

(3) Initiate the data copy by delivering the control block to a peripheral device or DMA engine.

(4) Wait for the DMA to complete. This may involve a context switch, interrupt, and/or polling.

(5) Check the status of the data copy to ensure that no errors are encountered.

Regardless of physical storage or memory media access times, the software alone for this process generally takes multiple microseconds to complete. Media, controller, and network related delays (if a remote peripheral device is involved) add to that. When using persistent memory technology, latencies enter a range that enables data to be exchanged between a processor register and PM in a single instruction.[2] Although the instructions to accomplish this are processor architecture specific, the software to do this has the characteristics of memory access which generally involves the following process:

(1) Identify a processor register containing the data.

(2) Execute a processor instruction such as a "Store" (ST) to copy the data from the register to a memory address.

(3) If no error exception occurs, continue execution with the next instruction.

---

[2]Actual latencies are technology specific. The RP impact of memory latency is explored in Chapter 9.

Although there are many variations on both of these processes, these two generally exhibit the following differences driven by the fact that the persistent memory access process forces the data copy to occur during a single processor instruction.

- The persistent memory access process is a compiler or interpreter generated instruction representing an assignment ("=") source code statement whereas the DMA process involves many lines of source code within an OS hardware abstraction layer or driver.

- Errors during the persistent memory access process can only be indicated by an exception which disrupts the flow of processor instructions. This makes the normal case more efficient; however, it also changes application error handling practices.

- After the persistent memory access, data may actually need to be flushed out of volatile CPU caches into physical persistent memory. There are many ways this can be accomplished including flush on demand triggered by applications, or flush on failure triggered by impending power loss.

- It is extremely inefficient to use the persistent memory access process when latencies approach or exceed 1 μS.[3]

In the following sections we will explore more specific data and process flow implications of PM and how they can impact Resource Proportionality. We will then quantify the Resource Proportionality implications of matching application practices to technology in software stacks including features such as atomicity and high availability.

8.2   System Implications of PM Technology

In this section we elaborate on the aspects of systems that are affected by new persistent memory technologies and can lead to resource proportionality issues.

8.2.1   Data flow implications

The differences in the way data flows during memory access as opposed to IO change several system data flow aspects. Workloads, non-functional requirements, and system conditions can impact the benefits of specific hardware and software feature implementations to a point where changing the way

---

[3]See Chapter 9.

software uses hardware improves resource proportionality with respect to performance. A number of examples of this type of interplay appear in Chapter 3. These situations create the need to manage enablement or deployment of hardware and software feature implementations at a granularity that reflects application requirements. Granularity may take the form of a memory address range or it may be more complex depending on the purpose of a particular data access and the system context.

8.2.1.1 Marshaling

Marshaling is the reformatting of data structures for the purpose of transmitting data between components within or across systems. This is a very common practice in networking and storage based on the following motivations.

- Many applications have historically intermingled persistent and temporary data elements in memory resident data structures to optimize both real time access and source code understandability. For example data structures in memory may include volatile indices that are reconstructed when an application restarts, or data may be stored in memory using sparsely populated data structures even though they are fully populated on disk. The persistent parts of such data structures must be separated from the temporary and reorganized in a well known or self-describing order for sequential transmission to storage components. This increases the efficiency of both capacity use and data transmission.

- When sending data to storage or to another system, the architecture of the receiver may be unknown. Conventions such as the way integers or floating point numbers are formatted or the way fields in data structures are delimited may differ from the interpretation of raw data by a differently architected system. For example data formats such as floating point numbers may be placed on disk in a canonical format but converted into a processor specific format when data is read. This same issue has led to the definition of the "Presentation Layer" in the OSI ISO network stack reference model[195] although some aspects are actually application layer considerations. The issue can arise if stored data is transported to, or remotely accessed by, a different system.

- Remote procedure call implementations must convert function calls with their parameters into messages for other systems that may have very different native function call implementations.

When data is made persistent in memory there is no need for marshaling because the native structure remains in memory as is. Some protocols[90] avoid marshaling by transmitting data without modification and requiring that the receiver translate the data if its interpretation differs. It has been demonstrated that marshaling can take anywhere from 100 μS to 100 mS depending on protocol and payload size. The bloat due to unnecessary marshaling can represent a very significant departure from a resource proportional design.

In many ways the overhead of marshaling represents the energy payoff, over and above media and low level protocol differences, for enabling applications to store variables directly in PM. This enablement requires changes to applications and introduces a number of new trade-offs. Marshaling remains a required overhead unless applications and libraries are reprogrammed to represent and manage data in native form. Applications that do not make this transition can use PM by emulating disk access. Even if the data copy at the core of disk emulation is very efficient, the overhead of marshaling is part of the way applications use disk storage.

8.2.1.2   Flushing and fencing

Good processor performance depends on volatile caches within processors. In order to survive power loss, data must be flushed out of volatile caches to persistent memory. Various processor architectures provide various methods of assuring non-volatility broadly represented by the following approaches.

- Flush and fence instructions[4] to ensure that data reaches memory on demand with the necessary ordering constraints.

- Non-cacheable memory regions, write-through caching, and/or instructions to ensure that some or all data in certain memory regions is always non-volatile by the time the processor instruction that manipulated it completes.

- System level solutions to flush volatile data to PM after power loss is detected. Such "flush on failure" solutions depend on some type of power hold time using capacitance or batteries, either of which may be associated with power supplies or individual components. Power hold time is the minimum duration that the components necessary to complete flushing are guaranteed to continue operation after the loss of external power input to the system.

The overhead of several flushing and fencing approaches has been evaluated and published for various workloads[56]. The following observations surface in that work.

- Flushing a cache line often consumes about 300 processor cycles regardless of whether it is dirty or not.

- Fine-grain write-intensive workloads increase flushing overhead leading to a cross-over point where it becomes advantageous to selectively avoid

---

[4]Flush instructions trigger processes that deliver data from CPU caches to RAM. Fence instructions ensure all flushes complete before any accesses to flushed data that appear before the fence instruction.

caching using techniques such as write-through instead of repeatedly flushing.

- For write-intensive workloads the difference between flushing on demand and cache avoidance may be significant (up to a factor of 2 time difference in micro-benchmark experiments).

These results call into question whether volatile caches in CPU's are beneficial features for some PM workloads. If not then those caches themselves may be adding unnecessary work (bloat) to the system. Still, volatile read cache is an immense benefit for mixed or read intensive workloads so the granularity of caching feature enablement is important.

Flush on failure may appear at first to be beneficial because it avoids real time flushing and fencing. Depending on overall reliability and availability requirements, however, there are several scenarios where flush on fail could be too risky.

- If the system is required to avoid all single points of failure, there may be failure modes such as failure of the volatile cache itself when there is no time to flush.

- Failure of the media to which volatile caches are flushed could cause a CPU to be unable to flush during power hold time which ends when CPU or memory chips do not have enough auxiliary or residual power to function. Intermittent failures may be most difficult to handle in that they create uncertainty about the best strategy for completing the flush. For example if flushing to alternate regions or redundant copies is implemented and failures occur randomly during a short period, flush on fail may be delayed and not have time to complete.

- The processing power that was assumed to be available for flush completion after failure may be inhibited by processor features such as throttling due to high temperature or power capping, so that once again, flush does not have time to complete.

If required, several of these risks are mitigated using cross-server redundancy which could be viewed as a non-local type of flushing and fencing with the added overhead of network communication. Even with modern protocols like RDMA and NVMe, remote access across servers often pushes latencies too high for memory access, so high availability features should only be used with critical data. This again raises the need to precisely manage the granularity of feature enablement.

8.2.1.3 Data recoverability

Obviously, persistent data must be recoverable after power failure or system reset. Unfortunately most processors do not guarantee that data stored just before a power failure or reset will be consistent afterwards. Data consistency is defined at the application level. Consistency means that the application is able to make sense of the data. The most primitive form of consistency applies to the contents of an integer or pointer type variable. After a processor instruction that updates an integer or pointer, write data flows through processor caches to persistent memory. For consistency the update to persistent memory, whenever it occurs, must result in either the prior value or the new value in its entirety. Updates with this all or nothing property are said to be atomic with respect to failure.

For most applications data recoverability requires a greater scope of consistency than a single pointer or integer. Analysis has shown that this requirement ultimately leads to transaction style data manipulation with well defined commit points[391]. Transaction constructs have been well studied for decades in the context of databases and applied in various ways to persistent memory [165],[131], and [393]. During recovery in most transaction implementations, committed data is driven into a consistent state representing the results of transactions while uncommitted data is driven back to its pre-transaction consistent state. Some data structures can be made intrinsically transactional while others require more formal transaction constructs[391].

If high availability is required, increased overhead appears in the data path. Various data recoverability requirements and approaches apply to persistent memory systems using RDMA[390]. Ultimately the cost of recoverability can be summarized by enumerating the various ways in which it forces additional PM writes to occur along with the necessary ordering constraints among them. We view this as a type of write amplification.

- Intrinsically transactional persistent memory data structures are most efficient. For example PM trees can be updated using subtree replacement accomplished with a commit point comprising a single pointer or integer update at the root of the subtree. Replacement subtrees containing new data are constructed in PM free space prior to the commit point. This temporarily used free space must be tracked by some means that is itself persistent and coordinated with the commit point of the subtree. Specifically, at the instant the root pointer is updated, previously free space is now part of the tree, and old tree space is free. Until the old space becomes part of a free list there is risk that the old space could be orphaned, creating a persistent memory leak should power fail. Intrinsically transactional data structures require at least partial ordering of PM updates which causes additional flushing and fencing in a flush on demand (as opposed to flush on failure) environment. In the tree example, the subtree must reach PM in its entirety before the root of the subtree is updated in PM.

- Any persistent data structure change that is larger than a pointer or integer and is not contained within a single intrinsically transactional data structure cannot use unprotected in place memory updates. Generally protection in this usage means capturing a partial pre-image or post-image of the data structure [131]. Either type of image requires additional writes to PM which again requires partial ordering of PM writes.

- If high availability is needed, additional writes to PM are explicitly required and they must be coordinated to assure recoverability.

Additional PM writes and ordering actions can cause resource proportionality issues in the form of write amplification if they are used when they are not needed. The most common scenarios of this type are those in which atomicity is assured independently for the same data in multiple layers of software or hardware. For example an OS might be performing additional PM writes to assure PM cache page atomicity while an application using the PM cache requires only that all writes in each transaction reach PM before a final write that commits the transaction. In this scenario an unnecessary PM cache page atomicity requirement creates a resource proportionality issue. Examples of this issue are quantified in Section 8.4

As another example, some applications may be able to afford some loss of recent work in the event of a failure as long as the state of the data after recovery is consistent. If no work loss can be tolerated then action must be taken to assure recoverability with local and possibly remote consistency before every data update is committed. If, on the other hand, limited work loss such as the loss of some transactions that actually completed before a failure can be tolerated, then some of the actions necessary to assure recoverability can take place less frequently. These actions may include the transmission of data between PM components to assure redundancy and the associated coordination of consistency. This is strongly analogous to remote copy features in enterprise storage solutions such as IBM storwiz[378]. In general, remote copy solutions describe the amount of work that can be lost during a failure as a "Recovery Point Objective" or RPO. RPO of 0 indicates that no effective work loss is tolerated. Non-0 RPO indicates the amount of work loss usually in terms of a time limit or a transaction count. Setting the right RPO for recoverability helps to control the bloat of excessive data transmission and coordination between components.

8.2.2 Process flow implications

Persistent memory also changes process flow surrounding the points in applications that previously would have involved IO to storage devices.

8.2.2.1 Context switch elimination

As described earlier in Section 8.1 the low latency of PM enables context switches and interrupts to be eliminated. Context switches are artifacts of IO protocols used with SSD's and HDD's. Memory protocols, on the other hand, do not involve context switches. One way to eliminate context switches is to use memory protocol instead of IO protocol. Another is to use polling instead of interrupts within an SSD style protocol.

PM and SSD protocols can be simulated in such a way that they play similar roles in a memory and storage hierarchy[383] so as to evaluate the effect of eliminating IO including command/response protocol, DMA orchestration, context switching, and interrupts.[5] For random workloads, direct access to PM without a file system experienced only 20-25% of the latency of SSD style access for media latencies up to 25 μS. Applying the equation 4.1 from Section 4.4 this means that the SSD protocol has a resource amplification of up to 5 relative to the optimal method of accessing PM on random workloads. Sequential write workloads got much lower benefit from PM protocol and there was no PM protocol benefit for sequential reads. This is because SSD style protocol overhead becomes inconsequential when a single command is used to stream large amounts of data. Note that these experiments did not include marshaling. Furthermore, memory style protocol eliminates virtually all of the software stack overhead that is present in SSD style IO protocol.

As another data point, context switching time has been measured independent of IO to average 3.8 μS on a 2.0 Ghz Intel Pentium Xeon[349]. This result does not tell the whole story in that while another process runs, it tends to evict data from processor caches that must then be fetched from memory again. Depending on workload this indirect cost is shown to escalate into the 1 mS range.

Unnecessary context switches could be viewed as a resource proportionality bloat issue; however, a total lack of context switches is generally undesirable even if they are not required for persistence. Systemwide factors such as process priority, time-slicing, and scheduling fairness should still occur and some of these may effectively share context switch overhead with IO. For example a process may start an IO and block when it was about to be preempted for time-slicing anyway.

In addition to the lack of context switches when directly accessing PM, applications that use disk emulation with persistent memory can create new resource proportionality issues especially with large data transfer sizes. "Smart Context Switching"[154] has been proposed for making decisions about whether to context switch or not when doing IO to low latency devices. Device type, latency, CPU Utilization, and IO size are considered in real time. Smart Context Switching serves as an example of a best practice for resource proportional system design.

---

[5]In order to evaluate protocols independent of media, trials comparing simulated memory access and IO protocols assume the same media latency for each in a given trial. This makes access method differences comparable only within trials.

8.2.2.2   Perturbation of CPU utilization

In a very real sense, the use of PM instead of storage converts IO bound applications into CPU bound applications. This creates a system perturbation that may significantly change the effectiveness of many existing applications. For example, extensive development has gone into the optimization of database queries based on an assumption that CPU's must wait for storage access. Many recent databases have moved to in-memory implementations that make these older optimizations unnecessary or even detrimental. To date, in-memory database optimizations have focused on read intensive rather than write intensive workloads. The use of PM creates opportunities to extend in-memory database optimization to write intensive workloads. This is likely to cause another wave of re-optimization.

To further explore the impact of PM on CPU utilization consider the simple model illustrated in Figure 8.1 showing systems that use IO to access storage.

An application running on a CPU core periodically generates IO's for storage access. The application itself receives requests at some rate. Each request requires the application to generate some number of IO's interspersed with its computation. This can be thought of as an alternating pattern of IO's and think times that result in a new arrival rate of IO's from the application. Although it is not shown, each (non-PM) IO may have a driver stack computation component and a wait time during which the CPU is idle.

In the context of this model, PM causes the following shifts in processor utilization.

- Elimination of IO related software overhead – All other types of storage access require software driver stacks that can be completely eliminated with PM. Driver stack overhead has been measured for SSD's at 45 μS or more[383].[6] The CPU utilization impact of this depends on the proportion of time previously spent in the IO stack relative to the think time of an application between storage or PM accesses and the application arrival rate. For example as a thought experiment, suppose an application using PM with an SSD style IO protocol stack receives and completes exactly one request per CPU core every mS where each request involves 20 IO's and each IO consumes 25 μS of CPU time with no wait time. In such a scenario the CPU may be spending half of its time in the storage stack.

- Elimination of idle time – Since PM latencies are much closer to instruction execution times than those of other types of storage, there is no reason for processes to block waiting for PM access to complete. For

---

[6]This measurement was taken using a conventional SSD driver in a memory mapped configuration on a 4 core 2 GHz PCIe gen 3 system without NVMe. Shorter times have been advertised more recently in other software configurations such as NVMe without memory mapping.

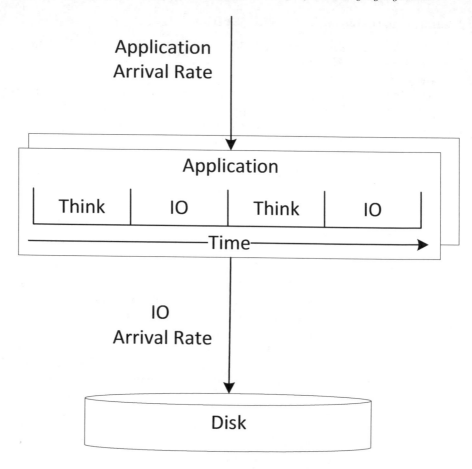

FIGURE 8.1: Simple application and IO arrival rate model.

example if the accumulated application think time to fulfill a request is equal to the inverse of the application arrival rate, then the application becomes entirely compute bound with PM (because PM accesses occur entirely during processor instructions) when it would have been IO bound with SSD's.

- Increase in CPU utilization for flush instructions – Although computation in the IO stack is eliminated with PM, flush and fence instructions may still be needed. Each of these instructions can consume significant CPU time without any wait time. This may increase the minimum time between context switches when CPU's are being time sliced among processes for scheduling fairness.

The impact of PM on processor utilization relates to resource proportionality in the following ways.

- The conversion of IO bound workloads into CPU bound workloads causes bottlenecks to shift. The system response to this may be positive if the new workload leaves CPU's less heavily utilized even though it is compute bound. This would only occur if the arrival rate to the application remained the same even though the time to complete each request was reduced. On the other hand, if processor utilization decreases, customers might modify system configurations to consolidate requests onto a smaller number of systems. When that happens, CPU's or networks are likely to become bottlenecks rather than storage. As a result, resource proportionality issues related to performance that were previously inconsequential may become significant.

- The elimination of IO bound workloads changes the motivation for concurrency. Historically it has been important for a CPU to have many threads of execution ready to run just to take advantage of available processor time during IO's. With the disappearance of IO bound workloads a given core can execute an application request from end to end without blocking as long as coordination with peers is limited. When this happens, applications are increasingly motivated to drive parallelism to the point where they can consume all available cores, or use higher throughput cores.

- Core motivated concurrency in turn increases the demand for cores that share memory regions so they can work on the same data. As memory scales, the latency to reach it becomes more varied due to traversals across processors or memory interconnect switches. This increases the impact of non-uniform memory access (NUMA) on software execution efficiency. Systems that manage memory locality efficiently will make more resource proportional use of processors by avoiding stalls in their memory pipelines. This topic is explored more deeply in Chapters 9 and 12.

It is difficult to anticipate the implications of IO bound workload disappearance. That said, applications that waste computation on storage optimizations that are no longer relevant, and those that cannot adapt to high parallelism in a NUMA environment, will be at a resource proportionality disadvantage.

8.2.3   Code reuse implications

The characteristics of PM are driving a significant transition in application architecture. Such transitions are protracted at best and in some cases never complete. An example in recent networking history is the IPv4 to IPv6 transition. The IPv4 to IPv6 transition is a renovation of the primary internet routing protocol that has been underway for over 20 years. Even though core infrastructure is now almost entirely upgraded, the new protocol is only available to about 25% of google users[403]. Bidirectional translation between IPv4 and IPv6 was built into the design, enabling the necessary long lived dual stack scenario in network infrastructure. The transition to PM has similar characteristics including the following.

- Some applications may never fully transition either because they do not get much benefit or they are legacy artifacts which, though still in use, have little or no ongoing support.

- While each technology (PM and disk) has a different stack, cross-stack translations between each application access approach and underlying technology can be accomplished in software. The dual native stacks can be cross-connected for inter-operability.

It is interesting to observe the role of memory mapping (the practice of assigning virtual memory addresses to data that permanently resides in non-volatile memory or storage) in the PM and MM disk paths. Memory mapping first appeared in the 1980's as a way of using DRAM to present disk contents as variables that can be manipulated directly by applications. Such use of memory mapping matches the "MM Disk Path" in Figure 8.2 below. With the emergence of PM, new implementations of memory mapping can be used in the native PM path without changing applications that already used memory mapping with disk drives. Both of these memory mapping implementations feed into the dual stack scenario created by the transition from disk drives to PM, illustrated in Figure 8.2.

The left side of the figure differentiates applications that use block structured IO from those that use MOV/LD/ST instructions[7] to access data. The right side of the figure differentiates block structured hard disk or flash technology from byte structured PM technology. The Disk path represents an application that uses IO to access hard disks or flash using disk drivers. This path is used pervasively to date for storage. The PM path at the bottom represents an application that has been re-written to be PM aware, using MOV, LD, or ST processor instructions to directly access memory mapped PM. The PM path uses the processor's memory access pipeline with additional flush and fence instructions to ensure persistence.

The RAM Disk path shows how applications that are not PM aware can still use PM. Depending on application requirements the RAM disk driver could be as simple as a single Mov instruction to copy data between an application buffer and a range of PM that is designated to be part of a RAM disk. More sophisticated RAM disk drivers such as the BTT driver[361] may provide additional block atomicity[8] guarantees needed by some applications.

---

[7]MOV instructions copy words from one memory location to another. LD instructions copy words from memory to processor registers. ST instructions copy words from processor registers to memory.

[8]Block atomicity assures that a given 4K disk block is never partially written even in the event of power loss or reset.

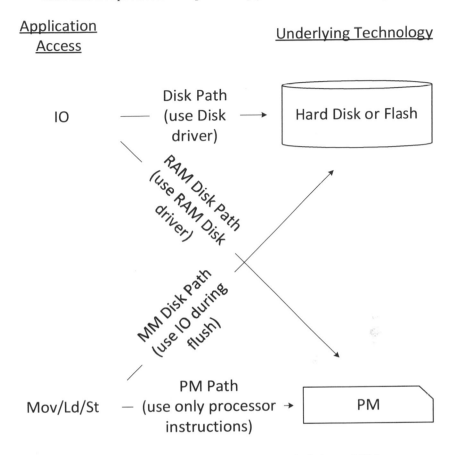

FIGURE 8.2: Dual stack scenario with disks and PM.

The Memory Mapped (MM) Disk path shows how PM aware applications can function in systems that do not have PM. For such systems the application accesses regular volatile RAM as if it were PM. When the RAM is initially memory mapped, data is read from hard disks or flash to populate the memory mapped area with the contents of storage. During flush or sync operations, modified portions of the RAM are written back to hard disks or flash to ensure persistence. This is the way existing *mmap* and *msync* actions work in Linux file systems as described by Linux man pages[245].

The Disk and RAM Disk paths allow excellent code reuse of existing applications while enabling some benefit from PM if it is present. The PM path is advantageous for many applications, but significant amounts of existing application code may not be reusable depending on its internal software architecture.

The least advantageous of the paths here is the MM Disk path. This path allows memory mapped application access but it does not offer much, if any, application performance improvement, due to the high latency of disk IO. This path should only be used when an application that depends on MOV/LD/ST access is forced to run on a legacy system without PM.

When PM is present, the choice between a RAM disk driver and processor instruction access to PM depends on the data access granularity and the transactional requirements of the application.

Code reuse across the disk-to-PM transition is the main motivation for the dual stack. It can also be a significant source of resource proportionality issues in the form of marshaling, access granularity conversions, and atomicity requirements that affect each path in a different way. In the following sections we will elaborate on access granularity and atomicity. Then in Section 8.3 we explore dual stack RP more quantitatively.

8.2.3.1   Data access granularity

Even applications that use only IO to access storage can only change data using processor instructions to manipulate memory. When using IO, applications make those updates in a RAM buffer and write the buffer to the disk. With PM the application can use processor instructions to directly manipulate persistent memory bytes. Although processor instructions appear to manipulate compiler recognized fundamental data elements such as integers or pointers, processors actually manipulate memory in processor cache line size units. In general, integers and pointers are smaller than cache lines and aligned to fall within cache line boundaries. When modified memory is written to disks, the memory cache line updated by the processor is embedded in a larger storage block. The same thing happens with a RAM disk.

Figure 8.3 illustrates an example of this pervasive hierarchy of granularity.

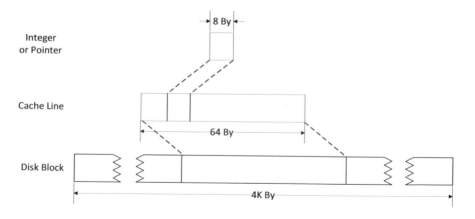

FIGURE 8.3: Granularity transitions between processors and disks.

Here we see an 8 byte pointer or integer embedded by the processor in a 64 byte cache line. We then see the cache line embedded in a 4K storage block. Although these sizes are very common in processor and storage architecture, the actual values are not critical to reasoning about resource proportionality as it relates to access granularity. Whenever a granularity mismatch occurs, system hardware and/or software must expend additional cycles to compensate. In particular, the larger region must be read, partially modified, then written in its entirety. This may cause the demand placed on memory or storage to be amplified over and above the actual workload observed from within the application. These variations in amplification create differences in resource proportionality with respect to performance, especially in workloads that involve small random accesses as opposed to large block or streaming access.

In Figure 8.2, the PM path traverses only one granularity boundary while all of the others traverse at least two. This is unavoidable on the Disk path but on the other two, the failure to align application needs and underlying technology creates an RP issue in the form of extra granularity boundaries that can affect both capacity efficiency and performance.

8.2.3.2 Atomicity

Section 8.2.1.3 described data recoverability with PM. Data recoverability is critical with block IO as well as PM. Data recoverability requires that the state of data after a power failure, hardware failure, or reset adheres to application specific consistency rules. Often the enforcement of these rules depends on the ability to create atomic (all or nothing) actions in the event of failure. If a piece of data is partially updated due to power loss it may inhibit the application's ability to re-establish consistency. There are only a few ways to assure atomicity with respect to failure. These can be broken down into three basic elements.

- Checksums – Records are written to disk or PM with checksums or other error detection code that allows incomplete or invalid records to be discovered. If a prior version of the record is still available, consistency can be re-established without loss of data.

- Atomic writes in hardware – A write to a disk block or a fundamental data type is guaranteed by hardware to complete once it has started. This is achieved using batteries or capacitance to complete the write after power loss has been detected.

- Transactions – Writes are tracked in such a way that they can be rolled forward or back during recovery from power loss or failure. The means of tracking writes accurately is based on one of the above elements (checksums or hardware). In many cases a log of data images is maintained using one of the above methods to assure log consistency which, in turn, is used to assure data consistency.

There are many well known ways of combining these elements with careful management of multiple data versions to achieve data recoverability after failure. At the same time it can be very difficult for one layer in a complex application and storage stack to know what atomicity guarantee is given by a lower layer. Therefore the simplest way to develop highly leverage-able code is often to implement one of these approaches in each layer. This is a very common source of bloat.

Considering the dual stack of Figure 8.2 once again, the granularity of atomic writes in hardware depends on the path through the diagram. Furthermore the granularity expected by the application does not match the granularity of hardware atomicity on the two middle paths. This is why the BTT RAM disk driver[361] is needed to emulate block IO atomicity using PM hardware atomicity. The MM Disk path is likely to contribute to avoidable bloat because the application may be written to manage PM hardware atomicity when in fact atomicity is being provided only by the disk drive at block granularity. If the application uses multiple flushes and fences to exploit PM hardware atomicity on that path, it creates many extra disk IO's that would have been avoided on the upper two paths in the dual stack diagram.

Suppose an application is layering transactions over the RAM disk path in a dual stack. The RAM disk adds considerable software overhead to provide atomicity on every write. At the same time the atomicity the application actually needs is derived from its transaction implementation which only requires ordered groups of non-atomic and atomic writes.

8.3   Storage Stack Bloat in the Dual Stack Scenario

With the example of Section 8.2.3.1 in mind, along with the storage and memory latency information from Table 8.1 we will now explore the performance and resource proportionality impact of the data flow, granularity conversions, and atomicity features implied by each path through the dual stack of Figure 8.2.

The following figure is a reference model showing data flows from processor cores to persistence (memory or storage) for the four paths that comprise the dual stack scenario.

The far left of the figure shows a label for each of the paths through the dual stack scenario. Each path is represented by a horizontal stripe. The native PM and Disk paths are lightly shaded while the paths that cross in Figure 8.2 are slightly more shaded. Next to the path labels, a CPU is shown with cores running applications and volatile caches (e.g., L1 - L3) to accelerate data access. Immediately to the right of the CPU there is workspace RAM where applications manipulate data using processor instructions such as LD, ST, and MOV. In the PM path the workspace is in the PM itself. This illustrates

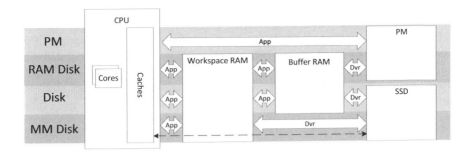

FIGURE 8.4: Data flow reference model for dual stack scenario.

a fundamental efficiency advantage that is not available in any of the other cases.

Although applications may be presented with a variety of data element sizes such as characters, integers, and pointers, CPU access to RAM or PM (any "App" arrow leaving the CPU in Figure 8.2) is constrained to cache line granularity as shown in Figure 8.3. The pervasive practice of embedding application data elements such as 8 byte pointers in cache lines is the result of a processor-to-memory interconnect performance trade-off that is effectively hardwired into processors. CPUs only read and write cache lines to memory to accommodate memory interleaving and predictable pre-fetch. Both of these practices help longer data transfers while meeting typical performance expectations for smaller elements. Some CPUs attempt to complete the writing of a cache line during power loss so as to provide atomicity with respect to power loss; however cache lines are not visible to the application except when delivered as byte, integer, or pointer atomicity. While some applications benefit from memory accesses smaller than a cache line, this is essentially a moot point as cache line granularity of memory access is embedded so deeply in CPU design.

To the right of workspace RAM the reference model illustrates buffer RAM where variables flowing to and from SSD's or HDD's are marshaled into disk blocks. The workspace RAM and the buffer RAM are assumed to use the same DRAM technology. The buffer RAM is analogous to the Linux buffer cache; however, in this model flushing is controlled entirely by the application so there is no cache algorithm. Other operating systems may have different buffer RAM analogs. Buffer RAM is where the second granularity shift of Figure 8.3 occurs on the Disk and RAM Disk paths. Marshaling is included in the RAM disk path because the application is not aware of any functional difference between disks and RAM disks. This illustrates the downside of the ease of integration that comes with RAM disks. The memory mapped disk case does not go through the buffer RAM because the layout of the application variables in the workspace is presumed to precisely match the data layout on the SSD.

The PM and SSD boxes to the far right represent non-volatile (or effectively non-volatile) devices.

Some modern processors allow processor caches to participate in DMA operations to improve IO performance [191], a forward looking capability that is likely to become more common over time. This means that processor cache contents left over from the transfer of data from a workspace to a buffer may be used instead of memory as the source of the subsequent DMA operation that transfers the workspace to a disk. The flow of data directly from processor cache to disk drive is represented by the dotted line at the bottom of the figure.

8.3.1   Analytic model of the dual stack scenario

For the purpose of evaluating dual stack scenario RP, we model an application workload comprising some number of reads, writes, or read-then-write updates[9] (A_R, A_W, A_U), all having a uniform size (S), randomly distributed over some total capacity (C). Writes (A_W) only model situations where applications are ingesting new data or generating fresh data that completely fills cache lines. Read-then-write (A_U) updates apply to any other application use case wherein partial cache lines might be updated. After a specified number of bytes written (S_C) the workload is defined to execute a commit. Note that S_C is constrained to be less than or equal to $A_W + A_U$ for a given workload trial. The activity that takes place during reads, writes, and commits depends on which of the four paths is used.

More formally, let

A_R, A_W, A_U = Number of Read, Write, and Update accesses, respectively

A_X = Any of $A_R, A_W, or A_U$

S = Size (Bytes) of all accesses (all are the same size in a given latency calculation)

C = Total capacity of PM or Disk media

S_C = Size of commit. Number of bytes written or updated before a commit occurs

The following assumptions apply to all paths.

- The flow of data between the CPU and any type of RAM or PM has cache line granularity. If the application read/write size S is smaller than a cache line then it is rounded up to a whole cache line. Cache line size is specified as a property of the memory components in a given system even though it is actually CPU specific.

- The number of cache lines accessed by a given workload is calculated by superimposing a uniform address distribution of all reads, writes, and

---

[9]Read-then-write is referenced here rather than read-modify-write because modification is not modeled. The modify is assumed to occur in the CPU cache so the time consumed is negligible in this model.

updates (A_R, A_W, A_U) having a single size S, on the collection of cache lines in the total capacity C. This is used to calculate a probability that the workload will access a given cache line which, in turn, is converted into the expected number of cache lines that the workload will access. This is tracked separately for reads, writes, and updates. The write that occurs during a given update accesses the same memory location as the read for that update.

- The transfer of data between workspace and buffer RAM is under processor control and uniformly randomly distributed due to marshaling.[10] The cache line access calculation described above is applied to workspace and buffer RAM.

- The number of blocks accessed on a disk or RAM disk is calculated in the same manner as the number of cache lines. The input to the calculation is the total number of cache lines accessed and the output is the total number of blocks accessed assuming a uniform distribution of cache lines across blocks. Block size is a property of a disk device.

- Processor cache flush times are assessed during commits where PM is the destination. It is assumed that the processor cache can hold all of the data written between commits so none of it was flushed in advance. Processor flush latencies are not assessed on disk IO because we assumed a processor implementation where DMA can access the processor cache[191].

Based on those assumptions, to compute the expected number of cache lines accessed in a workload, let

S_W = Word (pointer or integer) Size in Bytes
S_L = CPU Cache Line Size in Words
S_B = Disk or RAM Disk block size in Cache Lines

then

Probability that a given word is accessed $P_W(A_X) = \dfrac{\left\lceil \frac{A_X * S}{S_W} \right\rceil}{\left\lceil \frac{C}{S_W} \right\rceil}$

Probability of no words accessed in a given cache line

$$P_N(A_X) = (1 - P_W(A_X))^{(S_L)}$$

and

Expected cache lines accessed given a number of read, write, or update accesses in a workload

$$E(A_X) = \left\lceil (1 - P_N(A_X)) * \left\lceil \frac{C}{S_L * S_W} \right\rceil \right\rceil \tag{8.1}$$

---

[10]We assume that marshaling reorganizes all of the data between workspace RAM and buffer RAM.

8.3.1.1 PM path

On the PM path the application uses LD, ST, and MOV instructions to access S bytes at a time, although accesses are rounded up to cache lines when they reach PM. Commits flush the cache lines associated with S_C writes to PM. Flush latency is a property of the memory component.

For the PM path, let

LP_R = Cache line read latency
LP_W = Cache line write latency
L_F = Cache line flush latency
T_P = Total PM workload latency
then

$$
\begin{aligned}
T_P = (E(A_R) + E(A_U)) * LP_R + \\
(E(A_W) + E(A_U)) * LP_W + \\
E(S_C/S) * (\frac{(A_W + A_U) * S}{S_C}) * L_F
\end{aligned}
\tag{8.2}
$$

The first two terms of T_P account for accesses to PM due to A_R, A_W, and A_U where A_U generates one of each. The third term accounts for flushes due to S_C. Note that flush latencies are never probabilistically combined across commits so the expected number of cache lines per commit, $E(S_C/S)$, is multiplied by the number of commits ($\frac{(A_W + A_U) * S}{S_C}$) in the flush latency term.

8.3.1.2 RAM disk path

On the RAM Disk path the application uses LD, ST, and MOV instructions to access workspace RAM. The access pattern between the CPU and workspace RAM is the same as with PM, but workspace RAM latency may be different from PM latency as defined by parameters of the RAM component. The activity of a commit is quite different from PM because cache lines that were written since the last commit are copied one by one from random locations in workspace RAM to sequential locations in buffer RAM. After buffers have been composed they are copied sequentially from buffer RAM to PM in block size chunks. Each cache line write followed by an eventual commit generates a write and a read from workspace RAM, a write and a read from buffer RAM, and a write to PM. The write to PM is modeled as a single non-cached move from workspace RAM to PM to assure persistence as efficiently as possible.

For the RAM Disk path in addition to the above, let

LW_R = Cache line workspace RAM read latency
LW_W = Cache line workspace RAM write latency
LD_F = Flush latency for A_X disk blocks – accounts for memory copy from workspace to uncached PM destination
$E_D(A_X) = E(A_X)$ modified to compute expected blocks accessed (using S_B)

rather than cache lines
T_R = Total RAM Disk workload latency
then

$$
\begin{aligned}
T_R = \; & 2 * (E(A_R) + E(A_U)) * LW_R + \\
& 2 * (E(A_W) + E(A_U)) * LW_W + \\
& (E_D(A_R) + E_D(A_U)) * LP_R + \\
& E_D(S_C/S) * (\frac{(A_W + A_U) * S}{S_C}) * LD_F
\end{aligned}
\tag{8.3}
$$

The first term of T_R accounts for the reads from workspace RAM due to A_R and A_U. It is doubled because the workload also applies to the buffer RAM on this path. The second term accounts for writes to workspace RAM due to A_W and A_U, also doubled due to buffer RAM. The third term accounts for PM reads due to A_R and A_U. The last term accounts for the number of cache lines flushed to PM during commits due to S_C.

8.3.1.3 Disk path

On the Disk path the only flow difference relative to the RAM Disk path is that reads and commits perform DMA from buffer RAM to SSD instead of sequential cache line transfers. Of course SDD latency contribution is at least an order of magnitude more than PM.

For the Disk path in addition to the above, let
LD_R = Block read latency from disk to workspace
LD_W = Block write latency from workspace to disk
T_D = Total Disk workload latency
then

$$
\begin{aligned}
T_D = \; & 2 * (E(A_R) + E(A_U)) * LW_R + \\
& 2 * (E(A_W) + E(A_U)) * LW_W + \\
& (E_D(A_R) + E_D(A_U)) * LD_R + \\
& E_D(S_C/S) * (\frac{(A_W + A_U) * S}{S_C}) * LD_W
\end{aligned}
\tag{8.4}
$$

As with the RAM disk path, the first two terms of T_D account for reads and writes to workspace and buffer RAM. The third term accounts for reads from disk due to A_R and A_U. The last term accounts for writes to disk during commits due to S_C.

8.3.1.4 MM disk path

On the MM Disk path it is assumed that the data organization in workspace RAM matches the organization on the disk so the buffer RAM access and associated marshaling does not occur. During commits, DMA is used to transmit data directly from workspace RAM to the SSD.

For the MM Disk path in addition to the above, let
T_M = Total MM Disk workload latency
then

$$
\begin{aligned}
T_M = (E(A_R) + E(A_U)) * LW_R + \\
(E(A_W) + E(A_U)) * LW_W + \\
(E_D(A_R) + E_D(A_U)) * LD_R + \\
E_D(S_C/S) * (\frac{(A_W + A_U) * S}{S_C}) * LD_W
\end{aligned}
\tag{8.5}
$$

The first two terms of T_M account for reads and writes to buffer RAM. Unlike the RAM disk case these are not doubled because the buffer RAM is not involved in the MM Disk path. The third and fourth terms account for disk Reads and Writes using the same calculation as the RAM Disk path even though the workspace RAM is accessed instead of the buffer RAM.

8.3.1.5   Dual stack RP baseline

Although there are many potential patterns and distributions of memory and disk access, this workload model can be used to explore significant trade-offs across the paths through the reference model. The PM scenario is the resource proportionality baseline because all of its granularity translations are handled by the processor at memory speed. None of the other paths through the dual stack can be faster than this because all of them require memory access to buffers in support of the IO stack.

Note that both axes are log scale for all of the graphs in this chapter.

As a first analysis we plot latency vs. bytes written for each of the 4 paths, using component latencies described in Table 8.1. For this analysis we assumed single threaded accesses were grouped into 8 byte words with cache line size of 64 bytes, IO size of 4K, and total capacity of 2^{24} bytes. The number of bytes written before a commit (S_C) for this trial is 1K.

This chart clearly shows that PM latency is slightly better than RAM Disk which is in turn one order of magnitude faster than overlapping Baseline Disk and MM Disk results. Commits are 1K unless fewer total bytes were written. In the PM path, commit sized flushes contribute significantly to latency. As described above the RAM Disk path involves word-by-word marshaling of commits through 4K buffers. The Disk and MM Disk latencies are nearly identical because both they are dominated by disk latency.

Applying RP amplification equation 4.1 from Section 4.4 at 64 bytes accessed with PM as the specialized scenario, the RAM Disk amplification is 2.3 while the Disk amplification is 16. This is not the whole story, of course, because SSD cost/capacity is less than half that of PM.

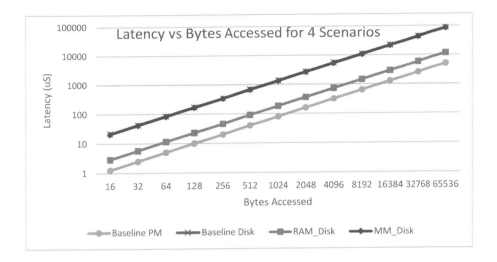

FIGURE 8.5: Latency vs. bytes accessed with 16 by write size and 1024 byte commit. The Baseline Disk and MM Disk lines coincide at about 7.3 times the latency of the RAM Disk. RAM Disk latencies are about 2.3 times PM latencies.

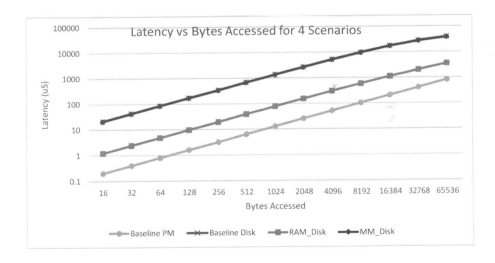

FIGURE 8.6: Latency vs. bytes accessed with 16 by read size. Here too the Baseline Disk and MM Disk lines coincide.

To explore workloads more broadly, consider the read graph in Figure 8.6. Read latencies in this model are very predictable with RAM disks at about 4x PM and the other disk workloads coinciding an order of magnitude above. Reads are simpler because there is no equivalent to the commit.

Several primary conclusions can be drawn from this very basic model.

- Using memory mapping with disk drives as in the MM Disk path yields no appreciable gain over raw disk performance. While this path does avoid marshaling, that overhead is much smaller than disk latency. While memory mapping may reduce software overhead slightly in the MM Disk case, it is bloat when it comes to performance improvement relative to disk. Note that although the buffer in Figure 8.4 is analogous to the Linux buffer cache, no caching is modeled here so its removal is largely inconsequential to the results of this model.

- The overhead of marshaling is visible in the write workloads as data movement through the buffer RAM is the main contributor to the difference between PM and RAM Disk curves in Figure 8.5. While it was very small on the scale of disk workloads it is observable on the scale of RAM Disk workloads, giving PM a moderate advantage over RAM disks. Once disk overheads are removed from the equation, marshaling becomes a more significant bloat factor.

- Since disks dominate the latency on the paths that include them, it is reasonable to construct systems with 10 or even 100 disks per processor. This would require a storage area network (not modeled here) and IO concurrency would need to be increased. In such systems marshaling may again become a significant bloat factor with respect to CPU utilization, so the MM Disk path may provide more benefit.

8.4 Multi-Layer Storage Stack Bloat Related to Atomicity

Section 8.2.3.2 described approaches to atomicity in support of application consistency requirements. This model assumes that the processor offers word atomicity which can be used with metadata to implement larger atomic actions such as all of the bytes in a commit or block. Atomicity can be viewed as an additional feature that was not accounted for in the Section 8.3 dual stack analysis. We will now analyze the latency impact of adding atomicity to PM and RAM disk paths. Additional atomicity related latency is modeled slightly differently for each of the two paths. Atomicity latency is always over and above baseline latency for the path being analyzed.

- On the PM path, memory accesses to a log in PM are added to record every S-sized (8 byte) write. Writes are sequentially packed into cache lines along with some metadata for tracking, which adds an additional 20 percent to the data that must be written to the log. Each cache line written to the log is flushed. During each commit, an additional cache line is written and flushed to represent the recording of the commit in the log, and the updates recorded in the log are applied to the PM image. This delayed image update during commit has a latency equivalent to the baseline PM model without atomicity. As with any log based transaction, the replaying of the log after failure provides the all-or-nothing property of atomicity.

- On the RAM disk path, each 4K block written is made atomic by writing it to a new block taken from a free pool. This is modeled as writes to 3 PM cache lines for metadata to track the block allocation, committing and freeing of the block. Each of the cache lines are flushed individually. The result of these writes is the same block atomicity property that is built into disk drives. In addition the model adds one atomic block write to each commit to record groups of blocks being committed together. It is assumed that the additional 3 cache lines of metadata per block write plus the one extra block write are sufficient to allow chains of blocks to be accumulated and manipulated together during the commit. This type of solution requires garbage collection of free space during failure recovery.

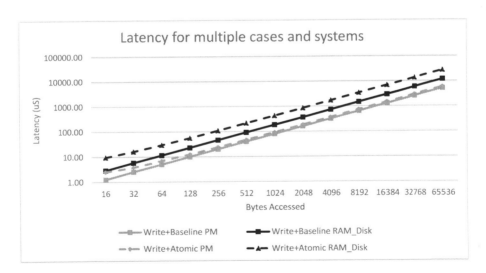

FIGURE 8.7: Latency for baseline and atomic PM paths. The baseline curves do not include the Atomicity feature. The Write+Baseline PM and the Write+Atomic PM curves converge for larger numbers of bytes accessed.

Figure 8.7 shows the baseline and atomicity latencies for each of the baseline PM and RAM disk paths. The distance between the solid and dotted gray lines shows the latency increase for PM atomicity including additional metadata writes to manage the log. The increase due to data duplication in the log is moderated by the packing of random words written into cache lines. That effect increases in significance as the number of bytes written approaches the maximum bytes per commit (C=1024) at which point it nearly converges with the baseline. The latency difference for the RAM disk is larger because it uses atomic block writes rather than a log based on atomic word writes. This illustrates a scenario where the feature of block atomicity implemented using word atomicity in metadata is wasted. Using equation 4.1 once again, for nontrivial amounts of data written, RAM Disk atomicity shows RP amplification factors between 4 and 7 relative to the direct use of PM's word atomicity in a log.

In a multi-layer storage stack such as a file system layered over a RAM disk, both the file system and the RAM disk might implement log based and block based atomicity independently. Perhaps the file system depends on block atomicity and does the same thing for both SSD or PM based RAM disks. This is a potential RP issue with respect to both performance and resources consumed because a word atomicity based log would be more efficient. Another possibility is that multiple clients of the block device have different atomicity requirements. For example a log may need atomicity from the block device while writes to the rest of PM may not because their atomicity is guaranteed by the log. The bloat in this scenario comes from sharing an implementation that is overkill for some users but not for others. Chapter 6 describes several approaches to removing this type of bloat.

8.5   Resource Proportional High Availability

High Availability (HA) is often an important feature of storage systems that also applies to PM. The primary costs of HA are the additional capacity for redundancy and writing additional copies in real time. Managing redundancy in storage is a well understood field encompassing RAID and more broadly Erasure Coding[309].

The simplest form of redundancy for analysis is mirroring, in which all data is written to two or more locations that are unlikely to fail at the same time. Failure independence is assured by avoiding single points of failure in which a single component failure of any type can cause all copies of data to be rendered inaccessible or lost. In addition, redundant copies must be consistent, meaning that they represent the same state of the data whenever needed, even if a power failure or reset occurs during a write. Section 7.4.1 describes approaches to this problem including two of specific interest here.

- Compare copies after the failure and force a secondary copy to align with a primary copy. This can involve comparison of large amounts of data unless metadata is maintained indicating what data was in flux concurrently with the failure. There are many well known approaches to this that account for various combinations of power loss, reset, and component failure.

- Maintain a third copy of data being updated so as to enable atomic commit points of updates. In some cases the third copy is maintained as a log, while other implementations involve temporary versioning of blocks to avoid updates in place. The third copy is used to restore consistency of the other two after a failure.

The first approach manages consistency of copies with each other but does not provide a data atomicity feature. The second approach manages consistency of copies by providing a type of data atomicity feature. The remainder of this section describes additions to the PM and RAM disk paths to implement data redundancy with or without the atomicity feature described in Section 8.4. The scenarios are summarized in Table 8.2.

TABLE 8.2:   Redundancy Feature Approaches

Technology	without atomicity	with atomicity
PM	Write each cache line to remote PM while maintaining lists of cache lines being committed	Transmit the atomicity feature log to remote PM and apply it in the background
RAM Disk	Write each block to remote PM while maintaining lists of blocks being committed	Perform atomic block writes on remote PM

In the PM model, if HA is enabled but not atomicity, every cache line is committed individually to remote persistent memory (RPM). The latency for the RPM component is increased to account for high speed networking such as RDMA since no networking was assumed to be present in the model of Section 8.3.1. Writes to RPM are also flushed at the remote node in the event that volatile caches appear in the path from the network through the remote CPU to its PM. Even though applications do not get an atomicity feature over and above the word size atomicity provided by some CPU's, consistency of the redundant copies with each other must be maintained. The model adds two words of metadata writes per cache line to record a list in PM containing the addresses of potentially inconsistent cache lines. This minimizes the number of cache lines that must be compared and reconciled after power loss. These additional words are packed into whole cache lines that are written to RPM during each commit. An additional cache line is written to RPM at the end of each commit to clear the list of potentially inconsistent cache lines.

On the other hand, if both HA and atomicity are enabled, the log that is used to implement the atomicity feature is transmitted to RPM during commit. The model assumes that the log contains the new contents of all cache lines that were updated since the last commit. The log is assumed to reside in a single range of cache lines that can be transmitted sequentially to RPM. Each commit also writes two additional cache lines to RPM indicating the start and the end of the remote commit log. After each commit the log is replayed at the remote node so that the RPM image is updated there. Any log replays that were not completed are re-attempted after power failure or reset. This background log replay process could become a bottleneck if commits occur too frequently. The model includes a scenario in which RPM is a steady state bottleneck, in which case remote write and flush latency is experienced during commits. The RPM bottleneck condition is illustrated separately as a worst case latency for PM with atomicity and redundancy enabled.

For redundancy without atomicity using a RAM Disk, RPM receives the same pattern of block writes that goes to local PM, plus 2 additional blocks per commit to track potential inconsistencies. The model assumes that a list of blocks written between commits can be accumulated within a single block of metadata. This is a valid assumption for the commit sizes illustrated here.

Finally if both HA and atomicity are enabled for the RAM Disk, the local PM workload is precisely duplicated on RPM. No additional block writes are added to commits. The block atomicity metadata is sufficient for redundancy consistency due to the lack of background log processing that appeared in the PM case.

8.6   HA and Atomicity Function Deployment Scenarios

Based on feature implementation combinations across PM, RAM disks, atomicity, and HA described thus far we can analyze some performance aspects to shed light on the sources and magnitude of non-RP behavior. Note again that while there are many potential workloads, implementations, and performance metrics, this analysis sheds some light on applications that randomly update variables stored in PM.

Beginning with the PM path, Figure 8.8 shows all four combinations created by the presence or absence of atomicity (Atomic) and HA features. A fifth curve is included to illustrate the case where RPM background activity becomes a bottleneck. All of the workload and component characteristics here are identical to those illustrated in Figures 8.5 and 8.7. The curves that show Atomic alone and Atomic plus HA converge to within 5 to 10 percent of each other so that their curves lie on top of each other when bytes written exceed commit size. For small numbers of writes the Atomic plus HA curve aligns with the bottleneck case that includes Atomic, HA, and remote image garbage collection updates (GC). As the number of writes increases, all but the bottleneck case shift to align with the latency of the baseline. The case with HA and not Atomic has the highest latency because it generates the largest volume of RPM access which, unlike a log, is fragmented and distributed across all of PM.

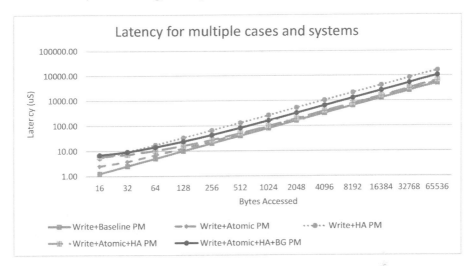

FIGURE 8.8: Latency for baseline atomic and redundant PM paths. The three HA curves coincide for small numbers of bytes accessed. The Write+Atomic+HA curve shifts to coincide with the Write+Baseline curve for large numbers of bytes accessed.

Figure 8.9 shows the RAM disk cases along with two of the PM curves that tend to bracket the PM path cases for reference. Note that the PM cases are usually lowest latency; however, the case with both Atomic and HA where RPM is the bottleneck exceeds the latency of the RAM Disk baseline for small numbers of accesses. The RAM Disk cases with Atomic alone and HA alone fall within 10 percent of each other across the board. This is because the overhead to manage block atomicity and the overhead to manage consistency of redundant copies is similar. Unlike the PM path, the RAM disk case with both Atomic and HA shows the highest latency of all. This is because the block atomicity implementation does not assist redundant copy consistency in the same way that the PM log does.

This case study illustrates several recurring phenomena in resource proportional multi-feature deployment.

- Feature combination analysis can expose implementation aspects that have common value across multiple features, such as the atomicity log in the PM path.

- Backward compatibility features become RP issues in the form of software bloat as upper layer feature implementations evolve to newer technologies as illustrated by the block atomicity feature in the RAM Disk path.

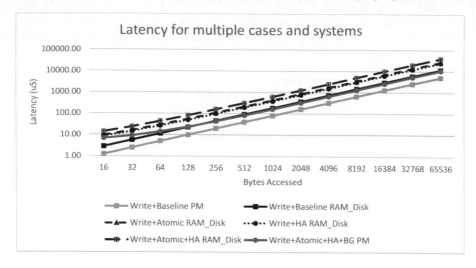

FIGURE 8.9: Latency for baseline, atomic, and redundant RAM disk paths. Write+HA and Write+Atomic lines coincide. The PM and baseline curves are shown for reference and behave as in Figure 8.7.

8.7　Keeping up with the Evolution of Persistent Memory

In this chapter we have described numerous potential resource proportionality impacts of PM, and quantified a case study that illustrates some basic RP trade-offs surrounding PM adoption. As with most RP issues, development time versus run time efficiency is the core consideration in navigating the emergence of PM technology. The system and application considerations involved can lead to several kinds of architectural and business disruption that should be considered in planning a response.

- Incremental disruption: Even without any application level change, the performance differences of a new technology may cause bottlenecks to shift. System provisioning decisions and design rules that were based on obsolete assumptions about bottlenecks become RP performance or software bloat issues. In particular the shift from SSD to PM places more stress on networking especially in scale out systems, so the evolution of higher bandwidth networks is important. Also, IO bound applications may shift to compute bound while at the same time spending less time in storage software and more time in application processing.

- Transformational disruption: The economics of a new technology impacting cost, capacity, and performance may create a new user experience that changes market dynamics. In the case of PM, the expansion of in memory database capacity and the ability to achieve persistence without IO may accelerate analytics. Suppose the advantage of accelerated analytics was leveraged to create faster and more accurate methods of anticipating future user actions. In this type of scenario there may be a multiplicative effect, creating a more radical user perceived change in application responsiveness than one would expect when considering the underlying technology alone. This kind of cascading effect can have transformational impact on user experience.

- Foundational disruption: Fundamentals of application or system architecture may need to change in order to get full advantage of a radical new technology. Persistence without IO is an example of a very disruptive change with a high return in increased RP. It ultimately leads to new data structures and increasingly pervasive transactional behavior that directly impact compilers and libraries, and raise the level of abstraction preferred by applications[390],[391].

The increasing rate of change and competitiveness in computer systems will likely continue to create high stakes RP challenges.

Chapter 9

Resource Proportionality in Memory Interconnects

Having explored the implications of PM latency in Chapter 8 we now turn attention to the interaction of memory with CPU cores across a memory interconnect. Memory interconnects are important to system resource proportionality because processor instruction execution can be delayed or stalled during memory accesses. That means memory access dynamics can cause the processor resource to be wasted. Wasting resources that are provisioned and claimed by applications is non-RP behavior.

Computer system architecture innovation related to memory interconnect technology has been accelerating over the last 5-10 years. Since the early 2000's the technologies that connect the integrated circuit chips on server printed circuit boards have evolved within three distinct roles all of which connect directly to processors.

- Memory modules – synchronous Double Data Rate (DDR) connections to Dynamic Inline Memory Modules (DIMMs)

- IO – Peripheral Component Interconnect (PCI) for connection to networks, storage, GPU's, and all peripherals

- Inter-processor – cache coherent interconnect between processors proprietary to processor vendors

Since 2010 a new role has emerged for a Memory Interconnect (MI) that overlaps the three incumbents. In some ways the new MI role is a trickle-down of large scale non-uniform memory access architecture already present in high end supercomputers. A number of factors have contributed to this turn of events, including the following.

- Decreasing cost/performance of serial interconnect technologies

- Economic feasibility of memory speed optical interconnect

- Increasing role of accelerators such as GPGPU's

- Increasing diversity of memory technologies such as PM described in Chapter 8

- Increasing demand for larger scale memory pools, disaggregated from compute components [305],[306]

In this section we will use instruction processing efficiency as a metric to explore the resource proportionality opportunities and challenges that accompany rapid change in memory interconnects and how they interact with processors.

9.1   Characteristics of Memory Interconnects

Table 9.1 describes characteristics of the current common processor interfaces plus networking and the new memory interconnect (MI).

Generally network interfaces are at the periphery of processor complexes so as to connect servers to the internet, to each other, and to external devices such as storage while electrically and logically isolating all three. Most storage related accesses require multiple network round-trips in practice so access latencies are in the 10 μS range for the fastest devices typically attached. Ethernet is the most pervasive network technology although InfiniBand (IB) is popular in high performance computing. Ethernet networks tend to be large scale so data routing uses the Internet Protocol (IP) which is the most sophisticated of all of the data routing approaches in this table. Typical overarching protocols for networks include TCP/IP, iWarp, and ROCE2. For storage networks, Fibre Channel (FC) and Serial Attached SCSI (SAS) are common. Remote Direct Memory Access (RDMA) is often used for cross-server memory access that may achieve lower latencies than other types of messaging depending on how they are measured.

TABLE 9.1: Interface Characteristics

Interface	Purpose	Access Latency	Routing	Protocol
Network	Connect servers to each other at rack, LAN or WAN scale	10 μS and up	IP address	TCP/IP, iWarp, ROCE2
PCIe	Connect CPU's to GPGPU's and IO cards at chassis scale	1 μS and up	Physical Memory/Register address	Custom register definitions, NVMe
MI	Connect memory, GPGPU's and IO cards at chassis to multi-rack scale	100 nS and up	Intermediate virtual memory address	Gen-Z, Open CAPI, CCIX
DDR	Connect processors to DIMMs	20 nS and up	Memory address	DDR5
Inter-Processor	Connect processors to each other	20 nS and up	Virtual memory address	Processor specific cache coherent

PCIe forms the next ring of electronics closer to processors. Its role is to allow a wide range to electronics including all types of IO, graphics processing, and general purpose processing accelerators. PCIe is also sometimes used to interconnect processors at small scale without cache coherency. Storage access latencies over PCIe are generally at least 1 μS for the fastest devices. PCIe uses memory mapped registers as a pervasive control plane so routing is primarily through memory address. All of the upper level protocols over PCIe come through these custom defined control registers which often refer to blocks of memory containing detailed command and/or data content.

The next closer interface to processors in this taxonomy is the class of new MI's which are the primary subject of this chapter. The new MI's can play the same role as PCIe; however, they are primarily targeted at memory and they are designed to expand to a larger scale than any of the other interfaces except for networks. Minimum MI latencies are between DDR and PCIe latencies although they can be much larger depending on devices attached and overall scale. MI routing uses a memory address space that is virtualized at the edge of the interconnect to allow each processor to have its own view of a memory pool that could otherwise be too large to address. MI address virtualization is implemented in a memory management unit (MMU) or an IO memory management unit (IOMMU) that is placed between the MI protocol implementation hardware and the CPU's virtual memory system. Additional address translation occurs in media controllers between a memory module's MI protocol implementation hardware and the memory itself.

Still closer to processors is the DDR interface which has gone through multiple generations of standardization, DDR through DDR5. DDR is a synchronous interface designed for memory access with latencies as low as 20 nS. It uses memory addressing to access ranges of memory that are limited to the capacity of a small number of DIMMs.

Finally, the interface used to connect processors to each other in a multi-socket complex is unique in that it is designed to support cache coherency across processors using proprietary protocols that are tied into the processor's virtual memory system.

Although the new MI has the potential to replace several other interfaces listed in Table 9.1 it appears that PCIe is the most obvious candidate with the addition of pooled persistent memory (PM) as a focal point. Figure 9.1 illustrates a 2 server reference model that includes all of the interfaces of Table 9.1 except PCIe.

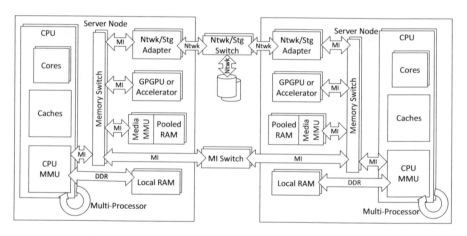

FIGURE 9.1: Memory Interconnect (MI) reference model.

At its extreme left and right edges Figure 9.1 shows a pair of CPU's, each having integral processing cores, caches and a memory management unit (MMU). The MMU's are connected to local RAM via a DDR interface. If there are multiple processors in one server they are connected to each other via a proprietary Multi-Processor interface. This part of the reference model aligns precisely with currently predominant processor architectures.

The reference model diverges from current convention where the new MI connects to the CPU's MMU. Although this connection is shown with the processor's integral MMU, early implementations use a separate chip containing an additional MMU connected to the inter-processor interface. The MI uses local switches to connect to a pool of PM components each with its own media MMU. General Purpose Graphics Processing Units (GPGPUs) and adapters such as Network Interface Cards (NICs) and storage Host Bus Adapters (HBAs) are also connected via the MI. Storage and network interfaces use technologies such as Ethernet, InfiniBand, FC, or SAS to connect to external switches that may in turn be attached to shared SSD's or HDD's. Finally the MI itself may connect to other server nodes through MI switches that are typically integrated into chassis and/or racks.

Many other permutations of the interfaces in Table 9.1 are feasible with strengths and weaknesses that align in various ways with the motivations listed in the introduction to this chapter. We will dive deeper into two aspects of MI architecture that are especially interesting in terms of RP.

- Separation of memory management related to memory media (Media MMU) from memory management related to the CPU's virtual memory system (CPU MMU). The two have historically been intermingled or tightly coupled in ways that, as we will see, are no longer beneficial.

- Enable a memory centric architecture paradigm[1] [392], [40].

The success of flash technologies has attracted considerable R&D investment in PM technologies including STT-MRAM, PCM, and RRAM [270]. Read and write latencies across these technologies range from less than 10 nS to more than 100 nS [96], any of which could be packaged as PM. While details of specific technologies are beyond the scope of this book, several of them involve complex technology specific media management for features to extend media lifetime while protecting data integrity. Various PCM write processes involve multiple phases, in some cases including reads to sense integrity while minimizing and distributing re-writes to extend life [219].

Many complex writing schemes have been proposed for various PM technologies. The net result is a situation where current memory controller architecture and the use of DDR creates the following types of RP issues related to performance and bloat, especially when multiple memory technologies are accessed from a single processor.

- DDR interface functions poorly in the face of unpredictable and asymmetric read and write latencies given the order of magnitude range of variation spanned within and across emerging technologies

- Media specific intervention in the path to physical memory

Section 9.2 examines these memory controller issues and the potential benefits of memory interconnect alternatives to DDR as illustrated by the Media MMUs in Figure 9.1. Section 9.5 further examines RP aspects of media

---

[1]Memory centric architecture promises new levels of performance and flexibility with lower power consumption, e.g., by leveraging the speed of persistent memory, hardware acceleration, and low-latency fabrics.

and storage related functionality in the context of large scale shared memory fabrics.

Several of the factors listed at the beginning of Chapter 9 have created increased interest in memory centric system architecture as illustrated by the connectivity of the MI in Figure 9.1. In particular, larger pools of shared PM and the growing importance of GPGPU's increase the attractiveness of memory centric computing systems wherein memory accesses paths that circumnavigate the CPU. The MI path between server nodes avoids slowing down PM with network (Ntwk) overheads that have historically applied to shared storage. The evolution of Graphics Processing Units from real time video or game rendering into general purpose vector processing for analytics (GPGPUs) has moved significant computation out of CPU's to the point where traversing the CPU, MMU, and DDR paths on the way to memory becomes an RP issue related to performance. Section 9.6 discusses RP aspects of memory centric system architecture.

9.2 Resource Proportionality and the Separation of Media Controllers from Memory Controllers

Memory performance primarily affects memory bound, as opposed to core or IO bound, applications. To analyze RP of memory interconnects and controllers we need a metric that quantifies their impact on core instruction execution rate (IER).

- The IER of a CPU core running an application entirely out of CPU cache (IER_C) is sensitive to processor clock speed, processor implementation, and application instruction usage.

- The IER of an application that accesses memory decreases IER_C by an amount that is sensitive to application and processor interaction characterized by a cache miss probability (P_M), and a stall probability (P_S) that represents the probability that the instruction pipeline will be blocked due to a memory access. Processor behaviors such as hyperthreading and write pipelining often allow processors to avoid blocking the pipeline during memory access.

- The decrease in IER relative to IER_C due to cache misses is sensitive to Memory Latency (L_M) including interconnect, memory controller, and media factors.

We will use Processing Efficiency (PE), defined to be the ratio of IER to IER_C, as the metric for the relative resource proportionality of memory systems. It has values between 0 and 1 with larger values being better.

Let

IER_C = Instruction Execution Rate (Instructions per μS) when running with all cache hits

P_M = Probability of a cache miss on one instruction

L_M = Miss latency (μS)

P_S = Probability that a miss will cause a processor stall

then

Instruction Execution Rate $IER = \dfrac{1}{\frac{1}{IER_C} + P_M * L_M * P_S}$

and

Processing Efficiency $PE = \dfrac{IER}{IER_C} = \dfrac{1}{1 + P_M * L_M * P_S * IER_C}$

All of these parameters except L_M are characteristics of a specific combination of application and processor. A similar model has been used to measure and analyze effective clock cycles per instruction (CPI) for Enterprise, Big Data, and HPC workloads [109] yielding parameters reproduced here as Table 9.2 (© 2015 IEEE. Reprinted, with permission, from "Quantifying the performance impact of memory latency and bandwidth for big data workloads." [109]). CPI is translated into IER_C assuming a 3.1 Ghz processor. The Blocking Factor and Misses per 1000 Instructions (MPKI) provide values for P_S and P_M, respectively. These measurements can be used to calculate PE's as a basis for comparing the RP of different memory system features and implementations across those workload categories.

TABLE 9.2: Workload Class Parameters

Workload Class	CPI (Cache Only)	Blocking Factor	Misses Per 1000 Instructions
Enterprise	1.47	.41	6.7
Big Data	.91	.21	5.5
HPC	.75	.07	26.7

9.2.1 Cost of asymmetric R/W latency with asynchronous MI

One of the new characteristics of PM compared with DRAM is that read and write latencies are up to an order of magnitude apart [270] with write latencies 2-10 times larger than read latencies.

To represent this non-uniformity we need to add a write probability P_W and differentiated latencies to the calculation of PE. Let

L_R = Read miss latency (μS)

L_W = Write miss latency (μS)

P_W = Probability that a miss will cause a cache line to be written to memory

then

$$PE = \frac{1}{1 + P_M * ((1 - P_W) * L_R + P_W * L_W) * P_S * IER_C)} \quad (9.1)$$

We can now plot PE for several workloads to illustrate the impact of DRAM and SCM media technology and interconnect choices on processing efficiency. The leftmost group of bars in Figure 9.2 shows efficiency calculation results using the parameters derived from Table 9.2 in equation 9.1. The baseline uses 200 nS latencies for both reads and writes representing typical performance of DDR3 memory in systems used for the measurements in Table 9.2 [109]. All of the other parameters of equation 9.1 vary significantly across workloads as per Table 9.2. Rather than compare these metrics with each other we will use them as independent baseline benchmarks for Enterprise, Big Data, and HPC workloads.

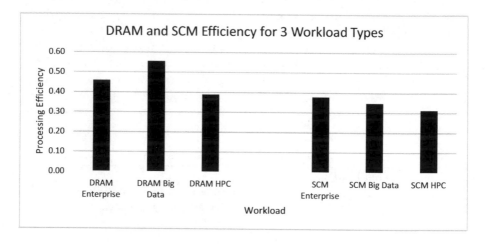

FIGURE 9.2: DRAM and SCM baseline for 3 workloads.

Examining the rightmost 3 bars in Figure 9.2 we see the impact of increasing write latency to 500 nS on each of the three workloads. This decreased efficiency by 8–20% depending on workload. The observation that all three landed in the 30–35% range is coincidental as each resulted from a different combination of the parameter values in equation 9.1.

While figure 9.2 accounts for the impact of media latency on IER it does not account for efficient use of interconnect bandwidth. Bandwidth analysis can shed light on a number of RP aspects related to technology and scalability. As a first example, consider the use of DDR style interconnects to media with non-uniform read and write latencies. DDR style interconnects are designed for media with uniform latency for reads and writes on a given DIMM. We can account for additional latency due to bandwidth utilization by dividing latency into a base component and a wait time component [109]. Equation 9.1 represents the base component. The wait time component $W_B(X)$ is a function of memory system bandwidth utilization measured at the CPU interface to DDR for the same system used in Figure 9.2.

Let

B_M = Maximum bandwidth of memory interconnect

S = Memory access size in bytes

$W_B(X)$ = wait time as a function of bandwidth utilization at the CPU memory interface

then

Bandwidth Demand $B_D = IER * P_M * S$

and

$$IER' = \frac{1}{\frac{1}{IER_C} + P_M * ((1 - P_W) * L_R + P_W * L_W + W_B(\frac{B_D}{B_M})) * P_S} \quad (9.2)$$

In equation 9.2, L_R and L_W are characteristics of the media while the W_B term is a function of the interconnect. IER' is calculated iteratively until it reaches a stable value which is used to calculate a processing efficiency (PE) that takes interconnect bandwidth into account. Figure 9.3 shows the result for 64 byte cache lines in a memory system with 3 GBps per core bandwidth. Compared to Figure 9.2, the Enterprise and Big Data workloads showed no impact because their bandwidth demand is low due to a small value of P_M relative to the HPC workload (see values of Misses per 1000 Instructions (MPKI) in Table 9.2). The DRAM HPC result, however, shows a 10% decrease in efficiency (from .39 to .35) due to high utilization of the DDR3 interconnect.

Comparing the height of SCM HPC bars in Figures 9.2 and 9.3 we observe that very little efficiency is lost due to wait time in the SCM HPC case. This is because the higher L_W in the SCM HPC case decreased IER relative to the DRAM HPC case even before bandwidth utilization was taken into account. As a result the SCM HPC case did not reach high enough bandwidth utilization to incur a significant wait time penalty.

While the rightmost bars are a reasonable baseline for the use of an asynchronous interconnect for SCM, they do not account for the behavior of the synchronous DDR interface in the face of non-uniform media read and write times. The DDR protocol is synchronous in that the number of clock cycles between a read or write request and data transmission is configured during system initialization. On writes, data transmission can be completed before the media is written; however, the DRAM DIMM may not be ready for the next write in time to allow its data to be accepted until data is written to the media. Although there is some timing flexibility between reads and writes, it is limited and the protocol functions best when the same time is used for

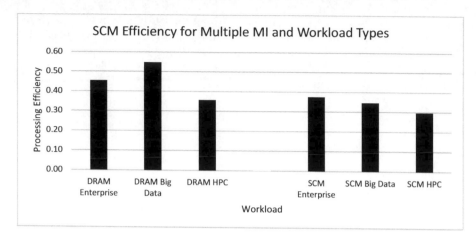

FIGURE 9.3: DRAM and SCM for 3 workloads with 3GBps per core BW.

both reads and writes. Due to these timing limitations use of DDR with non-uniform read and write times is likely to require one or both of the following.

- The transmission time for both reads and writes may be artificially increased to accommodate the largest expected media latency without holding onto the DDR interconnect during long media accesses.

- The system may need to tolerate frequent re-transmissions due to the loss of transmissions that collide with media accesses that hold onto the DDR interconnect for long periods of time.

Figure 9.4 models the efficiency impact of the bandwidth wasted by each of these phenomena compared with the behavior on an asynchronous interconnect. The leftmost group of bars duplicates the SCM bars from Figure 9.3. The middle group (labeled as SCM ... write) has both read and write latencies set to 500 nS to represent the case where the DDR interface is configured so that expected read latency is inflated to match write latency. The rightmost group (labeled SCM ... Retry) assumes that one read or write data transmission will be lost for every write, so writes transmit data twice.

Here we see that the retry approach had very little efficiency impact compared to the forced uniformity approach. This is because even with the high write probability of SSM HPC, the retry approach only caused interconnect bandwidth utilization to increase from 70% to 85% which is below the knee of the interconnect wait time function $W_B()^2$. Still, a system that is designed

---

[2]The graph of $Y = W_B(X)$ has a shape that is typical of response time vs. throughput curves. For low bandwidth utilization the curve is flat, but as utilization approaches 100%, wait time approaches an asymptote. The knee refers to the region of the curve where slope transitions from nearly horizontal to nearly vertical.

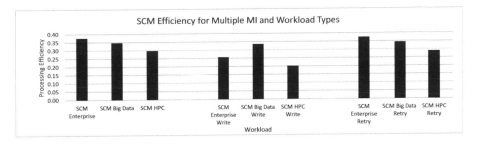

FIGURE 9.4: DRAM and SCM for 3 workloads with 3 GBps per core BW.

to have frequent retries tends to be less reliable because retries due to failing components may be obscured by expected retries.

In summary, ignoring memory interconnects the higher write time of SCM impacted instruction execution efficiency for the workloads modeled by 8-20%. Local interconnect bandwidth considerations reduced efficiency by an additional 5% but only on the HPC workload. These observations point out the performance impact of bloat if SCM is used for random access in situations where persistence is not needed. The choice between synchronous and asynchronous interconnects appears to have less performance impact in the scenarios analyzed so far.

9.3 Efficiency Model of Memory Fabric

To further analyze RP issues related to memory interconnect performance, consider a new reference model illustrated in Figure 9.5 which includes a subset of Figure 9.1 with elaboration on the MI fabric. Both reference models show the same processor constructs within server nodes. Within each server node the new reference model shows Media MMU's controlling multiple PM components connected by dedicated point to point links. These are viewed as separate from the MI fabric as they may be memory technology specific in nature. The "Memory Switch" components from Figure 9.1 are renamed to Node Switches 1 and 2, indicating that node memory switches are analyzed as part of the MI fabric. The other components from Figure 9.1 are omitted from this model so as to focus on the MI fabric.

While there are many ways of constructing and scaling fabrics, we choose a model with three layers of switches: Node, Cell, and Spine as a common datacenter network architecture ([72] figure 1). Our modeling approach can be extended to more complex architectures as needed.

Node and cell PM access miss rates are used to represent the probabilities that PM accesses will cross from node to cell or from cell to spine switches so as

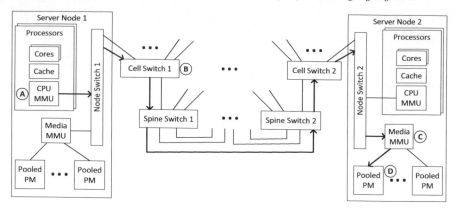

FIGURE 9.5: Memory interconnect fabric reference model.

to model locality within MI fabric layers. Figure 9.5 shows a limited view of a large scale system with ellipsis (...) representing points where the interconnect scales. For example even though only 2 nodes are shown and only 2 switches appear in the cell and spine layers, there may be many more of each. Spine switches are connected to each other to provide cross-sectional bandwidth as demanded by the traffic flow between cell and spine layers. Multiple cell switch ports may connect to the same node and/or spine switches as dictated by the total bandwidth of a systemwide workload along with are aforementioned node and cell miss rates.

There are several components in this model that may implement PM related functionality including CPU MMU's, interconnect fabric components, Media MMU's, and PM components. These are represented by circled letters "A" through "D". For purposes of illustration, one path from server node 1 CPU to server node 2 PM is illustrated. The primary input parameters to our model include the number of cores per node and the total number of nodes in the system. These are combined with the application and processor specific bandwidth demand per core (B_D in equation 9.2) to derive a total demand for the scaled-out system. Total demand is then combined with miss rate, port bandwidth, and port-per-switch parameters to determine the number of cell and spine switches required for the number of ports interconnecting each type of switch.

Depending on system scale, the number of required spine switches may vary from 0 to many. As a result accesses that flow from node into cell switches will traverse between 3 and 6 switches total. For the MI fabric model we used a utilization-to-wait time function typical of PCIe [329] as a proxy. All switches in the system are assumed to have the same port count, port bandwidth, and queueing delay characteristics. The latency added to a PM access includes a fixed per-switch component and an egress queue wait time that is a function of bandwidth.

Node port utilization is used to calculate a wait time that applies to node egress ports and node-facing cell egress ports. The wait time across all such ports is assumed to be identical given a steady state uniform traffic distribution between all node pairs in the system. A second wait time is calculated based on the bandwidth demand on spine-facing cell ports. This wait time is also applied to all cell-facing spine ports. A third wait time is calculated for intra-spine ports. The resulting MI fabric latencies are fed into equation 9.2 where the fixed per-switch part affects L_W and L_R while the bandwidth specific part affects B_D.

The following parameters apply to this MI fabric analysis. Let

CPN = Number of cores per node

$Nodes$ = number of nodes

PPS = ports per switch

B_P = Maximum bandwidth of a port (MBps)

PM_N = Probability of node miss (causes node-to-cell traffic)

PM_C = Probability of cell miss (causes cell-to-spine traffic)

L_S = switch latency (us)

M_O = over-provisioning factor[3]

M_T = round-trips multiplier

then

PPN = Ports per Node = $\lceil \frac{B_D * CPN * PM_N * (1 + M_O)}{B_P} \rceil$

N_C = Number of cell switches[4] = $\lceil \frac{PPN * Nodes * (1 + PM_C)}{PPS} \rceil$

N_S = Number of spine switches - $\lceil \frac{PPN * Nodes * PM_C * 2}{PPS} \rceil$

B_N = bandwidth demand at node port = $\frac{B_D * CPN * PM_N}{PPN}$

B_C = bandwidth demand at cell-to-spine port

$\quad = \frac{B_N * PPN * Nodes * PM_C}{\lceil (\frac{B_D * CPN * Nodes * PM_N * PM_C * (1 + M_O)}{B_P}) \rceil}$

B_S = bandwidth demand at inter-spine port = $B_C * (1 - \frac{1}{N_S})$

$Q_P(X)$ = queue depth as a function of bandwidth utilization for a port

and Total Fabric Wait Time

$$T_W = Hops * L_S + (Q_P(\frac{B_N}{B_P}) * 3 + Q_P(\frac{B_C}{B_P}) * 2 + Q_P(\frac{B_S}{B_P})) * \frac{S}{B_P} \quad (9.3)$$

where

Hops is a discrete function of N_S,

$B_C = 0$ in cases where N_S is 0,

$B_S = 0$ in cases where $N_S < 2$ and

$$IER' = \frac{1}{\frac{1}{IER_C} + P_M * ((1 - P_W) * L_R + P_W * L_W + W_B(\frac{B_D}{B_M}) + T_W) * P_S} \quad (9.4)$$

---

[3]Over-provisioning helps to avoid extreme network latency due to saturation.

[4]There must be enough cell switches to connect all node ports as well as the spine facing ports necessary to handle cell misses (PM_C).

9.4   Resource Proportional Capacity Scaling

The first use of the model is to illustrate the impact of MI scale on the same processors and workloads described in Section 9.2. For this purpose we assume that all memory accesses are remote to both the node and cell layers, so miss probabilities are 1. MI parameters are as follows for this analysis:

CPN (Number of cores per node) = 32
$Nodes$ (number of nodes) = variable
PPS (ports per switch) = 64
B_P (Maximum bandwidth of a port) = 25000 MBps
PM_N (Probability of node miss) = 1
PM_C (Probability of cell miss) = 1
L_S (switch latency) = .03 us
M_O (over-provisioning factor) = 0.1
M_T (round-trips multiplier) = 1

Figure 9.6 shows the processing efficiency (PE) resulting from MI scaling for the 3 workloads analyzed earlier. A 1 node data point was added to represent the original SCM workload efficiency baselines from Figure 9.2. The drop in efficiency from 1 to 4 nodes comes primarily from 2 switch latencies, one from a cell switch and a second from a remote node switch. For all but the solid line, each series shows a worst case for one of the Enterprise, Big Data, and HPC workloads because all PM accesses are remote. All of the lines show a series of efficiency plateaus as more switch layers are added for connectivity. Note that the x-axis is not linear as there are fewer datapoints towards the right of the chart.

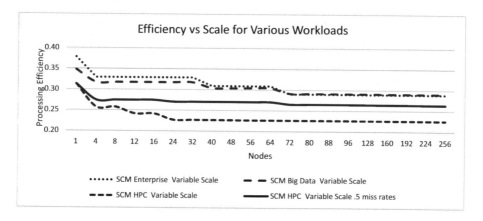

FIGURE 9.6: Processing efficiency vs. number of processing nodes (scale) for various workloads.

The assumption of all remote access creates a well known RP issue related to performance that occurs when reference locality is not driven into systems that have non-uniform memory access times. Since the HPC workload is the most demanding of network resources we further consider a case where node and cell miss probabilities are both .5 with that workload. This is represented by the solid line in Figure 9.6. For systems with 24 nodes or more this cuts the efficiency loss for the HPC workload in half. Since most scale out applications have some means of managing locality the solid line represents a much more realistic expectation. Similar improvements occur with other workloads when locality is introduced.

Given the processing efficiency impact of MI scaling it becomes important to reason about access density in concert with connectivity. Access density is the rate of reads and writes applied to a PM media component. Access density generally varies within a scale out system due to the following.

- Application demand for data that happens to be stored in a given media component. Load balancing of application demand across media components is a determining factor in access density. Load balancing may also account for the mix of heavily accessed data (working set) and other data in the same component.

- Media performance including the potential for tiers of data aligned with different media types.

- Processing efficiency decrease caused by MI fabric path length. This can artificially decrease access density by slowing down applications.

For example, consider a system where the total data capacity required mandates 256 nodes but only 10% of the capacity is heavily accessed over a time interval that is much larger (e.g., $>= 10^9 x$) than data access time. If the heavily accessed data is distributed across all components in the system then maximum media throughput is enabled, but network path lengths may be longer than necessary. The opposing forces of network scale and media load balancing create a complex RP trade-off, especially when media access times and MI fabric queuing delays are similar. This trade-off is generally managed today using a combination of data locality and system provisioning management decisions that often remain relatively static during the life of a system.

As another example, consider a system where some components have low access density. This could be because there is little demand for the data stored in those components, or because the components themselves have low performance. High demand on slow components should be resolved through multi-tier data placement algorithms. Low demand on high performance components may or may not be an RP issue related to data placement depending on workload dynamics over time. Access density trends should be monitored over time as high performance components that never experience high demand indicate an RP issue related to data placement. Inadequate availability or misuse of lower media tiers could be root causes of this issue.

The quantification of PE impact due to the issues described here should motivate attention to the sustainable alignment of demand, media performance, and MI fabric deployment. One increasingly viable approach to sustained resolution of MI fabric related issues is to apply analytics or ML driven by resource utilization trends across processing, interconnect, and media components over time. These trends are often application specific so when multiple applications are present, trends must be tracked for each and super-positioned.

9.5   PM Related Functionality Placement

The model described in Section 9.3 illustrates several positions in an MI fabric where media related functionality can be implemented. In this section we examine several types of media related functionality and analyze the RP implications of their placement around the MI fabric as shown in Figure 9.5.

9.5.1   PM related functions

PM technology is sufficiently leading edge that researchers are continuing to explore fundamental trade-offs between capacity, performance, and endurance, the latter being the rate of media wear due to access, especially writes. In addition, a number of existing storage and memory related features can be inter-mingled with more recent innovations. Here we examine several of these features and their MI implications.

9.5.1.1   Multi-phase write

Some power saving approaches to writes for Phase Change Memory (PCM), a type of PM, exploit a unique characteristic of PCM whereby setting a bit from 0 to 1 takes less power than resetting it from 1 to 0. As a result a multi-phase write approach is implemented to read the prior contents of a memory cell; then set or reset only those bits that need it. Exotic variations include encoding contents so as to incite more bit changes in the lower power direction.

Other approaches are needed to offset write variations in Resistive RAM (ReRAM) [436]. Compensating for ReRAM memory cell phenomena may require writing using increasing duration write cycles until the correct contents are assured using periodic calibration with error rate feedback.

In both of these examples, cell write times are increased by some multiple in order to compensate for undesirable media characteristics. The algorithms that determine or guide write timing could be located in any of the locations

A through D enumerated in Figure 9.5, although conventional wisdom would suggest location D in pooled PM components.

9.5.1.2   Atomic rewrite in place

Most PM technologies do not intrinsically guarantee write atomicity as discussed in Section 8.2.3.2. If it is important that a media component provide this function then an implementation such as those described in 8.4 can be implemented in software or in a memory controller. This feature creates multiphase write behavior similar to, but separate from, that described in Section 9.5.1.1.

9.5.1.3   Memory interleave

Memory interleaving is a very common main memory performance improvement technique whereby bytes in a cache line are spread across multiple memory media modules that can all be accessed at the same time. The resulting parallelism increases cache line access rate. Today, interleaving is generally orchestrated by CPU MMU's. With an MI fabric, interleaving could occur at any of the functionality placement locations in Figure 9.5. The interleave function translates a single cache line access into multiple media facing reads or writes. In the part of the MI fabric that sits on the media side of the interleave function, multiple traversals of the MI for each cache line increase the number of switch hops involved in a cache line read or write, but payloads are smaller and they occur in parallel on different network paths rather than serially.

9.5.1.4   Redundancy

Section 8.5 analyzed redundancy functionality for high availability from a software perspective. Similar functionality can be implemented in and around the MI fabric. The net result is an increase in round-trips and media accesses between the point where redundancy is implemented and the media.

9.5.2   Functionality placement given split vs. monolithic memory controllers

Since the functions listed in Section 9.5.1 all involve changes in media access time, MI fabric round-trips, or both, we can gain some initial insight by examining the impact of these two perturbations on processing efficiency. Figure 9.7 shows the permutation of 1x, 2x, and 3x media access times and 1, 2 or 4 MI fabric round-trips.

Here we see that as media access time increases by 2x or 3x without adding MI round-trips, efficiency decreases by 20% and 33% using efficiency of .24 as a baseline. Note that this baseline is the same as the 256 node efficiency for SCM HPC Variable Scale in Figure 9.6. Increasing round-trips by 2x or 4x

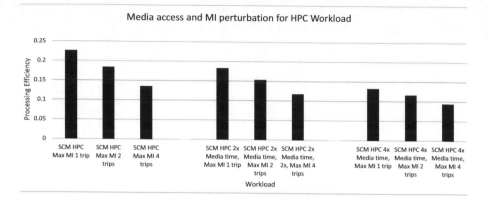

FIGURE 9.7: Media access and MI perturbation for HPC workloads.

decreases efficiency by another 12-18% for each media access time. Generally, media access time has more impact than round-trips but both are significant.

We can now estimate efficiency impact based on the way each function perturbs media times and trips. In the following table the functions are those described in Section 9.5 and the locations are the following, taken from Figure 9.5.

- A – CPU MMU

- B – Cell Switch

- C – Media MMU

- D – Pooled PM media component

TABLE 9.3: Split Controller Function Placement and Efficiency

Function	Location	Media Mult	MI Trips	Efficiency
Multiphase Write	D	2	1	.18
Write In Place	C	2	1	.18
Interleave	C	1	1	.23
Redundancy	B	2	2	.19
All	B-D	4	2	.12

In Table 9.3 the media multiplier indicates how much the media access time is increased (multiplied) given the indicated function placed at the indicated location. The MI trips indicate how many additional times the MI is traversed given the indicated function placed at the indicated location. The media multiplier and the MI trips parameters allow us to choose a combination in Figure 9.7 that approximates the PE of function placement. Here we see that interleave is the least disruptive function because it does not impact media time and it does not change MI fabric traffic. Placing multi-phase write and write in place on the media side of the MI fabric avoids disruption of the MI fabric, but both double write times. Since the redundancy function was placed in the cell switch, it does not impact traffic on the first node switch. The model was modified just for this case so it is not reflected in Figure 9.7. The "All" row in Table 9.3 indicates the efficiency impact when all of the functions above are deployed in the same system.

Now let's consider function placement with monolithic (as opposed to split) memory controllers. In this analysis, the MI fabric still exists for media scalability; however, locations B and C do not. All functions must be in locations A or D.

TABLE 9.4: Monolithic Controller Function Placement and Efficiency

Function	Location	Media Mult	MI Hops	Efficiency
Multiphase Write	D	2	1	.18
Write In Place	A	2	2	.15
Interleave	A	1	1	.22
Redundancy	A	2	2	.15
All	A-D	4	4	.09

In Table 9.4 we see additional efficiency impact from all but multi-phase write. When all functions are implemented, efficiency decreases by 25% relative to split controller function placement.

In several ways this is worst case illustration. MI bandwidth was set to 25 GBps and miss rates were 1 in both nodes and cells. As an alternative, suppose MI bandwidth is increased to 100 GBps with a node miss probability of .5 and cell miss probability of 0.1. This allows us to scale up the MI interconnect from 256 nodes to 10000 nodes without overstepping switch connectivity. The reason for the large scale difference is that the cell-to-spine port utilization was often the limiting factor in Figure 9.7. The difference between node and cell miss probabilities was chosen to model a cell centric locality management approach in which data is often spread across nodes in a cell but seldom across cells. All of the other system parameters remained the same.

Figure 9.8 shows the results of media write time and MI round-trip count perturbations for that larger system with increased locality running the HPC workload. The impact of write time perturbation is essentially unchanged, but the impact of round-trip perturbation is reduced to 5-10%. The point here is that there are many ways to provision networks and manage locality, and choosing between them depends on application specific behaviors.

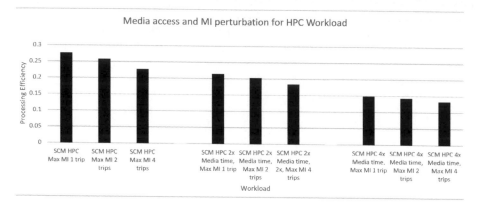

FIGURE 9.8: Media access and MI perturbation for HPC workloads on larger systems with locality.

In this analysis, instruction processing efficiency (PE) has proven to be an interesting RP metric for PM systems driven by cache line access (as opposed to bulk DMA) workloads. We have shown how PE can be used to characterize performance trade-offs between CPU and memory interconnect provisioning in scale out systems. At first glance it might seem that over-provisioning MI's is an issue because we never approach full utilization of MI resources. On the other hand, when CPU's drive memory interconnects towards saturation, queuing delays reduce the PE of the system, causing it to settle into a non-RP operating point where neither network nor CPU are well utilized. This conflict is mitigated by data locality which must be driven into the system by applications, automated caching, or, more often, both.

Once locality management is added, memory capacity provisioning for performance (in addition to capacity) becomes a first order factor in system RP. If node or cell level capacity is not sufficient to keep an application's miss probabilities low, then the MI vs. CPU provisioning dilemma is once again exposed. The interplay between CPU, memory, and MI provisioning could be analyzed and managed from a whole system perspective by using RPD to prioritize provisioning and function placement based on resource utilization and miss rate monitoring. This type of function placement management could involve the creation and use of tools that use the logic represented in Table 9.3. It could be extended to include the consolidation of multiple applications in the same system.

Throughout this analysis and our response to it we see an application of the "Identify - Quantify - Transform - Solve" process described in the conclusion of Chapter 1 applied to systems. System RP management also presents an opportunity for system analytics as described in Section 12.4.

9.6 Resource Proportionality and Memory Centric System Architecture

Once again considering the MI reference model, Figures 9.9 and 9.10 show the original model from Figure 9.1 with the following paths added to illustrate how memory centric architecture characteristics introduced in Section 9.1 can improve system RP related to performance.

- A – GPGPU access to local pooled RAM (solid gray line) – this path has the same characteristics as CPU access to SCM described in Section 9.2, but the data consumer is the GPGPU rather than the CPU.

- B – GPGPU access to local DRAM or pooled RAM (dotted gray line) – this path has the characteristics of the minimal scale system described in Figure 9.6. The CPU MMU acts somewhat like a memory switch, albeit not as low latency. In today's systems, PCIe replaces the memory switch on the path from the GPGPU to the CPU so this path represents the norm.

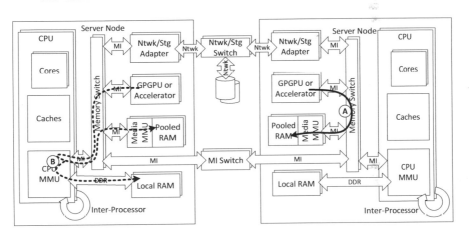

FIGURE 9.9: GPU local memory access paths.

- C – GPGPU access to another node's pooled RAM (dashed gray line) – this path has the characteristics described in Section 9.4. Path C has higher latency than path A because path A has fewer switch hop equivalents.

- D – Network access to memory (double gray line) – This path is included to illustrate that networks such as Ethernet can get the same benefits from memory centric system architecture as GPGPU. In fact the two can communicate directly with each other. For small accesses the MI Fabric is more efficient than the network but for bulk data transfer with an efficient RDMA implementation the two paths may be similar.

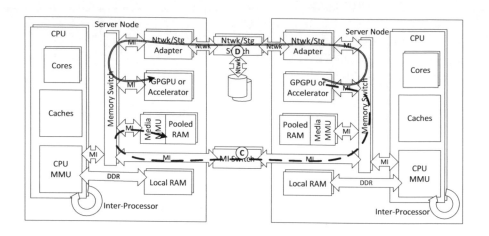

FIGURE 9.10: GPU remote memory access paths.

One key theme of memory centric architecture is caching and cache coherency. Section 9.5 concluded that locality management is critical to RP behavior of MI fabrics. Applications can take charge of locality management by allocating memory from pools that are near the core(s) during a computation. Today this is often done by parallelizing an application into multiple threads, each allocating local DRAM and moving shared data to local DRAM. Sharing is controlled by application lock management. If the application thread remains pinned to a single core or processor then this sort of application-managed-cache approach is effective.

If, on the other hand, threads move around or if the application does not manage locality then automated caching takes over, causing coordination of CPU MMU's in a multi-processor system using a cache coherency protocol. Numerous such protocols exist, and all involve additional communication among cache managers to maintain state information about copy validity. While these protocols generally run over the inter-processor link near the bottom of Figure 9.10 they can be extended to the MI fabric. Cache coherency adds state to MMU tables stored in high speed SRAM. Performance wise, it is reasonable to assume that an extra round-trip and/or multi-cast of cache state information is required as a result of shared data modification. Such communication may occur prior to the modification or before the next time the modified data is accessed by another processor depending on the protocol. The PE impact of this depends heavily on the read/write ratio of application memory access.

Driven by earlier trends that created symmetric multi-processing, cache coherency is generally enabled in all modern multi-processor systems whether the application is managing locality on its own or not. This can lead to RP issues related to bloat when cache coherency overhead is incurred in spite of application locality management. Furthermore, the need for applications to manage data consistency using techniques such as locking does not go away when automated caching is used. Earlier in Section 8.2.2 it was observed that PM impacts process flow by eliminating context switches and making more processes compute bound. This phenomenon may eliminate the need for thread mobility among cores, thus weakening the value of cache coherency. Finally, cache coherency state is difficult to maintain in a fault tolerant system, where node loss may compromise cache state information in unrecoverable ways. This limits the scope of cache coherency in an MI fabric. Perhaps symmetric multi-processing has run its course.

On the other hand, the virtual address space maintained by CPU MMU's is very valuable for flexible memory resource management and inter-process security. This value could be expanded by making the virtual address space pervasive across the entire MI fabric, creating a higher order hardware abstraction that enables function placement independence and data reference resolution. Maintenance of a pervasive virtual address space is much lower overhead than cache coherency so its RP impact would likely be very positive. It would improve the ability of sophisticated locality management and function placement services to increase system RP without adding a significant RP burden itself. This same logic applies to GPGPU's and networks (especially when using RDMA) as well as CPU's.

Chapter 10

Applying Resource Proportional Design Principles to a Deeply Layered or Complex Software Stack

10.1 Introduction

In the previous two chapters, memory design has been discussed in the context of newer developments of persistent memory and memory interconnects in the past decade. These developments have fueled what has been termed the "killer microsecond problem" and point to the deeper issue of optimization across the full hardware and software stack[50]. Newer low-latency IO devices such as datacenter networking, accelerators and non-volatile memories have latencies of microseconds rather than milliseconds. There have been many optimizations at the nanosecond level (e.g., in computer architecture) or at the millisecond level (e.g., in operating systems). A synchronous paradigm is suitable for delays at the nanosecond level while an asynchronous model may often be suitable for the millisecond level. However, the intermediate microseconds level has not seen sustained optimization efforts either due to the increased number of heterogenous entities at that level (requiring multiple coherent modeling efforts), or programming concerns such as the possible unsuitability of both pure synchronous or asynchronous models at such delay values. A good example where such issues arise is in the design of RAMCloud[291] storage system; here, software overheads need to be minimized and software stack modified to take advantage of hardware accelerators now necessary or present in such systems.

If we apply Resource Proportional Design principles to a deeply layered software stack or to a modern software stack, the choices for redesign are thus much larger as many parameters are likely to be involved and simple models (such as those resulting from assumptions that disks are the bottleneck) are not sufficient. In general, one high level model for designing large complex systems could be the following: Assume that each subsystem can be separated into a control plane and a data plane. Note that the data plane, for analysis, can be approximated not only using traditional performance models but also newer ones such as DNNs and approximate models, including even analog models, where meaningful. The data plane should ideally be as non-mediated as possible to prevent interference with the attempted high speed computation, high speed networking, or other efficient functionality. In addition, there should be a provision for performance counters, preferably in hardware, that collect information about past behavior in a RP manner. Some triggers using these counters/state can be used to directly pass information to other control planes of other subsystems or to a managerial plane to control other control/data planes. The control plane across the subsystems can be used to attempt k-RPD in each subsystem or across the system, using control-theoretic[398, 311] or other means (see Section 7.1.3). It can also attempt to discover bloat across the subsystems using, for example, utilization as a metric (Chapter 3).

The control planes across subsystems can, in addition to performance tracking, also now be used to track direct or covert attacks. This can be seen as the beginnings of a model-driven approach[197]; here we have a simple model of the system split into control and data planes. It is likely that these planes are different for performance and for security analysis due to the different needs. Given such an approach, the abstracted systemwide data and control planes can now be tested both for performance and security. The interconnections of the control planes of the subsystems along with an optional systemwide managerial plane can be tested with functional tests that are derived from specifications where available or from design considerations. Similar analysis is possible with the data planes also, but now bloat and RP aspects can also be checked, for example, using some types of bottleneck analyses[68].

The data plane performance counters should ideally be in hardware or in fast components such as L1 caches while control plane information may be in slower memory such as LLC (last level cache), memory, or even in SSD. Note that technological changes will affect each of these differently and redesign may therefore be appropriate at different epochs.

A related approach is the one proposed in the roofline or gables models for performance modeling and design, and also for diagnosing anomalies[407, 185]. Here, performance parameters are modeled, to a first approximation, as one or more linear functions with each function saturating after a control point in the input space. Essentially, the aggregate control plane in such models is the set of all the slopes of the linear parts along with the control points where saturation occurs.

Also, when we consider much larger "universes" as part of our design, the RP heuristic may interact with structural constraints that may require rethinking of this principle. The RP principle can sometimes thus be overridden, ignored, or modulated in some cases. We will discuss some simple examples first and then we will take a quick detour to discuss some examples where resource proportionality has to be co-designed with other wider aspects in the design, with some of them from the past and some more recent.

10.1.1 Simple examples of RPD in systems design

A desirable systems goal is the ability to write some computational logic and have it executed on any substrate that can support that computation. Operating systems provided this early on by using a process model (process as "container") that uses a runtime system and kernel services ("orchestration") and some specific hardware. Independence of hardware has also been achieved through virtualization by creating virtual machines (VMs). To avoid mistranslations or underapproximations that can compromise systems, Java and related languages have reimagined this goal using a linguistic framework, including bytecode verification. From a systems perspective, the same goal has now been attempted with a cloud infrastructure using VMs as hardware, containers as applications, and orchestration through middleware (such as Kubernetes[5]). Such a design has multiple levels of abstractions and multiple levels of translations; hence, investigation of resource proportionality is warranted.

How congruent is cloud-based design with RPD? The cost of a typical design and deployment is the hardware procurement cost + hardware cost itself + analogous software costs + maintenance costs and operational costs. Procurement costs are fixed costs encountered in the beginning and when failures occur; these are difficult to amortize. In a cloud environment that serves many users, such fixed costs borne by the service provider become part of the variable cost as seen by the end user. In the non-cloud context, as the intensity of usage increases, the variable costs approach and finally go past the fixed cost. As long as that (high) usage does not materialize, one can argue that a cloud solution is closer to a RPD design and a solution a mature economy would reward. Thus in many designs, solutions based on aggregation or multiplexing (such as cloud solutions) become more cost effective when fixed costs are higher than variable costs.

10.1.1.1   Layering costs

Consider the differing rationales in the design of filesystems (FS) such as for scale and reliability in the Google filesystem (GFS) and for commercial high availability (HA) filesystems. The HA designs are well engineered systems but with high cost ("fixed cost high") and failures are carefully contained typically at the level where they happen but some may be handled at adjacent or immediately higher levels. It has also been found in the design of a large RAM based storage[291] that reducing complexity of fault tolerance in the design requires minimizing the number of distinct error cases that must be handled; it is best if they are handled at the very high level or at a low level. GFS uses standard computer systems and tolerates failures, as a part of the design, mostly at much higher levels of the system. In a large enough system, the failure handling part can be a (small) establishment whose (variable) cost could be smaller than the extra fixed cost of the HA system. In such systems, a failure-tolerant design such as GFS is likely to be a RPD candidate.

Another good example is the ZFS filesystem that integrates traditional volume manager functionality into a filesystem. Integrating layers makes for a more complex design and is not worthwhile unless it also effectively solves a problem not feasible before. In both GFS[1] and ZFS, the failure aware integrated filesystem design provides economical solutions for certain problems that are much more difficult if attempted with multiple layers. In GFS, for example, replication is provided through the use of ext3 lower level filesystems on networked nodes and not, for example, at the physical disk level. This enables the use of commodity high speed networks rather than the more costly specialized backplanes. Failures can now be managed by the upper level GFS filesystem which can also provide consistency models for storage (such as the eventual consistency model) as needed by the applications. The ZFS design, on the other hand, provides better atomicity guarantees in the context of failures.

---

[1]Note that further developments have taken place with time in the design of successor systems to GFS as the data has increased orders of magnitude. See http://www.pdsw.org/pdsw-discs17/keynote.shtml for a recent snapshot; a "recursive" design (Colossus) has emerged where GFS is now used for metadata only with a new storage design for data alone.

A similar analysis is meaningful for the SPARK framework[429] that uses Resilient Distributed Datasets (RDDs). In a standard (HA) software, one provides redundancy all through the stack; hence a substantial fixed cost. In RDDs, the attempt is to recover the computation only when failures occur; this may be RP in a well designed system. With failures, however, latency variance can go up in a RDD-based design.

Is it possible to combine the two? At the lower levels of the stack, redundancy is better (for example, ability to fix the errors at close to hardware speeds) but, at higher levels of the stack, RDD-like designs are likely to be better.

We can also clearly see this in the design of ethernet-based networks. Consider ethernet originally. Normal wired ethernet design assumes, as a feature, that the error rate is smaller than some threshold bit error rate (BER) and hence there is no retransmission feature in the design. Suppose we now consider a new design feature where the error rate is much higher (say, by 3 or 4 orders as in wireless networks). In such a case, retransmission is fine as long as it is local and (local) signal levels are not permanently "bad". (If neither is true, then the no retransmission option is better.) As time progresses, if the radios become sufficiently good (using, for example, MIMO), the effective BER decreases again. We now have a design that may be "bloated" as it is retransmitting even when it may not be helpful (for example, in one or more hops the signal-to-noise ratio (SNR) is so low that only higher level insight, such as TCP timeout, can stop retransmissions).

In 1Gb and 10Gb ethernets, flow control and drops are handled where possible at the physical/data link (DL) layer itself (including use of redundancy) as utilization of a link goes down if higher level protocols have to be used to handle errors due to the longer timeouts at higher layers. Errors in upper level protocols such as TCP and SCTP across transport links, if handled at that level, are likely to be RP as they are based on actual losses at the flow level. As the cost of handling losses is quite high at higher levels, it is important to design mechanisms to mitigate them; for example, at the physical layer, PAUSE frames are available in Gb ethernet to control buffer overruns, and to prevent head-of-line losses in the transport layer protocol SCTP, multiple lanes are provided. With RDMA over Converged Ethernet (RoCE), Priority Flow Control (PFC) is used to reduce losses at the lower level DL layer.

At a deeper level, a cross layer design that informs the upper layer about the exact physical conditions it encountered would be helpful rather than the decoder just decoding the signal into bits[395]. Instead of using either frame receipts or SNR estimates to select bit rates, the confidence value (actually, the log likelihood ratio) calculated by the physical layer on decoding each bit can be exported to higher layers, which can then use it to estimate the BER and send the BER to the sender. The sender can use this BER estimate to select suitable bit rates. Such a design is likely to be RP as fading and interference can be distinguished in this model and inappropriate responses avoided.

In addition, co-design of adjacent layers is often very helpful. For example, in the context of scheduling of user level threads multiplexed to kernel threads, if there is a communication channel between the kernel and the userspace thread library, any thread getting blocked in the kernel due to an I/O event can be reported ("upcalled") to the userspace thread library manager. The latter can then take meaningful steps to achieve some goal such as a minimum throughput (by starting another thread to handle requests)[33]. In recent years, similar designs have reappeared in the context of PaaS (platform as a service) models in cloud systems: instead of an agnostic layer that manages the "elasticity" of computational needs without knowing the specificity of application needs, an application aware layer can spawn the required parts of the ecosystem as needed. Such designs can be seen in frameworks such as Kubernetes[5]. However, "spill free" designs (see Chapter 13.3 for further details) are not compatible with such a coupled or integrated design as opaque pointers used in those designs are problematic as they need the services of the providers of these pointers for operations such as creating and comparing pointers. One option is to use copy-by-value to provide information across interfaces rather than copy-by-reference and possibly resample now and then (with the sampling frequency being dependent on the system dynamics). Or, we can use weak pointers as in Rust (Chapter 13) and manage them explicitly.

With "generalized" pointers involving persistence and/or involving multiple address spaces, a resource proportional design needs to avoid scanning/traversing the full "memory" space. Some operations in a system are systemwide; for example, garbage collection (GC) or fsck ("file system check") as their "pointers" can point anywhere in the system. If such pointers can instead be allowed to point only in some specific part of the whole, such operations can be faster and also RP. We discuss here one such pointer design for GC with wrappers for non-local accesses. In the context of dynamically typed languages such as Javascript, GC is needed. Since Javascript is heavily used in web browsers, and many browser tabs are often used, GC can involve memory across all the browser tabs due to the Javascript model, a clearly non-RPD (as partial GC or local GC is not possible). Using compartments or multiheaps[396], however, objects within a heap can be GC'ed as usual. Direct pointers across heaps are not allowed; any communication between compartments must go through special wrapper objects. With this design, it is easy to see that local GC is possible, a RP design. A similar RP fsck/dirty region logging for disks and SSDs can be designed for recovery from a crash or at initialization as needed in some storage designs.

Careful design is also needed in automatic and scalable memory reclamation[27][2] or, more generally, for data structure traversals without any synchronization, i.e., read sequences with either no memory updates (therefore, no memory fences, contention, or cache pollution) by utilizing the semantics of the given data structure, or with only a few updates (e.g., RCU)[313].

---

[2]The problem of how a deallocating thread verifies that "no other concurrent threads hold references to a memory block being deallocated."

A systematic effort in substantially automating debloating of containers [327] is needed due to the increased number of layers. Given information about interferences between applications by the user, one approach[327] uses dynamic analysis (for code coverage) to discover resource usage by application in the container and then partitions the container while satisfying the constraints. It also uses techniques such as partial evaluation, symbolic analysis, compiler assisted specialization as well as OS kernel debloating.

10.1.1.2 Resource rate mismatch costs

Given a large system, and since we can often only optimize two out of three aspects such as latency, bandwidth, and cost[375],[3] or read latency, update latency, and memory size[42],[4] there will be many subsystems with different bandwidth and latency characteristics. One can expect to see "bloat" in the system if resource rate matching is not present across the subsystems.

We now discuss a concrete situation from a bandwidth perspective. In newer memory designs[152] using eDRAM, the HBM ("High Bandwidth Memory" units) performance on the CPU package can show "counterintuitive" behavior: increasing cache hit rates on such memory side caches may not necessarily increase the "effective memory" bandwidth (that of cache and memory together) available to the processor. If the system is designed to maximize cache hit rates, the bandwidth available at the main memory is underutilized. A lower hit rate needs to be "learned"[5] so that the system operates at the combined bandwidth of all paths available. However, such a design might still not be effective if evaluated from a different perspective such as energy consumption. Similarly, analysis is needed[287] to move specific accesses that are in the critical path to be promoted to the L1 cache rather than increasing the number of cache levels. Here too, any attempt to also optimize non-critical paths (due to design simplicity) may result in bloat in energy or for some other metric.

In an OS, a disk access may be followed by another access on the same cylinder or nearby one. It makes sense to examine non-work-conserving models in such cases where moving to a request "far away" has costs. Here, throughput

---

[3]If cache density is increased, cost decreases but latency increases; we can see this in the design of L1, L2, ...Ln caches.

[4]A heuristic says that one can design optimized data structures for read (R) latency, or for update (U) latency while trying to minimize memory (M) need but not all three are possible[42]. For example, to minimize the cost of updating data, differential (or, delta) structures can be used so that updates of many queries can be combined avoiding the cost of reorganizing data. However, this can increase the space overhead: read costs also go up as the deltas have to be merged before responding. Or, read costs can be decreased by storing data in different formats, with each format specialized for a particular workload. However, update and space costs become higher. Thus, when we fix any two of the three (R, U, M), there is a lower bound for the other one that cannot be reduced. Also, note that if improper update strategies are followed, they can cause difficulties such as COW storms[337] (see Chapter 7.4.1).

[5]This approach has similarities to a non-work-conserving approach when that is useful.

is attempted to be maximized rather than latency as such, though it may show improvement also.

In traffic studies, it has been shown that adding a new link can slow down the system. Such game-theoretic models show the conflict between individual perceived gain vs. system gain and is seen in the above examples too (such as in disk scheduling and cache bandwidth optimization).

10.1.2   Copy elimination in RDMA based storage stacks

For better performance in large complex systems, attempts can be made to reduce execution path lengths. In this exercise, one can assume that no additional information is available, or that some information such as statistical distribution of some inputs or execution paths is known or, in the other extreme, complete information is known about execution paths and inputs. Alternatively, one can explore (negative) interactions: for example, check where the execution paths increase due to lack of some critical information.

We illustrate the possibilities through an example involving a storage stack. Suppose we provide a new feature RDMA[8], i.e., the ability to write to memory of a different node ("remote memory") securely in a distributed system. Let us consider a specific example of introducing user level 0-copy transfers in such a system with NVMe/PCIe interfaces[7]. Consider NVMe and RDMA together. Both use a doorbell architecture[15] where there is a submission queue and a completion queue along with a doorbell.

If a RDMA RNIC sends a packet, the arriving packet data is stored in the host's memory using the RDMA protocol by the local RNIC. If this data has to be written to the NVMe, the host submits a new request into the NVMe controller's submission queue and it is written to the NVMe using the NVMe protocol. Note that the data traverses the PCIe interface twice as well as incurring a write and a read of the data into/from memory, thus requiring twice as much bandwidth to/from memory.

The multiple operations to memory through PCIe (similar to the situation of memory being a hotspot due to multiple IP (intellectual property) components in mobile devices[276]) can be avoided if we can use RDMA to write to destination directly (for example, from fabric to NVMe SSD).

A simple optimization would be for the local RNIC to write the data into its local buffer and transfer it through PCIe to the NVMe controller. Here, the double bandwidth required in the previous approach is not necessary. However, there is still a write and a read to/from the local buffer. Another optimization would then be for the RNIC to write to the NVMe controller memory buffer directly but still use control commands that involve the RDMA RNIC. We can also use a TCP offload engine to further offload CPU processing.

In this series of refinements, the multiple data indirections are optimized based on accesses. If there are also correlated accesses with certainty or high probability (for example, if a write from RNIC to memory is always followed by a write from memory to NVMe SSD), then one can propose shortening the paths traversed in the system.

Some of the negative interactions could be that

- the system is no longer fully mediated by some kernel component for security purposes.

- it is not clear where error control, flow control, and the like are to be handled.

Essentially, it is not clear how to handle errors or security issues. If the cost of handling errors becomes high in the highly optimized versions (due to new monitors introduced or due to handling these issues at higher levels of the system), it may be better to "go back" in the design (i.e., keep some "superfluous" indirections).

10.1.3 High level RP systems design

While deeply layered software stacks are quite common, "serverless" computing has also become common in cloud frameworks in recent times. Here, we will investigate the connection between such ideas and RPD and how they can be used for realizing a RPD.

The most general model of computation is a graphical notation such as dataflow[388] or Petri nets.[6] A useful and widely used dataflow model is MapReduce[116] and its cousins such as Dryads[193] that use DAGs instead of trees; other realizations are microservices[31] and "serverless" computing. Such models have gained importance post-2000's due to wide availability of computing devices at lower cost; when this was not the case, say between 1980–2000, layered systems were the only norm. The webstack of the last decade or so, for example, was often a 3 layered one (frontend, business logic, backend being the (L)AMP stack: (Linux/)Apache/MySQL/PHP).

Layering makes the interactions between components very structured and the design has to handle dependencies across layers carefully; these can become very complex, for example, in the context of locking and persistence. Due to this, a request may require multiple traversals of the stack and may result in non-RPD behavior.[7] While layering is important to deal with complexity, latency, for example, can increase. One strategy is to find the common case and introduce as few layers as possible to service requests for this case (the "fast" path) while other requests traverse the full set of layers.

For example, a user might be interested in browsing (and not purchasing) the products available at a website; if the database is in the backend, the request has to still go through the other layers, adding to latency. A better

---

[6]While Petri nets may not be Turing complete, they can be useful in certain contexts.

[7]The STREAMS framework in Unix SVR3, while elegant, suffered from performance problems due to its highly layered design. In a different and unrelated context, linear blockchains may not be as performant as the DAG chains of IOTA design[192].

design could be a (browsing) product catalog with a simpler design better integrated with the frontend (performance/security wise) and a purchase catalog with the backend (with higher security and with additional interfaces such as with banks). This may be a more responsive system but with some additional cost or even sometimes complexity.

If a design is thus split into its various functionalities (e.g., browsing vs. purchase),[8] a natural decomposition of the architecture may result except that there is a need for allocating computing and other resources as needed to execute the business logic and storage functions. With cloud computing and VMs, this job of allocation could be the responsibility of the cloud provider and, as long as this aspect can be done effectively, we may see the outlines of a RPD design. Serverless computing[29] is based on this insight; it decouples application logic from resource management. Note that the splitting of functionalities is also possible in the context of a hardware design[276, 43, 78] with data flows being taken as a starting point.

In the serverless computing model, users structure applications as collections of functions that are instantiated based on user requests or calls by other functions. A function can be specified using any language along with a specific runtime environment, including, for example, specific versions of interpreters and libraries. However, the function may be instantiated either as a physical machine, a virtual machine, or a container as the runtime system decides it. The cloud platform is responsible for managing where the function is placed and its scheduling, and starting new function invocations on demand. This requires application state to be independent from that of the functions and stored in shared storage (e.g., a filesystem, key-value store, or database).[9] This makes it possible for all function invocations to have access to the state independent of their physical location in the cloud. In some recent design explorations for the 5G control plane, serverless models are seen as a good fit as the control plane is broken down into services communicating via HTTP. However, asynchrony is needed as otherwise a control plane function can interact with, for example, a long data plane function and get blocked. If failures are also to be handled, combining asynchrony with external storage needs deeper programming abstractions such as continuations.

Note that the Infrastructure-as-a-Service (IaaS) cloud model requires over-provisioning to meet bursty traffic demands and hence can be non-RP. Using managed services and a serverless platform therefore presents a good way to get cloud benefits such as scalability and cost-effectiveness.

In the context of such a serverless platform, a serverless function is a natural unit of information flow tracking also but we may need a termination-sensitive non-interference (TSNI)[29] model; see Chapter 13.3.2 for further

---

[8]This is termed as a "Command Query Responsibility Segregation" (CQRS) microservices pattern[332].

[9]Interestingly, an idealistic school of philosophy (the Vignānavāda school of Buddhism) posits multiple external stores to model reality across sentient beings while perceptions are explained as projections of the percipient mind.

details. However, the store can become a bottleneck just as memory can become a bottleneck as discussed in Section 10.2.1. There are also some other issues in this model such as what model of execution is supported in the context of failures and non-idempotent operations: is it at most once semantics or exactly once semantics? Furthermore, if functions are composed, is there a possibility of double billing[48], once at the leaf node and then higher up?

One model of safety in the context of serverless computing ("ST-safe") requires the following "trilemma"[48]: (i) functions should be thought of as black boxes (cannot assume that all written in same language), (ii) the composition should obey the substitution principle (so asynchronous models may not be fine as callbacks may be involved), (iii) invocations should not be double-billed (which can happen if a function cost is counted as part of its parent as well as in its own right). Embedding the internal scheduling of function invocations in a continuation passing style, a serverless runtime can affect the sequential composition of functions that satisfies the trilemma at the same time[48].

A related complementary model, at a practitioner's level, is proposed in ZeroMQ[17]. Here, a "topology" is defined as the computational structure (encompassing many nodes, reachable from any other, connected through many data channels) that has a uniform command set and wire format all across.[10] The data channel also has uniformly either a REQ/RESP or PUB/SUBSCRIBE model. The goals are not only scalability of the topology but also two properties called:

- Uniformity: "Uniformity principle states that it should not matter to which node in the topology you connect your application to. The service provided should be the same."

- Interjection principle: "Inserting an intermediary node into the topology should not change the behavior at the endpoints."

But are such resulting designs RPD? This is unlikely as these highly general principles can cause extra computations at the interfaces (including costs such as marshaling) that could make the system non-RP, especially if the granularity of the functions is low. Instead, processing-in-memory (PIM) (embedding logic in memory device) is one way of offloading computation and reducing data movement[78], especially due to the emergence of the newer 3D-stacked DRAM architectures. Reducing data movement is also important for other reasons: consider an example analysis[11] where two functions that account for a large part of the system energy are packing/unpacking and quantization. The first one is for minimizing cache misses during matrix multiplication through reordering the elements of matrices. The second one, quantization, converts 32b real and integer values into 8b integers to decrease the runtime and energy use for inference. In spite of that, there is still a large movement

---

[10]This might also be seen as a generalization of control and data plane models.

[11]In the Tensorflow Mobile library.

of data and part of energy savings gets lost. The net gain depends on the system relationships and the RP aspects as critical hardware parameters are changed.

Another design approach avoids "state spilling"[75] across interfaces by decoupling entities based on possibility of state spill between them by eschewing encapsulation (as code and data being bundled often results in state spill)[12] and mandating stateless communication (as state across entities often creates (spilled) dependencies between them). Instead, after any interaction, the server returns to the client an opaque handle that can be presented in the next interaction. We discuss this in more detail in Chapter 13.3 including RP aspects of such resulting designs.

However, in the context of concurrency and transactions, abstract datatypes (which do emphasize encapsulation) have been shown to have higher performance[184] through the use of datatype semantics, reduction of bookkeeping, limiting false conflicts, and enabling efficient concurrency control.

10.1.3.1   Proportional design at different levels of the stack

While we have given many examples of RPD that are at a systems processing level, there are many other levels at which the RPD can be said to take place:

- at an information-theoretic level[57] by reducing redundancies (for example, compress the time series during small time steps, and deduplicate across longer time steps). Some other examples in this domain are as follows:

  - at an error correcting level using coding theory which has been widely studied. In newer areas such as quantum computing too, managing errors due to decoherence and other quantum noise is critical.

  - at an error significance level (such as in approximate computing or in reliability): The training costs of a DL model can vary substantially depending on the accuracy attempted by 3-4 orders of magnitude[372]. Similarly, the effort in going from a system reliability of three 9's to four 9's or even higher is highly disproportionate; the cost-effectiveness is an important consideration. We can also look at such issues from other angles such as at the language level. We discuss this in some detail in Chapter 13.2.

  - at an error logging level. From a systems perspective, we can view the problem as locating root-causes. For example, using lightweight

---

[12] If there is an interaction between a src entity and a dst entity, the resulting state should be in src but opaque and not in dst; otherwise, an entity internally stores the progress that other entities have made when interacting with it.

triggers, one can identify the first time a problem occurs and then to use its recurrences to effect "blame-proportional logging"[246]. On the first occurrence of a problem, a "blame rank" can be assigned to methods in the application based on their likelihood of being relevant to the root cause of the problem. One can then enable heavyweight logging on highly-ranked methods for short periods of time. "Over a period of time, logs generated by a method is proportional to how often it is blamed for various misbehaviors, allowing developers to quickly find the root cause of the problem"[246].

– at an error recovery level. We have already discussed RDDs. Let us consider resource proportional recovery for synchronization in pointer-based systems when failures occur. Current synchronization mechanisms such as RCU[258, 85], hazard pointers[258], Stack-Track, ForkScan, and ThreadScan[27] handle this problem in different ways but we can broadly classify them into two major approaches. In the epoch based reclamation (EBR)[258, 85], when a quiescent state is reported by a thread or CPU, this means that it has no outstanding reference to any object that is using the synchronization mechanism. When all report this state, no such references exist and the objects can all be freed (in principle) in a single batch operation. EBR, used by RCU for example, offers high performance as it has minimal to no tracking involved but, with a thread or CPU fault, can cause accumulation of large amounts of unused memory leading to application failure. Reference Counting (used by hazard pointers and ThreadScan for example) offers a precise way to finding out which parts of memory can be reclaimed at any point in time but it can lead to low performance due to extensive tracking of references. Reference counting (RC) is a costly operation when used in non-blocking concurrent programs loss due to the usage of costly read-modify-write operations. Approaches like ThreadScan do not track pointers and use RC only during failure situations by scanning the stack for pointers at that time. Memory reclamation is important in synchronization mechanisms like RCU that create garbage on every update, unlike other approaches where garbage is created only on delete. As the rate of updates is usually higher than the deletions, the rate of garbage creation in RCU and similar optimistic synchronization mechanisms is much higher.

Clearly, for EBR, we need a useful RP solution in spite of an actual failure of threads or CPUs, as they cannot report a quiescent state. Instead, we can treat the memory reclamation problem as a fault isolation problem. Instead of actively tracking references all the time as in precise reclamation, or not taking any action on failure and let the application or system die as in EBR, we can identify when precise reclamation is required (e.g., just before memory exhaustion due to likely failure of some thread or CPU) and which threads require it. At any point in time, all normal threads use high-performance EBR-like approach but slow threads that seem to elongate the epochs can switch to precise memory reclamation after receiving a signal followed by scanning the stack.

- at a circuit level: The energy consumption of some digital systems is often high. In a study[173] of the sources of performance and energy overheads in general-purpose processing systems, the overheads of a 720p HD H.264 encoder have been quantified. Using SIMD or VLIW techniques improves performance but still cannot match the ASICs in performance or energy as the instruction fetch and data fetch stages of the instruction execution pipeline still occupy 90% of area with only 10% of the area devoted to actual work. Only by introducing "magic" instructions that effectively increase the number of operations that are executed per instruction by 2-3 orders of magnitude and also reduce the pixel bits from 8 to 5 without decreasing quality that one can come to a factor of 2-3 in performance/energy to ASICs. If done in software, the overhead is equivalent to repeated read and writes from memory instead of a design that integrates computation and storage better. Tensor processing units[202] are a recent example of a design that is designed with DNNs in mind to reduce power consumption and increase performance at the same time.

At a even finer level, instruction execution time is bounded linearly even with pipeline stalls. But with anti-dependencies in multi-scalar designs, the execution time may not scale linearly. To avoid this, we need register forwarding, register renaming, etc., but these now require non-RP $O(n^2)$ circuits and power. To reduce such situations, designs like EDGE[358] use limited data flow with registers across hyperblocks only, with limited fanout in the broadcast bus, using lazy predication instead of eager, and also schedule code blocks intelligently.

With data prefetching, there is a trade-off between coverage (potential candidates) and accuracy (how many are actually useful). If accuracy is low, prefetching unneeded data can result in non-RP behavior. If coverage is throttled, the IPC is also affected. A better approach is to consider the potential candidates without decreasing coverage and use a cost-effective learning algorithm based on some features (such as page address) to learn useful candidates and then attempt prefetch[60]. Such an approach is also useful in other areas such as storage processing where determining the parts of data that provide useful insights and processing only those can be a RPD (Chapter 11).

Each of the above can be an issue at any layer in a layered design; if resource proportionality is not feasible or designed in at some intermediate level only, the whole design may become non-RP.

10.1.4 RPD by mixing analog and digital components then and now

A feature can be small or as substantial as a new technology option (for, e.g., introduction of digital components in electronics post-1960's, or SSD devices in computer design post-2010's). Such a feature may be very compelling not only because it is cheaper but due to the accuracy and other aspects that are not possible with the previous designs.

Let us consider accuracy as a feature. If originally the system was based on analog designs, then to obtain higher accuracy digital models may have to be considered.[13] Analog computing in the '50s and '60s provided results at close to speed of light in some cases but accuracy or controllability was poor.[14] Similarly, software (SPICE) simulation for circuits is orders of magnitude slower than live electronic speeds. The same situation obtains with modeling through wind tunnels or using computational fluid dynamics (CFD) software[226] but CFD may provide higher flexibility. In all these cases, digital representations and computations based on such representations became necessary to provide some functional guarantees of repeatability and accuracy but there are some situations where computational or energy requirements are an issue.

Moving from an analog design to digital fully may be a few orders of magnitude costlier in terms of energy and sometimes may be some orders of magnitude slower, though with an increase in accuracy.[15] If the system interfaces with human beings in the loop or other slow systems, the slowdowns in the speed are not consequential. But if it interfaces with other fast systems, the costs may be substantial: one may incur the costs of a bloated design (even though often tolerated as the very feasibility of the new functionality itself is surprising).

An analog (VLSI) design could also be preferred due to power considerations as the transistors are operated in the linear region and do not switch between the "0" and "1" states that requires much larger amounts of power. Carver Mead and his student Misha Mahowald created low power "retina chips" between 1980's-early 1990's. The information from the retina to the ganglion cells is carried by the spike trains in the signal that model the "digital" all-or-none activation of neuron but they also use the timing of the spikes

---

[13]Consider the move to digital in aerospace and other defense industries in the '60s and '70s and even later.

[14]Note that even to get an analog amplifier to work with a reliable gain needed large capacitances that slowed down the system.

[15]An instructive example is in the (cinematographic) film preservation studies. Data preservation needs energy to move data from one format to the next new generation format; this has be repeated every few years. Due to the high cost of these moves, many digital movies have been stored in the analog form as a backup in case the migration is not successful. The US Academy of Motion Picture Arts and Sciences's study of digital archiving in the movie business (2007) says that to store a digital master record of a movie costs about $12,514 a year versus the $1,059 if kept as a conventional film master stored in a salt mine or a limestone mine. Another good example is the use of analog means (measuring current) to detect collisions in the early Ethernet systems.

themselves to carry the "analog" information needed to model the firing of neurons together.

A RPD may therefore be partly analog (for example, to conserve energy) and partly digital (for accuracy). Analog here could refer to systems without quantization, whether electronic, optical, chemical (such as DNA-based computing or storage), or, in the even longer timeframe, quantum-based systems. In communication systems, storage systems, and the like, it is now a widespread practice to use fiber optics that uses light ("fast" photons) for transfer of information ("optical") while using electrons ("slow") for computation as it needs bistable states ("electronic").

While the digital revolution has been strong the last six to seven decades, as it gives phenomenal accuracy advantages limited only by the speed and resolution of analog-to-digital circuits (ADCs), it is becoming clear that a new relook is necessary as some designs may need hybrid models for effectiveness with respect to energy or other parameters. For example, a well-designed RP system may have some parts optical and some electronic, not all digital: consider a recent hybrid design with optical components that works close to speed of light and that interfaces with electronic/digital components for control or other processing[95]. Here complex computations needed in a digital design are done simply optically using negligible power.[16] In this design, the feature interaction is simple: the frontend is optical and the backend digital but we can imagine more complex layered or even recurrent designs. We can think of this as a design that is counter to the common design where the "digital" feature has permeated (for accuracy?) all through the system instead of where it needs to be. A RP design, on the other hand, is tailored for the accuracy needed and hence optical convolution may be appropriate instead of a digital one[17] in the convolutional neural nets (CNNs) used. Given that digital circuits can be produced in the current mature ecosystem with significant advantages with respect to size, optical-based CNNs may still not be meaningful in many contexts. Note that as multimedia, sensing, and cyber physical systems (CPS) become the more common workloads, many CNN-based architectures (and

---

[16]See also[134]: "The optical computer operates by physically preprocessing image data, filtering it in multiple ways that an electronic computer would otherwise have to do mathematically. Since the filtering happens naturally as light passes through the custom optics, this layer operates with zero input power. This saves the hybrid system a lot of time and energy that would otherwise be consumed by computation." In a sense we can think of it as a camera that takes multiple views of the same scene but with each variation taken through a specially designed filter. These views are however captured optically, just like a digital photograph; otherwise each of the views would have to be extracted computationally using digital circuits. With the optical route, we have much fewer calculations and much fewer memory accesses but also taking much less time. As there are no preprocessing steps, the rest of the computer analysis has a considerable advantage.

[17]Essentially, a solution for a particular problem in the vision area can be found either by "bit hacking" (using, say, deep learning) or by "photon hacking" (understanding the physics of how light interacts with matter and designing optical systems), or a combination of both (a subject of much current research[14]). Each has its own strengths and weaknesses; for example, bit hacking uses much more energy than photon hacking but is more flexible.

others like recurrent neural nets) are more appropriate but the question is whether it should be digital based tensor processing units (TPUs)[202], or optical systems (even if the latter are still experimental as of now). Without TPUs, the cost can be high[372]. Note that even a fully optical design can be meaningful as described in [241].

In many computations, bloat is possible (due to the higher-than-needed accuracy of the computations) if the receiver (say, the eye) is not modeled. In the context of newer designs for large displays such as those used in smart-phones and where energy efficiency is critical, it has been shown[369] that "color transforms can reduce display power dissipation by over 66% while producing images that remain visually acceptable to users. The measured whole-system power reduction is approximately 50%."

Similarly, one can add a second mode to the standard camera pipeline (for display and archival with JPEG, for example) so that a different processing is effected for inputs to (human) vision algorithms[88] as such algorithms are "highly tolerant to approximate image capture." Again, working with fully "lossless" JPEG may not be RP.

In addition to multiple modes, adaptive modes may also be in order. In the biological world, it has been reported[426] that the retinal detectors that code for "up" motion in an animal start responding to all directions under low light levels. One can argue that the neural networks for "up" motion has now been reprogrammed to be effective for any motion under low light conditions.

10.1.5 Blockchains and the proportional heuristic

Consider next newer technologies such as blockchain. As one component of a consciously proportional aspect of the system design in the first blockchain system, the Bitcoin has a notion of *proof of work*.[18] asymptotically, based on the work put in, a miner should have proportional success in being able to commit transactions. This generally requires careful design across the whole stack but Bitcoin achieves this proportionality trivially by making mining very costly compared with the rest of the costs (just like the disk is often the bottleneck in disk-based systems, and other software overheads can be ignored). In other designs where mining is not the predominant cost (such as permissioned ones or those using proof-of-stake), the blockchain stack needs to be carefully designed. *Proof of work* assumes that success accrues proportional to work attempted but the "arms race" in hashing power has resulted in

---

[18]An early precursor is the "Hashcash" anti-spam system devised by Adam Back in 1996[44] that used a SHA256-based proof of work; it requires all emails to show evidence of proof-of-work so that spammers find it costly to send mass emails but allows individuals to send messages to each other. Here, in an email header that has recipient's email address, the date of the message along with a counter is used as the string to be hashed; a valid message is one with an email header that has a counter value so that the header's hash has k leading bits zero. k decides the cost of proof-of-work; in '96, 20 bits was considered enough. Bitcoin has a similar proof-of-work model with k increasing with time; it has increased from 32 in 2008 to now 74 in early 2019.

mining pools and very specialized hardware, not to speak of huge electricity consumption.[19]

Furthermore, these systems should be resilient to some subtle strategies (such as dynamic adversaries) that can be employed by adversaries as mutual trust is not assumed. In such systems, an important question is whether we can simultaneously scale storage efficiency, security, and throughput? Sharding has been one approach[247] that can be used to divide the network into an almost linear number of segments that scales with computation capacity; due to the fixed size of the segments, the message complexity within a segment for secure consensus is also fixed and the overall message complexity scales linearly with the size of the full network. The transaction throughput is also proportional with the computational power of the network. However, in the presence of a dynamic adversary, compromising as few as half the nodes in a shard will corrupt the shard and therefore the chain; in a non-sharded chain it requires a similar fraction of the total number of nodes which is much higher.

To provide much better security, coding strategies have been proposed[238] so that storage efficiency (through sharding), security (through coding), and throughput (through sharding) can be scaled independently. However, coding increases computation as well as communication; we now have to ensure that subsystems that provide these capabilities do not come in the way. As the designs are in the context of permissionless blockchains,[20] these are moot as the mining costs are sufficiently large that they are not an issue. However, in the context of permissioned blockchains (or other systems discussed below such as *proof of stake* or *proof of storage*), the mining costs are not dominant and hence coding and other costs can make a proportional design non-trivial; this may also depend on the bloat present in each of the subsystems (see Chapter 3).

Other systems based on *proof of stake*,[21] *proof of storage, proof of retrievability, proof of space, proof of replication*,[22] and many others have been designed and deployed. Because these systems work in the context of users who are possibly not mutually trusting, each of such principles requires a validation in terms of other features of the system that need to be also well grounded. For example, with respect to proof of replication of storage (which additionally produces a useful economic good in contrast to merely energy consuming Bitcoin mining), if servers and clients of storage services collude, timing based methods that check if the right amount of storage has indeed been stored (due to substantial copy costs) can be subverted as there is no theoretical model that relates real time and computation. In principle, custom hardware can

---

[19]This is an example where the proportional heuristic can become problematic; we discuss this further in Chapter 13.5.

[20]Any interested party can participate without any need to be part of a defined group.

[21]"Proportion of some resource such as money (or, may be even virtual money) staked relative to the market cap represents the proportional chance an individual has of committing the block and receiving the transactions fees contained in it". In this model, Sybil attacks that are based on manufacturing large numbers of ids are not possible.

[22]Success of these depends on resources such as storage created.

be deployed to break the thresholds for detecting the serving of (nominally replicated) storage using only a single copy if based on elapsed real time. A solution for this problem is possible[115] if we assume multiple servers with at least one honest server. Similarly, *proof of stake* can degenerate into "nothing at stake" as there is no computational effort, and can encourage "aggressive" mining on every competitive chain as there is a reward no matter what chain wins. There are also long range attacks such as forking at some older block which is easy if there is no serious computational effort needed. Both these can even prevent consensus from developing. Hence blockchain designs such as Ethereum reward failures also but at a fraction of that for a win.

10.1.6 Anti-RP designs

Very often, we need explicitly non-RP or anti-RP designs, for security purposes (especially against DoS attacks) or in the context of a game-theoretic mechanism design (see also Section 7.3). Examples:

- Memory hardness: The Ethereum Ethash hash algorithm slows down hashing through the memory bandwidth bottleneck to prevent runaway hardware acceleration (using ASICs, for example) that has happened with Bitcoin. Similarly, the DES encryption and decryption standard from the mid'70's is designed specifically to avoid competitive software based solutions.

- DoS-resistance is designed in by making potential attackers solve computationally hard puzzles before allowing entry into the system as in the Hashcash system[44].

- Side-channels are avoided by "mixing" specific types of noise to reduce S/N (signal-to-noise ratio) for the attacker.

10.1.7 Some newer issues

We discuss a few newer issues that are likely to be of importance in the future.

Security Proofs as Part of Protocols: Due to stricter notions of security and privacy considerations, a new development is the need to prove some property of interest in a running system with the proof being preferably at most linear in size of the current system if not smaller. Often, interactive or non-interactive proofs have to be devised (some also needing to be "zero knowledge"); the time or complexity of this proof itself is now an important component in systems design. Consider the design of restricted spaces (such as high security military installations) where an electronic or computing device (including cyber physical systems such as drones) can be operated only with a proof that its code observes the restrictions imposed in the restricted space.

If some of the code to be checked is dynamically generated, then the proof checking is in the critical path. Proof carrying code[280] has been used to ensure correctness of network filter code in an OS kernel (such as in Berkeley Packet Filter) or more generally to check conditions or invariants, and/or to get programmed visibility during kernel code execution in frameworks such as DTRACE[166], or for mobile code.

Goedel's speedup theorem (1936) says that proofs can be shortened by using more powerful logics[213]. Unfortunately powerful logics cannot be mechanized easily and the time taken for some of the steps may have high complexity. Interestingly, the celebrated PCP theorem[36] states that every decision problem in the NP complexity class (an important class that underlies many practical optimization problems) has probabilistically checkable proofs (PCPs) with the proofs having only a constant query complexity (the number of queries that the (randomized) verifier can pose to the prover) and using only a logarithmic number of random bits. The implications of this result and later developments (such as linear PCP and interactive versions of PCP) are increasingly driving research in zero knowledge proofs and blockchain technology (see below).

Note that proof of these properties is needed at every level of the stack in spite of using end-to-end arguments, as failures and subtle dependencies can invalidate some property of interest (typically, availability). At lower levels of the system, simpler models such as finite automatons are typically used and the proof systems needed will be correspondingly simpler. At higher levels of the system, deeper models such as logics that correspond to context free languages are needed.

While these may seem theoretical, there is now a definite need to ensure that some properties are invariant in certain systems; for example, those that employ blockchains or those systems that ingest new code, possibly untrusted, at runtime. Bitcoin, for example, needs to guarantee that money is not created from any of its user transactions: the input bitcoins should be equal to the output bitcoins plus any fees for the miner. One aspect of this problem is that if the amounts transacted on the public blockchains have to have privacy, some commitment scheme[23] has to be used to hide the amounts along with homomorphic (Chapter 13.4.2) properties. In addition, it has to be shown that the amounts in this scheme are within some ranges (range proofs) and that there are no overflows. As of 2017, from its 22 million transactions, using a 52-bit representation of bitcoin that can cover all amounts from lowest value up to 21 million bitcoins, this results in roughly 160GB of range proof data using the current proof systems[93]. Some zero-knowledge succinct non-interactive arguments of knowledge (zk-SNARKs) have been designed but they require a trusted setup. Some proof systems such as Bulletproofs[93] have been designed for use in blockchain applications so that they are compact, with fast verifica-

---

[23]This allows one to commit to a chosen value while keeping it a secret and reveal the committed value later.

tion (to enable the property "all parties should verify all proofs") but without any trusted setup; such a system has been shown to have an order of magnitude reduction in size. Note that this system suffices for Bitcoin as it has a simple scripting language. In comparison, Ethereum's language[3] is Turing complete as that capability is needed for the more powerful smart contracts system in Ethereum.

Speculation: If speculation is taken as a feature, limited speculation might be appropriate in the context of covert channels as seen in attacks such as Spectre/Meltdown[186]. Essentially, "run away" speculation can extract non-RP costs if speculation has to be modeled correctly to avoid security exposure; not modeling the cache state properly has given rise to, for example, the Spectre attacks. A new RP design may be necessary to take care of costs of covert channels; for example, a large number of "milli" cores (just like GPUs) rather than many powerful (superscalar) cores. Some architects have commented that the speculation feature may require completely new thinking, namely Computer Architecture 2.0 instead of 1.0[186]; for example, this may require a new perspective in the design where the bits leaked in covert channels are quantified and optimized for minimization along with performance, energy/power, area, etc., or, new developments in program analysis that model speculation explicitly.

Memory Layout/Design: For powering down (or increasing the number of) unused DRAM banks to save power, we need a mapping of user memory to physical memory, so that consolidation is possible[296]. Similarly, in rowhammer attacks[274], the issue is that logical vs. physical nearness may be known to the attacker and therefore, by repeated accesses to memory locations in certain technologies, the attacker can possibly flip needed bits for the attack. In general, we need some notion of nearness of logical objects not only in the logical space but also in physical space, or in some intermediate virtual spaces, or across these spaces. A more generalized idea is that of using a TCAM[64] that instantiates nearness properties based on need; this can be extended by having optional pointers that are instantiated on demand.

In the future, newer memory technologies may suffer read and write disturbances due to the very high density of bits being stored. This requires very conservative designs,[24] or access to information on the I/O patterns of the workload to mitigate the adverse effects of dense memory technologies such as storing multiple bits per transistor. With access to I/O patterns, the writes and reads may be rearranged within and across requests and rescheduled to avoid deleterious disturbances. Dense SSDs may thus require using safe program (i.e., writing) sequences that take advantage of the I/O request patterns[37]. For very large reads, for example, one can rearrange the requests so that all the bit planes of a transistor (that stores multiple bits in some

---

[24] For example, to avoid rowhammer attacks, the DRAM refresh rate has been doubled in many laptop designs.

technology) are read together, even after data is laid out in a different order on the wordlines to satisfy safe program sequences.

At a higher level in the stack, we may need sketches or quick summaries of memory accesses across GPU-CPUs for coordinating the populating of caches or preloading TLBs. In general, we may need, in the long term, linguistic notions for describing order of data writes (which could be rearranged using compiler transformations), or even just temporal or spatial patterns of accesses during certain windows of time.

10.2   Some High-level Recurring Patterns in RPD

We now discuss some recurring principles in systems design.

- Perform optimization across flows and also for each important flow separately: e.g., hyperblocks in compilers, and chains across large function blocks (see Section 10.2.1). But there can be duplication of code or hardware in this approach. It avoids a centralized (star) design as the center (star) can become a bottleneck. Furthermore, such a strategy can avoid the "data center" tax[204] as the data formats can be harmonized across a flow.

- Do not layer strictly: keep the design mostly layered (for simplicity in development) but also effect crosslayer optimizations to make crosslayer interactions efficient.

- Avoid computations where possible by using lazy/deferred models; also reuse computations when possible by memoizing. Similarly avoid excessive generality by tailoring the configuration space for the class of problems being run.

- If the system has to tolerate uncertainty, try statistical multiplexing such as through virtualization.

We now give some examples below.

10.2.1   Chains

It has been noticed[276] that many intellectual property (IP) blocks in smartphones repeatedly read from memory and write to shared memory and this memory can become a bottleneck or increase power consumption. Instead, it has been shown[276] that tracking the flows of data can enable transmission of data from where produced to the right submodule instead of repeatedly "dumping" the data into shared memory. Due to the sourcing of different IPs in a module, there may also be considerable marshaling/de-marshaling of data when produced by one IP and consumed by another. In general, we may need to use virtual chains of subapplication flow patterns to short-circuit memory traffic, not only in handheld platforms but across large systems with many subsystems with disparate IPs.

When consolidating functionalities into a shared resource such as a server, scheduling them requires care as the call chains may also need to be considered. In the telecom area, for example, the various network functions (NFs) require scheduling and chain management when run on Network Function Virtualization (NFV) platforms[229] using rate proportional scheduling and notions of backpressure to drop loads early when necessary.

Another example is the processing in memory (PIM) approach[78] where flows are modeled to analyze the energy and performance impact of data movement for several widely-used web consumer workloads.[25] With such an approach, PIM can reduce data movement for these workloads, by effecting part of the computation close to memory.

Otherwise, bloat can result as memory (or similar structure) is being used as the junction for all transfers across subsystems. This is similar to the bloat in the "date" string case with "strings" being the lowest common denominator[63] (Chapter 2) as various translations are effected across different formats.

10.2.2   Crosslayer optimization

Across deeply layered stacks (for example, network, storage, application stacks), crosslayer optimization has been one approach to reduce "useless" work. Some interesting case studies are

- (at the network layer): use by upper layers of the confidence value (actually, the log likelihood ratio) calculated by the physical layer on decoding each bit rather than SNR[395]. In the case of TCP-based networks, there has been interesting work at an analytical level[104].

- (at the storage layer): how to work around the problem of different "block" sizes in different layers in a RAID5 storage design[161].

- (in the Linux kernel): abstracting a minimal core of base features in the kernel has been discussed in Section 5.1.1.

Often, information in one layer or subsystem needs to be accessible elsewhere for an effective design. For example, if movability of kernel pages is not possible (as in the Linux kernel), anti-fragmentation and compaction suffer at the physical memory level[297]; a better design therefore keeps these in mind when allocating kernel pages.

---

[25]Specifically, the Chrome web browser, TensorFlow Mobile (Google's machine learning framework), video playback, and video capture, both of which are used in many video services such as YouTube and Google Hangouts.

10.2.3   Memoization, checkpointing, and lazy/deferred designs

To reduce the amount of unnecessary work, especially in the context of failure, it is useful to have computational objects that encode the effect of program execution after some checkpoint. If a failure occurs, such objects can be used to resume computation without starting from the beginning, for example, by the use of lineage information in SPARK[429] vs. the more costly replication approach.

However, there can be serious negative interactions sometimes. Consider a deferred free design, the Read Copy Update (RCU) mechanism[258] in the Linux kernel, for read-heavy shared objects. Updates in this design result in a new copy of the object and the old objects that are still being read are deferred for release once all the earlier readers exit. However, this means that there is a "bloat" in terms of memory and if this synchronization mechanism is not integrated with the memory allocation, out of memory situations can result[313].

10.2.4   Managing configuration space

With different feature sets, we may need different configurations of the system that are appropriate. How is the mapping discovered and how is it enforced? Given a deep stack, how does one select configurations of each layer of stack, for example, at the map-reduce (MR) layer in Hadoop? This may require stochastic optimization techniques such as SPSA[230].

10.2.5   Using virtualization

This can be used to stochastically balance variations in utilizations of subsystems across multiple applications. However, for map-reduce types of problems with barriers that all have to synchronize with, the "stragglers" problem can be acute and it is important to reduce variance, using possibly ML[424].

10.2.6   Applying the 80% 20% rule where possible

Often, a well designed system may have around 80% of the traces (or flows) that are typical while non-typical ones may be around 20% of the traces. If that is the case, a RPD is more cost-effective if applied first to the typical ones. For example, after collecting subsystem activity traces (e.g., in a mobile phone with multiple IP cores[276]), important flows can be identified and optimization of such flows effected (e.g., pipelining) and also avoid bloat (e.g., repeated copy to memory and read from memory).

The above approach is generally useful but the presence of "long tail" can conflict with RPD of selected portions of the flows.

10.2.7 Some theoretical insights useful for RPD

- To avoid centralized functionality wherever possible, simple but effective distributed algorithms can use randomization for scalability[291].

- c-competitive algorithms are often simpler as the costs are bounded to c times the "best" unrealizable algorithm that has full knowledge of the future. If a disk is likely to idle, when is it appropriate to do so? This depends on the cost of stopping the disk and its restart (say, s cost) as well as the cost of disk idling (say, p per unit time). A simple 2-competitive algorithm would wait for time $= s/p$ and then idle. While such a simple 2-competitive algorithm for the adaptive disk spindown[180] is sufficient in many cases, something better is possible. If spindown cost is s as before, Karlin et al.[207] give an expected $e/(e-1)$ (1.58) competitive algorithm[26] if timeout is chosen at random from $[0,s]$ with density function $Pr(timeout = x) = e^{(x/s)}/(e-1)$ as it is optimal in the following sense: every other distribution of timeouts has an idle time for which the distribution's expected competitive ratio is $> e/(e-1)$ [180].

- With increasing complexity of systems, cooperative and non-cooperative models of behavior across subsystems are meaningful. Hence game theory and mechanism design can give insights on how to design systems, especially in the context security and privacy. Higher costs may accrue to the system as a whole because of the "Price of Anarchy"[225] that is usually due to lack of information about strategies likely to be chosen by the various parties in the game. Hence, there are game-theoretic "overhead" costs in many systems; these are more especially obvious in the domain of security and fault tolerant systems. A detailed survey on the use of game theory for cyber security and privacy is available in [124]; we summarize it briefly here by giving examples for three types of game-theoretic solutions.

  - Use of mixed strategies such as cycling through a set of strategies as in the RPS (rock-paper-scissor) example. A good example could be the cycling between strategies that use ECC or retransmit facility when errors are high, and removing them when errors become negligible; this could be done adaptively when one moves across spatially when the signal's S/N varies widely, or across generations of technologies (as in wired and wireless communication).

  - Coalition games can be a useful model for studying selfishness in packet forwarding[23]. Since a mobile node directly experiences whether its packets are being forwarded by other mobile nodes nearby, it can use its observations to make Bayesian estimates of

---

[26]Here e is the constant 2.71828...

the types of these nodes. Based on these estimates, it can form coalitions based on its own payoff and the payoffs of other players in the current coalition as well as new coalitions.

– Study of the dynamics of the system to check if evolutionarily stable strategies (ESS) are possible. For packet forwarding in vehicular ad hoc networks, using both evolutionary game theory and public goods game to study cooperation, it has been shown[352] that the spread of cooperation across unknown users cannot be forced but evolves due to its dependence on networking conditions such as the average path length and user mobility.

• Modeling lack of information (due to stochastic events) or its opposite (the "excess" information due to, for example, covert channels) or similarity between two contexts as a cost or potential benefit can be used to develop information-theoretic models with explicit entropic costs. For example, entropy models may be useful to compute the informational distance between two replicas and thus the energy required to make them up to date.

With such a model, a resource proportional property can be redefined in terms of costs that accrue from physical resources, or from entropy in the input space, and/or from (covert) information loss from the space of intermediate and final values.

10.3   General Design Principles and Observations for RPD

A useful development in systems design is the discovery of patterns that can help us in the systems design and analysis. In the software engineering world, many software design patterns have been proposed[149]. Because of their generality (such as the *strategy, facade, proxy,* and *memento* design patterns), they are useful for crafting complex systems by composing the design patterns together (see, for example, [164] for an interesting middleware design that is based on patterns for a control systems design for mobility) but their problematic interactions also need discussion.

Given the above discussion, we now discuss some general observations that may be useful in the design of RP systems.

• FEATURE: LARGE SYSTEM

If large subsystems are designed independently, marshaling costs can increase as information is exchanged across subsystems. A significant fraction of total datacenter cycles is due to remote procedure calls, protocol buffer serialization, and compression, dubbed as "datacenter tax"[204]. Bloat can be an issue, especially if the system has many reentrant paths. Control-plane and data-plane separation may be helpful in reducing the overhead to enable systemwide view but this may be at the cost of future changes.

- **FEATURE: LARGE NUMBER OF FLOWS/PATHS**

  Knowledge of critical paths in the system is important for effective analysis; optimizing parts that do not happen to be on those paths may be futile. There can be substantial differential impacts with respect to power consumption and bloat given widely different subsystems with differing load and power consumption characteristics[68]. If bloat or slack is present on non-critical paths, it may actually be beneficial (as the lower utilization may result in lower power consumption, or a similar desirable situation for some other metric). Furthermore, if the design is "very tight," and there are many critical paths that have the same or similar figure of merit (energy, time, etc.) due to very careful optimizations, then any small change in any input will impact a different critical path each time and the interpretability of the system becomes poor.

- **FEATURE: HIGHLY OPTIMIZED FLOWS**

  If flows are optimized extensively, error/flow control may be compromised. For example, if RDMA is used, the "kernel" (or more generally, a management structure) may not be in the new optimized path and we may lose the ability to monitor flows. Special structures have to be created to handle failure; this may result in "design" bloat where many such structures may be reintroduced in different places due to the optimizations carried out.

- **FEATURE: MONITORING INFORMATION FLOWS**

  If information flows are to be monitored, either centralization (checking labels) can be one strategy, or we have to tolerate increase in cryptographic operations as the payload has to carry authenticating information along. If there is a check every time, it may be more than what is necessary ("bloat"). Checking should be based on the needed redundancy to catch any non-authorized flows rather than every time but this is non-trivial in the general case.

- **FEATURE: TRACKING COVERT FLOWS**

  If covert flows are to be monitored, side channels have to be neutralized by, for example, additional events that act as noise. Again, it needs to be done *where* necessary rather than everywhere. This needs a learning component.

- **FEATURE: HIGHLY LAYERED SYSTEM**

  Any inefficiency or slack at lower levels in the stack has increasing (often multiplicative) impact at higher levels. Good examples are trap handling or address translation in nested virtual machines. Similarly, doing computations at higher level of accuracy than needed (e.g., 8b vs. 32b) or with more generality than needed (e.g., IEEE FP semantics) across the stack can impact performance, or equivalently seen as "bloat" in the system.

- FEATURE: NEGOTIATION FOR FUTURE CHANGES

Either there can be incremental changes without new features (domain of optimization but this may also improve resource proportionality) or we add new features (domain of RP design). If we consider only the latter, when new features are added, it can impact many of the decisions discussed here: some positive, some negative. Depending upon the varied perspectives (or "interests") across complex subsystems modeled as independent actors, Arrows' theorem[38] says there is no way to decide what is best across the subsystems, assuming some desirable or "natural" relationships[38]).[27] Hence the developmental trajectory of a large system is not easy to predict.

For example, for performance (such as caching, speculation, or worst case execution timing analysis), or for security (such as information flow analysis), or for power (or resource) consumption, or for approximability, different subsets of relevant (time-varying) data are important for analysis and optimization (some of which may overlap partially). Unless there is a deep underlying theory between these somewhat disparate aspects of a program, any single generic design will be found wanting in some dimension or the other.

- FEATURE: MULTIPLE RESOURCE RATES and RATE MISMATCH

System balance is an important goal in large systems. Consider memory in a system that can be consumed and released by different entities in a system at different rates. In a large system, given different requirements of subsystems, different memory structures (e.g., deferred free memory) or types of memory (e.g., kernel or user memory) may be needed. To keep these memory demands balanced, there is a need to track use and release of memories across these disparate structures. If this is not done, the system may run out of memory, or can show significant "bloat" if there is no way to convert from one to another type. We may need active "market" structures to balance these stochastic demands. A good example is the study of RCU synchronization mechanism and its impact on memory allocation[313]. Similarly, the management of regular and

---

[27]The conditions required are "unrestricted domain, non-dictatorship, Pareto efficiency, and independence of irrelevant alternatives." For example, consider 3 parameters A, B, C for which 7 subsystems vote A > B > C (here, > can be read as "is preferable to"), 6 subsystems vote B > C > A, and 5 subsystems vote C > A > B. So A > B desired by 7+5 subsystems (12) vs. B > A by 6 subsystems. Similarly, B > C voted in by 7+6 (13) vs. C>B (5). And C > A voted in by 6+5 (11) vs. A>C (7). So, majorities prefer A>B, B>C, C>A which together cannot be satisfied!

large pages[297], if not properly integrated with the allocation of kernel pages (which are mostly, if not all, non-movable), can impact compaction significantly and hinder allocation of large pages and hence result in higher TLB misses.

- FEATURE: REENTRANT SYSTEMS

  Analysis is often simple if we assume flat organization of subsystems. However, there can be layering and multiple types of feedback vertically and horizontally. We thus have to deal with multilevel structures with multilevel feedback. Introducing feedback requires some temporal component; if done too frequently or the opposite, the system can be said to be "bloated." In a complex system, this may have to be learned. This learning can itself introduce lags and thus keep alive "bloat."

  The feedback itself may be in the form of queues across subsystems. If the size of queues are configured to be too large, "bufferbloat"[286] is likely. Oscillations in the size of queues can be a sign of "bloat." For layered (networked) systems, a "principled" design may be possible, for example, as discussed by Low et al.[104].

- FEATURE: CONCURRENCY

  Once concurrency is present in the system, locking can be an issue. Here again, critical paths in the system have to be examined. Critical paths may need to be examined carefully as a result of detailed modeling, for example, of memory models, locking, transactions, etc. Abstract data types can be helpful in concurrency control to avoid non-scalable checks for conflicts across the whole memory[184]. "Non-intuitive" temporary elevation in priority for certain threads ("priority inheritance") may be needed to reduce critical paths[344].

- FEATURE: HIGH STATE SPILL

  Due to extensive caching, high state spill[75] is possible (Chapter 13.3). Considerable amounts of state of software entities can propagate across functional boundaries; since these are directional flows (without explicit tracking) any attempt to encapsulate, migrate, or passivate is difficult without consistency problems. If this is not designed properly, runtime overheads of scanning from "root" may be necessary sometimes to track the flows. Instead, a good design principle can be to avoid state spill[75].

- FEATURE: CACHING, FAULT TOLERANCE, AND ADMISSION CONTROL

  If caching is employed, on failure or disablement of "faulty" caches, the system can become unrecoverable as the system is designed for admission control with caching in place[138].[28] If bandwidth matching is not in

---

[28] Analysis of a 2.5 hr Facebook Outage of Sep 23, 2010 (summarized from [138]): "Caused by an automated system to check for invalid configuration values in cache and replace them

place, "bloat" can also result (see Section 10.3). If multiple subsystems cache memory, "bloat" can result due to mismatched rates of production and consumption of memory[313].

- FEATURE: HIGH LEVEL LINGUISTIC ABSTRACTIONS (or first-class citizens)

If good abstractions that are useful for analysis can be discovered and used, static analysis based on these abstractions can be used to simplify or eliminate some wasteful aspects of the computation. Fundamentally, the set of abstractions required to express a program concisely but at the same time expose enough structure for analysis is the critical question. The basic issue is to decide on the "reality" or ontology of concepts that needs to be expressed in a linguistic formalism. Good examples are the liveness annotations in Rust, specifying error bands in approximate computing, or expressing probabilities or distributions in, say, Bayesian inference.

At a simpler level, dynamic correlations can be investigated using dynamic program analysis in a formally expressed linguistic notation. The related traces can be approximated through structures such as, for some examples, DNNs, RNNs[159], and the like but some abstractions may not be human-interpretable. Hence linguistic reformulation may be not easy and therefore such "features" not available for analysis; we may thus miss out on some opportunities for reducing "bloat." This tension between "Being/Becoming" of features (i.e., explicit in a linguistic sense vs. implicit in code) is reflected in our ability to recognize bloat. For example, if approximation is not available as a first class feature, it is not amenable for analysis (as in current programming languages). Similarly, in probabilistic computation[162], probability distributions have to be explicitly provided in a linguistic formalism for deeper analysis (such as for probabilistic inference).

- FEATURE: LIBRARY SUPPORT

If explicit linguistic representation for an abstraction is not possible, library realization is one possibility. C++ implementations "obey" the zero-overhead principle: "What you don't use, you don't pay for"[371]. The C++ language is widely said to be powerful enough to allow the defining of abstractions like RTTI[29] as library code, and opt out of

---

with updated values from the persistent store. Works well for a transient problem with cache, but it doesn't work when the persistent store is invalid. Somebody made an 'invalid change' to persistent configuration values. Each client attempted to fix the problem: it had to query a cluster database that was not scalable (=>1000's queries per sec). Also deleted cache key. Now queries do not succeed in the cache after the fix. Each new request had to go to a (non-scalable) database again. Queues built up rapidly without any hope of being processed in time. Only solution: had to stop all requests to cluster database to recover; site down."

[29]Run Time Type Information.

using it as a language feature (as LLVM does). The dichotomy between language support for a feature or its implementation through a library is the quintessential question of how to avoid bloat. If libraries are used, they may be callable from codes not all written in the same language as long as standard linkage conventions are used. Similarly library use with complex objects may require copy elimination analysis across the user/library interfaces, and if concurrency is used, thread-safety analysis. Such analyses may not be possible across multiple languages (unless translated to a common intermediate representation like LLVM) whereas a single language design may permit many sophisticated analyses.

- FEATURE: LEARNING

  If machine learning or deep learning is used, is the learning cost bearable in terms of time or resources such as energy? The computational cost can be quite high; a recent study suggests that the carbon footprint[30] can be substantial[372]. To reduce such costs, specialized Tensor chips may be necessary along with newer systems designed for such workloads. Another issue is whether the data provided is necessary and useful for the learning attempted in the system (discussed in Chapter 11). To avoid mistraining or defeat adversarial training, how much additional data has to be provisioned?

10.4   What Is Feasible Theoretically?

Computational complexity theory has an interesting "Speedup Theorem" first shown by Blum[400]: "There exists a total computable predicate P such that for any algorithm computing P(x) with running time T(x), there exists another algorithm computing P(x) with computation time $O(\ln T(x))$."[400]. Hence, "bloat" is in the eye of the beholder for this predicate as it can be sped up arbitrarily!

However, the theorem can be further generalized[74]. If f is any recursive function, it can be shown that "there is no effective procedure for going from an algorithm for f to another algorithm for f that is significantly faster on

---

[30]The model training process for a natural-language processing (NLP) framework "Transformer" with 213M parameters has been estimated to be about 192 lbs of CO_2 footprint (201kWh) based on a reported Jun 2017 work. Attempting to increase its accuracy by a "neural architecture search" makes it jump to as much as 626,155 lbs CO_2 (656,347 kWh) based on a reported Jan 2019 work. The cloud cost for the training has been estimated to be about a few hundreds of dollars for the first case but rising sharply to between \$1M to \$3M for the latter[372]. Note that such costs are still critical in the context of very large scale data centers even when deep learning itself has reduced the PUE (power usage effectiveness) overhead by 15% as reported by Deepmind in 2016. If PUE is 1.1, for example, it can be reduced to 1.085, a reduction in power consumption only by 1.5%.

all but a finite number of inputs"; essentially, it says that there is no general algorithm for optimization. On the other hand, for a large class of functions f, "one can go effectively from any algorithm for f to one that is faster on at least infinitely many integers"; note that this does not contradict the previous statement as there may be another infinity of integers on which it is not faster. Checking if computational bloat is present is thus equivalent to checking an infinity of inputs. Also, "if one has an algorithm for a given function f, and if there is an algorithm which is faster on all but a finite number of inputs, then even though one cannot get this faster algorithm effectively" (as there is no general algorithm), one can still obtain a pseudo-speedup, i.e., "a very fast algorithm which computes a variant of the function but differing from the original function on a finite number of inputs" [74]. Thus some approximations are possible but not necessarily in the complexity class of interest (for example, unless P=NP, it is hard to approximate problems such as MAX-CUT).

While this result may be of theoretical interest (for example, "non-constructive" techniques such as diagonalization are used in the proof), it still indicates that there may be no universal complexity or resource measures. While bloat is usually seen as a performance or resource measure (call it X) gone "bad," it is possible to use X as a security condition (for example, $X < $ value). Since in general a security property is in the co-RE class, bloat detection is also co-RE (co-recursive enumerable): refutation of membership in a predicate or property (such as security) is easy but proving the predicate or property itself may not be easy (or terminate).

Even if a program has finite number of paths, it may be computationally infeasible to investigate all the paths if the number of paths is huge; if there are a huge number of paths, only sampling of these paths may be feasible unless quantum properties are exploited. At a foundational level, since "nature" is quantum mechanical (vide Feynman), any problem that involves search or superposition of states may be realizable in a quantum algorithm, possibly in a RP manner as powerful quantum properties are used in the search.[31] Any "conventional" algorithm for such problems is likely or will be highly inefficient[321, 1].

Researchers from Google and others have shown (2019) that a processor with 53 qubits ("quantum superposition bits") takes, for a specific computational task, about 200 secs compared to an estimated 10,000 years with a current supercomputer[39] this has obvious implications for RPD. While larger amounts of storage, when available, can make the quantum advantage smaller as pointed out by IBM researchers [302], the overall picture is clear that solutions based on quantum phenomena can have a substantial edge in specific contexts. This is arguably the first quantum computation that cannot reasonably be emulated on a classical computer.

It is also noteworthy that this experimental result challenges the extended Church-Turing thesis that states a "classical" computer can efficiently implement any "reasonable" model of computation.

---

[31] Just as with accuracy in analog systems, managing error in quantum systems is currently the limiting factor in deployment.

Recently, however, using "quantum" algorithm ideas, a classical recommendation system algorithm has been described[376] whose runtime is exponentially faster than the best-known in the literature.

Furthermore, moving now to a more practical plane, a recurring theme in modern designs is that it should be adaptive (or, self-reorganizing using feedback/sensors) and data driven. But is this possible in a RP way, even theoretically? Even if possible, is the design resource proportional with respect to desired parameters?

It might be first useful to review some recent "general" results. For example, Wolpert[307, 411] proves that "there is always something that an inference device cannot predict, something that it cannot remember, and something that it cannot observe." In particular, Wolpert proves that "in any such system of universes, quantities exist that cannot be ascertained by any inference device inside the system. Thus, the demon hypothesized by Laplace in the early 1800s (give the demon the exact positions and velocities of every particle in the universe, and it will compute the future state of the universe) is stymied if the demon must be a part of the universe."

There is thus the "breakdown theory of software development"[409]: there is no way to design or engineer a *complex system* perfectly from the beginning. As breakdowns of the software occur, the best that can be done, simply put, is to study them and patch them to handle newer problematic cases or newer problematic interactions not suspected before.

Interestingly, one can see examples of this issue in the working of even some of the simpler program profilers[275] where these often disagree on the identity of hot methods. First, samples used for profiling should be as far as possible non-correlated (i.e., if a sample is chosen, another sample elsewhere that has a high correlation with it should not be chosen); this needs information on the structure of the program in the general case. Secondly, probing or introducing sampling code needs to be handled carefully; incorrect execution is possible otherwise (for a very simple case, consider time-sensitive code). Often, such code has to be placed at "safe" locations (for example, consider GC interactions); this itself can make the profiling non-meaningful.

However, systems can still be engineered to have self-adaptive properties using workload modeling and prediction. Some examples are recent work such as using similarity detection in storage-system centric designs[206, 57], workload detection, i.e., discovering an application and thereby its characteristics through its history of accesses (memory or storage, for example) [423, 308]. The issue is whether the adaptive design is RP. Essentially, the data volume required for training or prediction needs to be taken into account.[32] Here approximations are necessary for effective application-level de-duplication (for

---

[32]DNNs are the most general with high accuracy, but costly in terms of resources. Typically they are an order of magnitude more costly in computational terms over linear models like LR (logistic regression) and SVM (support vector machines). The three approaches (LR, SVM, DNN) correspond to models with increasing numbers of free parameters. Using labeled input examples, these models must be trained by optimizing predictive accuracy[178].

example, across video), or ability to ignore certain types of "noise" (such as timestamps in IoT data streams) using application level knowledge.

One issue that needs attention is the relative lack of runtime information that a design has to cope with: it can be extreme uncertainty (zero information) vs. bounded uncertainty. For example, in some cases, inputs may be from a distribution (here prediction, ML and AI help) or there might be no information at all. In the latter case, use of c-competitive algorithms or game theory might be appropriate.

To discuss the trade-off between data collection latency and analysis accuracy, consider the problem of using performance diagnostics in cellular Radio Access Networks (RANs) to do root cause analysis for anomalies[292]. Applying standard techniques of ML is not sufficient here. Building a model on a per-station basis may not be enough due to limited data but waiting for longer periods may not be useful at all as the learned model may not be valid in the immediate future due to the highly varying RF environment. To solve this conundrum, multi-task learning has been proposed so that multiple related models are learned in parallel by using the commonality across such models. In the DARPA Grand Challenge[148] for fully autonomous driverless vehicles, the vision module had to maintain an accurate classifier for identifying drivable and non-drivable regions in the image stream. For this task to be done reliably, an *adaptive* learning approach was however necessary due to the "many changing and not easily measurable factors such as surface material, lighting conditions or dust or dirt on the camera itself that affect the target concept."

In general, finding anomalies can be seen to be the same as detecting concept drift[148]: concept drift primarily refers to "an online supervised learning scenario when the relation between the input data and the target variable changes over time." In a sense, there can be a "learning" bloat! For example, if the learning rate is almost the same as the model churn rate, the learning is plainly useless. In effect, we can have consistent "blindspots" over large state spaces.

10.5   Conclusions

Due to the high complexity of computer systems, now and likely more so in the future, and the pressing need for efficiency, predictable aspects of the systems will increasingly get factored into the design of the systems. However, the end result of this is that each subsystem will increasingly interact with essentially a stochastic environment as any predictability is already factored into the models of the subsystems. In such contexts, reinforcement learning is one approach that can be helpful, game theory and mechanism design another. In the future, systems design is likely to not only have a core structural component but also a critical learning component in a stochastic environment. Design of autonomous cars is one example.

Chapter 11

Data Centric Resource Proportional System Design

In previous chapters we discussed resource proportional responses to technology shifts such as persistent memory and low latency fabrics. In addition to these technology shifts, emerging system designs are also motivated by a shift in the nature of application characteristics, such as increased data centricity and insight orientation, driven by the rising adoption of machine learning and artificial intelligence. In this chapter we explore how these characteristics bring in an additional dimension of "data-centric" resource proportionality trade-offs compared to previous generation workloads.

There are two ways in which resource proportionality could be addressed in a data centric software ecosystem. The first approach is to program or modify analytics application software to explicitly use techniques (such as approximate computing or early stage data reduction) that consume resources in proportion to the utility achieved. The second approach is to introduce system software modifications that make the analytics infrastructure resource proportional (RP-aware memory and storage). This could include mechanisms that enable a system to learn and apply RP techniques on behalf of applications and also make system level resource optimization decisions across applications.

To understand these nuances in depth, let us study some interesting characteristics of data centric applications, how they impact resource proportionality and then delve into case studies of three different classes of data centric applications: graph analytics, machine learning driven data mining, and streaming IoT analytics.

11.1 Characteristics of Data Centric Workloads

As it gets easier to generate (and capture) data from physical, human, and machine sources, application decisions can be increasingly derived (learned) from the large volume of data available instead of only using hard coded business logic. Thus workloads are evolving to become data intensive. When we measure resource proportionality in these environments, utility or function accomplished depends not only on what the code does (software features implemented) but also the nature and volume of data processed (data features inferred).

We observe two distinguishing characteristics of such applications that are relevant for RPD.

11.1.1 Data intensive rather than compute intensive

The first key characteristic is the focus on data intensive processing instead of pure computation or application logic operating on small amounts of input data. Data-intensive processing could encompass various kinds of analysis involving large volumes of data, such as map-reduce, stream processing, or deep learning applications. In some scenarios, such as those involving genomic data, astronomy data, seismic models (or other scientific data), compute intensive machine learning, complex large graph network analysis, and medical image analysis, a data-centric workload could include deep computing in addition to being data intensive. The latter scenarios have motivated cross-pollination between big data and HPC techniques, both in terms of DISC (Data-Intensive Scalable Computing)[86] for scientific applications and HPDA (High Performance Data Analytics)[233] for big data applications requiring HPC resources.

One of the implications of data intensiveness is that any non-resource proportionality (or bloat) present in software components, as exhibited in data structures and processing steps, may be repeatedly incurred for the data elements in a big dataset. These costs may be further amplified through compounded effects induced by the algorithmic complexity of data processing involved. This observation has motivated techniques geared at bounding or containing these overheads. For example, Facade[284] and Yak[283] support big data friendly garbage collection by splitting the JVM managed heap area into a control space with generational GC and a data space with region based memory management. Facade achieves this using a compiler transformation and runtime system for optimizing the data path in Big Data applications based on a user provided list of data classes (refactored for clear separation from the control path), to ensure the number of heap objects is almost statically bounded. Yak implements a full blown Big Data friendly garbage collector for JVM based languages that relies on annotations of epochs in the program without any additional manual refactoring.

11.1.2   Analytics oriented: Emphasis on insight derivation rather than data serving

The second key characteristic is insight-orientation. Typically, the primary purpose of data centric processing is the generation of insights, such as analyzing or discovering patterns, learning models, predicting events, guiding decisions, or triggering timely actions. Consider the following classes of data intensive application case studies discussed later in this chapter:

- Graph analytics

- Map-reduce based batch analytics

- IoT Streaming analytics

The value or function accomplished by these applications is determined by the quality (effectiveness) of essential insight derived. Distilling valuable insight does not always require precise transactions on exact data values for every single piece of input data, because *not all data or all the computation contribute proportionately to the desired outcome*. In some situations, processing more data than necessary may even introduce over-fitting or noise that degrades the quality of insight. Further, a certain level of accuracy may be sufficient (acceptable) for correct decision-making and any resources expended to achieve higher accuracy could also be perceived as a form of bloat. One interesting implication of this characteristic is the opportunity to apply a variety of data reduction techniques and approximate computing optimization strategies[269] to manage the accuracy vs. efficiency trade-off.

11.2   Data Centric Frameworks and Stack Evolution

The evolving landscape of data intensive applications illustrates a recurring tension between productivity, flexibility, and efficiency. These trade-offs expose a rich variety of RPD challenges along each of the three dimensions depicted in Figure 1.4. For example, during the exploratory phase of any data analytics scenario, there is a high focus on development productivity. This favors the use of tools, languages (python, R, matlab), and libraries that enable fast experimentation with new analysis, preferably on representative data. Once a suitable analysis approach is identified, scaling data sizes become increasingly important for most practical use cases. At this point, focus shifts to efficiency and scalability, such as the development of specialized algorithms and optimizations that exploit emerging hardware technologies and distributed computing for particular analysis types, e.g., graph analytics. However, when deploying analytics in production for wide usage, operational productivity and flexibility become critical, including fault-tolerance and resource management. This has encouraged the rise of standardized (large scale distributed) general analytics frameworks, which provide certain core primitives, e.g., Hadoop map-reduce for compute and Hadoop distributed filesystem (HDFS) for storage, often layered over commodity infrastructure because of cost or portability considerations.

With the emergence of more demanding and sophisticated analytics techniques as well as changing hardware trade-offs, new frameworks such as Apache Spark [363] have evolved, introducing core primitives that enhance both flexibility and efficiency, resulting in order of magnitude faster results in some applications using in-memory computation. For example, Apache Spark introduced a core abstraction called resilient distributed datasets (RDDs). RDDs record the lineage of computation on data represented as a direct acyclic graph (DAG) of operations, which allows efficient fault-tolerance for in-memory iterative computations very typical in machine learning. The framework provides a unified engine flexible enough to efficiently support and combine different types of analytics within the same application, e.g., ETL[1] using dataframes (SparkSQL), machine learning (SparkML), and graph analytics

---

[1] Extract transform load.

(GraphX), exchanging data between phases directly through RDDs or higher level abstractions such as dataframes. The ability to combine batch and streaming analytics is supported by discretizing streams as small batches of RDDs. However, as low latency streaming demands increase, alternate frameworks such as Apache Flink have opted for a different core primitive directly addressing streaming dataflows, where batch analytics could be treated as a special case of streaming replays, rather than the other way around. Meanwhile, the rise of deep learning has spurred a fresh set of frameworks, such as TensorFlow, and the associated challenges of managing, retraining, and serving models in production.

As this history shows, it is hard to anticipate all new opportunities and advancements that would present themselves in the future, no matter how stretchable the core primitives chosen are. Yet reuse, flexibility, and standardization are important for development and operational productivity. A systematic use of RPD principles can help characterize and balance these trade-offs.

11.3   Sources and Impact of Non-resource Proportionality

Resource proportionality in data centric workloads can be studied along the three dimensions described in Chapter 1 (Figure 1.4):

1. proportion of program features utilized vs. provisioned (as discussed in Chapter 5 and 6)

2. perturbation due to technology change or system architecture of deployment environment (as discussed in Chapter 8 and 9)

3. proportion of data (and computation) contributing to useful (acceptable) insight (new to this chapter)

While applying RPD requires addressing each of these dimensions systematically as shown in Figure 1.4, let us focus the discussion in this chapter on illustrating nuances that are very characteristic of data centric applications.

11.3.1   Characteristic resource cost amplifiers in data centric applications

Large scale data centric applications incur some resource expenses for data movement and fault tolerance which can amplify the impact of non-resource proportionality.

11.3.1.1 Data movement expense

One the challenges in data intensive workloads is that the scale of data that needs to be processed is so large that it must often be distributed across multiple compute nodes and also across multiple levels of memory and storage hierarchy. Moving volumes of data around (across nodes or between different media levels in a storage or memory hierarchy) is expensive and a potential source of non-resource-proportional behavior when data that is moved does not contribute proportionately to the analytics outcome.

Further, the need for serialization-deserialization when exchanging objects across distributed big data systems on different computing nodes incurs a high cost when a large number of data elements have to be transferred. Approaches such as Skyway[282] minimize this overhead by enabling object graphs to be directly moved across remote heaps without serialization-deserialization.

Shipping functions close to the data can help improve RP to an extent, but consolidating insights across the dataset requires communication and data exchange. Various data partitioning and distribution strategies have been developed to optimize data movement (typically as part of the framework control plane) in distributed analytics workflow systems. Another approach is to leverage system architecture advancements, such as fabric attached persistent memory and memory driven computing[139], that completely alters the RP trade-offs as discussed in the previous chapter. For example, Sparkle[218] optimizes the shuffle primitive in Apache Spark to directly use this shared (fabric attached) memory instead of network data transfers and associated serialization-deserialization.

Similar data movement RP challenges also arise in edge to core or hybrid cloud analytics environments where the source of data may be at a far edge which lacks enough computation resources or context for delegation of processing responsibility. We discuss this with a case study in Section 11.6.

11.3.1.2 Fault tolerance, check-pointing, and lineage

Data centric application frameworks are designed with fault tolerance capabilities to avoid losing the valuable insights derived by expending significant computation resources in processing and distilling data as long running analytics workflow progresses. Common approaches for fault tolerance include checkpointing state or saving redundant copies of intermediate outputs (e.g., HDFS typically keeps three copies of data across nodes in the cluster[2]). However, maintaining copies of already huge datasets across nodes can introduce significant overhead for in-memory analytics processing. A common RP optimization in frameworks such as Apache Spark[363] and Alluxio[28] is the ability to track lineage of data transformations in a way that allows recomputation of steps since the last saved copies available upon a node failure.

---

[2]Recent Hadoop versions (3.0) support more space efficient redundancy schemes such as erasure coding[146].

This enables applications to exploit a trade-off between fault tolerance cost vs. recovery cost (and RTO[3]) without compromising RPO.[4]

Notice how non-resource proportionality due to irrelevant data gets amplified as it incurs additional costs for fault tolerance.

11.3.2   Characteristic sources of overheads in data centric applications

There are three characteristic sources of non-resource proportionality in data centric applications: data locality impedance mismatch, data that does not contribute much insight, and needless synchronization.

11.3.2.1   Impedance mismatch between workload locality patterns and page locality of the system

Data movement across different levels in the memory and storage hierarchy within a single node has traditionally been automatically optimized by sophisticated processor caches, memory management units, and operating system virtual memory mechanisms. These optimizations are typically transparent to the application, assuming spatial and temporal locality of data reference by workloads, both at the granularity of cacheline sizes and memory page sizes (4KB-64KB). However, such assumptions may not hold true for many data centric workloads, where the data layouts may exhibit entirely different patterns, e.g., a graph traversal is likely to have very little sequentiality. This results in an impedence mismatch between the system level assumptions and application specific caching and prefetching requirements where needless data paging manifests as non-resource proportional behavior. Even optimized virtual memory translation using TLBs have been found to be an overhead (i.e., source of bloat!) for large graph analytics workloads[52] which do not exhibit spatial locality (as neighboring vertices in a graph may not be spatially contiguous in memory or in-storage). Basu et al.[52] showed that turning off address translation (using direct segments instead of 4KB pages and TLBs) can lead to a 51% improvement in the Graph500 workload. However, turning off a feature reduces the versatility of the system, while a more resource proportional solution would need to retain the power of this feature but implement it in a way so that it does not introduce an overhead under such conditions.

11.3.2.2   Computation on data that does not produce additional insight

While the abundance of data accessible is a boon for analytics, not all data produced is equally informative for a given application. In practice, an overwhelming volume of data that does not add much new information is

---

[3]Recovery Time Objective.
[4]Recovery Point Objective.

continuously generated by sensors, cameras, human activities, and computer systems, which can lead to a disproportionate utilization of resources to derive analysis insights. Any data or operation that does not contribute insight can thus be viewed as a source of bloat resulting in a resource overhead. This additional dimension of resource proportionality can be expressed as follows.

The data resource proportionality (amplification factor) is the ratio of the resource usage $R^{alldata}$ of an application in a deployment scenario when it uses the entire data set provided compared to $R^{reduced}$, the resources expended by an application that uses (and operates on) only the reduced informative data that is sufficient for deriving the insight required in that scenario:

$$R^{alldata(S_j)} = R^{D_{1..n}}(S_j)$$
$$R^{reduced(S_j)} = R^{D_j}(S_j)$$

Resource overhead $R^{overhead(S_j)} = R^{alldata(S_j)} - R^{reduced(S_j)}$

Data centric resource amplification $\alpha_d(S_j) = \frac{R^{alldata(S_j)}}{R^{reduced(S_j)}}$

Data bloat propensity $b_d(S_j) = \frac{R^{alldata(S_j)} - R^{reduced(S_j)}}{R^{reduced(S_j)}} = \alpha_d(S_j) - 1$

This type of bloat is hard to eliminate because of the intrinsic difficulty of predicting what data is useful and what isn't. Making this determination typically requires some computation on the data, usually as part of the early stages of an analytics workflow. For example, several data reduction strategies are often used to tackle the data overload, including dimensionality reduction (and feature selection), data aggregation, compression (similarity based de-correlation, transformation, and encoding), discretization, approximation (including sampling), abstraction (data numerosity reduction using compact representations such as parametric models or concept hierarchy generation)[175]. Different strategies incur differing levels of computation resources and accuracy trade-offs. These strategies may be applied to input data or intermediate data at any stage of the workflow.

Let $R^{select(S_j)} = R^{D_{1..n} \Rightarrow D_j}(S_j)$ be the resource cost incurred to perform the data reduction required to arrive at the informative data that is sufficient for scenario S_j.

An implementation (system + application stack) that is designed (staged) to reduce the data to this informative set and then analyze only the reduced data set would incur the following resource cost:

$$R^{alldata(S_j)}_{staged} = R^{select(S_j)} + R^{reduced(S_j)} = R^{select(S_j)} + R^{reduced(S_j)}_{transfer} + R^{reduced(S_j)}_{process}$$

where $R_{transfer}$ refers to the data movement resource expense incurred and $R_{process}$ the resources expended for the actual analysis on the reduced data.

$$\alpha_d(S_j) = 1 + \frac{R^{select(S_j)}}{R^{reduced(S_j)}}$$

In practice, it is too simplistic to assume an explicit separation between R^{select} and $R^{reduced}$. Typical (big) data processing workflows tend to be implemented as a series of transformations (e.g., map-reduce operations that filter the data set) and analysis iterations (e.g., principal component analysis, non-negative matrix factorization) that reduce and refine the data in stages to distill relevant information. The accuracy vs. efficiency trade-off varies depending on the staging approach and techniques adopted. Reducing the data set down to its intrinsic dimensionality may even enhance accuracy and understanding by eliminating noise and avoiding model over-fitting.

Carrying irrelevant data can tax system resources in many ways in addition to the extra analytics computation costs, e.g., flooding networks and precious faster tiers of storage, memory, and caches with low value content and also increases (degrades) the time to value for critical insight.

11.3.2.3 Unnecessary synchronization

Data centric applications are typically designed to scale by leveraging data parallelism at different levels, such as:

- leveraging multiple cores within each node using multiple computation threads and processes

- utilizing large scale distributed analytics clusters (e.g., Hadoop, Spark)

- performing distributed learning across remote edges connected through WANs.[5]

As in any multi-threaded or distributed application, this architectural approach incurs the overhead of synchronization costs (e.g., locking, barriers, data consolidation) for consistency and coordination. However, strict synchronization of data structures and operations across parallel execution threads may not be essential for satisfying the primary data processing purpose of insight generation. For example, researchers have demonstrated that an approximation strategy of relaxed synchronization[330] can produce acceptable results with a significant speedup for several common analytics algorithms such as KMeans clustering and Breadth First Search (BFS) of graphs. Similarly distributed deep learning with approximate synchronization of parameters can significantly reduce communication costs[278]. Thus synchronization operations that are not necessary for deriving acceptable insight can be viewed as sources of non-resource proportional behavior.

11.4 Graph Analytics Case Study

Graph analytics is a popular category of (big) data processing. Graphs associate context with data through a rich knowledge representation that

---

[5]Wide Area Networks.

expresses relationships between a variety of data elements. Patterns in both the connectivity structure and content properties can be analyzed to derive insights using a diverse set of techniques. Graph analytics approaches can be characterized by the choice of programming model, algorithm, and execution model. The choices appropriate for a given scenario vary based on the nature of the graph (structure, content, update pattern), the type of analytics questions, the data model, and the underlying system architecture layout. For example:

- A couple of commonly used graph data models are (i) RDF, resource description framework (a labeled directed multi-graph expressed as subject-predicate-object triples) and (ii) property graphs[336] (a directed attributed multi-graph that encodes both connectivity structure and properties with each vertex and edge)

- The type of graph analytics could range from transactional graph database queries (typically involving short hops from initial matching vertices) to whole graph analytics (e.g., betweenness centrality,[6] page rank,[7] SSSP,[8] graph coloring, connected components, loopy belief propagation) and could require any path (reachability, WCC[9]) or all path (shortest path) exploration between source and destination vertices

- The system architecture environment could range from single system environments, where the graph placement could be entirely in-memory or out of core (disk/SSD based), to distributed graph processing across a large cluster. Both environments could also potentially leverage GPUs or other hardware accelerators.

Graph programming models could include vertex centric, matrix, or subgraph centric approaches [425]. The vertex centric approach (with or without block centric optimization) is particularly popular and has many variations. The variations may support edge based communication (single hop) or ID based communication (which can cover multiple hops through pointer jumping or path doubling) and may use either message passing or shared memory. Execution of iterations could be designed to be synchronous (e.g., BSP : bulk synchronous processing) or asynchronous depending or even be adaptively switched between the two using approaches such as PowerSwitch[413].

In all these cases, a common opportunity for improving data resource proportionality is to focus more attention on the informative portion of (graph) data that contributes analytics insight, both by (a) reducing processing of

---

[6]The extent to which a vertex lies in the shortest path between other vertices, a measure of the influence it has on information flow in the graph.

[7]Algorithm that measures the transitive influence of nodes in a graph, originally used to rank websites in Google.

[8]Single source shortest path.

[9]Weakly connected components.

irrelevant data and (b) improving the impedance match of system level mechanisms for graph data loading and data movement at each processing iteration.

11.4.1 Reducing the size of the input graph data processed

Determining what data is informative and sufficient for a given analytics scenario can be especially challenging with large irregular graph structures with complex interactions. Techniques previously discussed (Section 11.3.2.2) for scalar or multidimensional datasets are not enough. An effective reduced graph abstraction must jointly preserve key structural (connectivity) properties and attribute information relevant for the kind of analytics queries presented by a given deployment scenario, while avoiding less relevant detail. To ensure resource proportionality, the computation needed to perform the reduction needs to be scalable, simpler, and significantly less resource intensive than the analytics computation on the full graph.

A survey of graph summarization approaches from Liu et al. discusses some of these challenges and abstraction methods such as (a) aggregation via vertex or edge grouping (combining vertices into supervertex, edges into super edges, or compressor virtual nodes) and (b) simplification or sparsification (skipping edges or vertices). The aggregation or sparsification heuristic could be prioritized based on connectivity or attributes associated vertices or edges. Suitable choices and their impact can vary significantly based on the type of analytics query and nature of input data.

For example, it has been shown that a 40%-50% input graph reduction obtained by applying a sequence of local, non-interfering input graph abstraction transformations can speed up iterative vertex centric algorithms by 1.5x[231] when both the original and reduced graphs can fit in memory. A two phase processing model was used which first runs the original algorithm on the reduced graph and then converts the results into precise answers using the original graph. GraphQ[397] and Wonderland[432] adopt an iterative approach of abstraction (skipping edges) and refinement (adding edges) using two very different schemes that progress with iterations of the analytics query across partitions for out of core graphs on a single system. GraphQ is geared at analytics questions of the form "find n entities from the graph with a given quantitative property" (such as page rank, single source shortest path, community detection, connected components) and incorporates a notion of budget-awareness, i.e., a memory budget that is resource proportional to the extent of query answering capability supported. Wonderland can speed up convergence of all path and any path analytics queries where the abstraction is guided by an application specific property of interest using custom edge selection and prioritization of edge blocks (e.g., keeping edges with the smallest weights for SSSP).

Applying generalized approximate computing techniques to graph analytics is non-trivial. In particular code centric approaches can run into issues as the workload characteristics depend heavily on the nature of the input graph which tends to be irregular (e.g., the structure may exhibit a power law, where a few vertices have a very high degree). Ongoing research on approximate graph analytics such as GAP [194] (Graph Analytics by Proximation) explores graph sparsifiers that leverage spectral graph theory along with a machine learning model to learn and predict the amount of sparsifiation needed for a given scenario and resource budget.

11.4.2 Enhancing the proportion of relevant data paged in by the system

Specializing graph processing algorithms around a given set of memory and storage hierarchy trade-offs limits resource proportionality with respect to technology change and system architecture deployment choices. For many traditional workloads, system level mechanisms for paging and caching data have allowed applications to remain transparent to low level technology and system architecture characteristics. However, because spatial contiguity of access may not hold for graph data intensive applications (Section 11.3.2.1), only a small proportion of data paged in automatically by the system may be useful for the next few processing steps. Thus a graph database workload, for example, effectively perceives the processor MMU, virtual address translation, and OS virtual memory management as bloat. The architecture community continues to research optimizations that could reduce complex translation overheads when avoidable, such as direct mapped sections, de-virtualization, and do it yourself virtual memory[26].

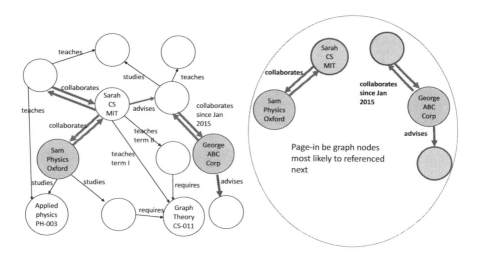

FIGURE 11.1: Sample property graph with logical paging of nodes and vertices that are likely to be referenced together next (e.g., the people Sam and George collaborate with or advise).

RPD could address the heart of this impedance mismatch problem in virtual address translation, caching, and paging by recognizing that not just locality patterns but even the very shape of what constitutes a page should be dynamic based on application associations (e.g., vertex edge relationships and common properties) rather than purely by spatial contiguity.

For example,

- In a property graph (Figure 11.1), we may observe that vertices that have a common property (e.g., companies which build a certain class of products, professors who teach a certain subject) and all the edges connected to those vertices labeled with a common property (e.g., customers using those products for more than a year, researchers they collaborated with in the last five years), tend to be referenced together.

- Or, during an iterative vertex centric application, only a subset of edges in a given partition may have new contribution during a given iteration and needs to be paged in[394].

- Or, everything but an abstract graph may be paged out to start with and its refinements paged in as iterations progress.

Can these patterns of what needs to be paged in or paged out together be expressed to or learned automatically by a virtual memory control plane? With application aligned paging policies, there is less wastage due to paging unnecessary content. The trade-off is the cost incurred in achieving the application aligned decision (which is typically more complex than spatial locality based fixed size paging). Hardware accelerators may be considered to optimize this cost, just as TLBs help cache and accelerate traditional address translation. For example, ternary content addressable memories (TCAMs) can be used to optimize certain types of pattern searches or associative lookups. A more radical approach to software defined virtual memory is to implement a content addressable virtual memory hierarchy as proposed in [64].

Sometimes just paging in relevant data may not be effective if the graph processing logic still traverses less relevant data. For example, consider, "Load the edges you need"[394], an optimization for out of core vertex centric algorithms that avoids loading edges (in the partition) which do not propagate any new values. This requires a corresponding change in the graph processing logic to avoid the need to access these edges. The optimized algorithm (i) uses incremental instead of absolute computation which accumulates deltas from incoming edges[10] to ensure unchanged incoming edges need not be referenced and (ii) delays propagation to out-of-core edges through shadow iterations.

---

[10] Assuming a distributive graph algorithm.

11.5 Data Mining Case Study (Map-reduce/Spark)

The rise of large scale distributed analytics, deep learning, and AI (using computation platforms such as Hadoop, Spark, and TensorFlow) reflects a steady shift from domain expertise guided traditional data mining approaches (modeling, collecting, and indexing data in a way that is pre-optimized for a specific class of analytics queries) to automatic discovery oriented analytics approaches. These approaches allow greater power and flexibility by leveraging machine learning to effectively learn programs (analytic models and even features of interest) from data[125, 126] instead of programming fixed rules to process data.

In the parlance of the data science and machine learning community a *feature* is "an individual measurable property or characteristic of a phenomenon being observed"[404]. A machine learning model (program) can be used to explain, classify, predict, prescribe, or control outcomes of interest based on features of available data instances.

Just as framework based software tends to over-provision program features in order to prepare for a wide range of deployment needs, these data centric systems tend to over-provision data processed (and features analyzed) in order to address a wide range of analytics possibilities, because:

- it is hard to anticipate upfront what questions would be interesting as we may later discover something important to ask

- even when we know the questions, we cannot always predict what data and features are needed to answer those questions lest we introduce bias or miss discovering a surprising insight

- even if we identify what is needed, given the sheer volume of raw data, we still must traverse or index some data that is not needed just to reach relevant data (e.g., to locate data matching feature ranges needed)

Fortunately, despite these unknowns, in practice analysis needs, data instances, and their features tend to exhibit some predictable characteristics with enough temporal stability for RPD. Standard data reduction techniques including correlation analysis and similarity based summaries help focus more processing attention on relevant data when designing data analytics workflows. Such pre-processing steps incur the cost of extra pass(es) on the entire data, but are often worthwhile when it is necessary to reduce the amount of data processed by a subsequent compute intensive analytics phase. Remembering the results of these passes is one way to amortize the cost when multiple analysis (applications) operate on the same data. For analytics steps which tend to be repetitive, computation saving can be achieved through result caching (including computation de-duplication across applications) [385, 377, 171] and approximate computing.

Data centric workloads can involve an analysis pipeline with many different types of data processing. One may explore resource proportionality of each such stage independently and then that of the entire workflow (including data transfer stages). Further, multiple analysis pipelines (ETL, batch mapreduce, deep learning) may operate on a common repository containing data accumulated from multiple data sources. Applying RPD to this entire system of datasets and applications (hosted in an analytics cloud environment, for example) helps amortize the trade-off between resources spent in discovering more informative parts of data vs. resources wasted due to processing less informative parts of data (especially costly when non-linear, e.g., joins where processing vs. data curve is steep). The key to optimizing the trade-off lies in persisting this learning (about relative data relevance) and improving on it over time.

11.5.1 RPD aware storage systems for insight-centric applications

There is a close analogy between human memory and a storage system that automatically sifts and abstracts information to focus attention on relevant data based on previously learned insights. Strategies for persistently organizing large data sets that make it easy to sift (grade) irrelevant or less informative data could go a long way in improving data resource proportionality.

For example:

1. Correlated data[11] sources or features may not add much new information and could be de-prioritized. This needs to be done carefully as in some situations, collecting correlated information helps reveal and measure small deviations with important implications (e.g., use of multiple cameras to gauge depth, use of multiple data sources to detect inconsistencies).

2. Remembering the results of (common) computation steps expended on data makes it more informative, i.e., distilled data could be graded higher. On the flip side, computing and preserving excess derivations hurts resource proportionality and should be graded lower.

3. Repeating the same computation may not contribute new insight and could be de-prioritized when previously learned analyses are available for reuse across analytics applications that operate on common data or run common processing steps repeatedly (to reduce recurring overheads).

---

[11] High mutual information.

4. Similar data may not add much new information and could be de-
prioritized or bypassed (by leveraging previous insights) [206]. At the
same time, any similarity detection policy needs to handle situations
where small differences matter and could lead to different analytics outcomes.

5. A representative sample of the data could be processed using approximate computing techniques to obtain analytics results within an acceptable accuracy range compared to that obtained with the full data. This
representative subset can be prioritized over the remaining data. However, some trial and error is typically required in order to learn a suitable
approximation for a given situation.

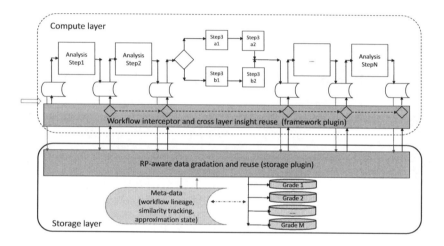

FIGURE 11.2: An RPD aware storage system with an application layer workflow
interceptor and resource proportional data gradation.

The following are examples of some techniques that could be employed to
create an RPD aware storage system (Figure 11.2):

11.5.1.1   Cross-layer insight reuse

Cross-layer information exchange between the computation layer and storage layer enables memory of past computation on data (i.e., derived insights)
to be preserved, reused, and possibly generalized to similar data and analysis scenarios. For example, consider an Apache Spark application, where the
workflow interceptor shown in Figure 11.2 could pass workflow lineage (from
the RDD) to the storage layer, enabling it to remember data and lineage
associations so that it can detect opportunities to reuse past insights if the

lineage matches an earlier computation. Along similar lines, Nectar[169] provided a distributed analytics framework service that unifies data management and computation and replaces repeated computation with previously computed results if available. UNIC[377] is a cache service proposed for secure deduplication of general computations on identical data. SEeSAW[206] extends this idea to bypass computations that can be replaced with results previously computed on (statistically or semantically) similar data. Potluck[171] provides cross application computation deduplication for mobile applications, while FoggyCache[170] goes a step further to allow approximate computation reuse across mobile devices.

11.5.1.2 Semantic similarity detection

Data that leads to similar analysis outcomes can be considered semantically similar. Predicting semantic similarity in advance can help save expensive analytics computations and storage resources. However similarity detection in raw data can be difficult and expensive due to the presence of irrelevant features and statistical variation. SEeSAW[206] detects statistical or semantic similarity of the intermediate outputs at a point in an analytics workflow, preferably after pre-processing steps have distilled features (fields) of relevance. Similarity in data patterns such as frequency or trends may be detected by comparing coefficients of a Discrete Wavelet Transform of an input signal; similarity in documents may use TF-IDF,[12] latent semantic analysis. If the intermediate outputs are found to be similar to a previous instance for the same workflow lineage, then corresponding raw data for both instances may be (de-prioritized and) marked as similar and subsequent analytics computation steps can be bypassed by reusing previously computed results.

11.5.1.3 Approximation reuse

Approximate computing [334, 45, 279] is based on the observation that there are several problems (e.g., in search, analytics, and media applications) where approximate results are sufficient and appropriate but computational resources are still needlessly expended on ensuring exact and accurate answers. A variety of mechanisms to support acceptable accuracy vs. efficiency trade-offs have been devised at different levels of the hardware, architecture, and software stack [354, 341, 135]. These include compiler transformations for relaxing accuracy while provably maintaining suitable acceptability tolerance bounds [260, 97]. There has been a surge in research interest on approximate computing based optimizations in the context of analytics workloads as they are naturally tolerant of inexactness [269]. BlinkDB [21] provides approximate answers to SQL queries and ApproxHadoop [157] provides input data sampling and online aggregation of partial intermediate results for Hadoop applications.

---

[12]Term frequency - inverse document frequency.

Such approaches can be viewed as exploring resource proportionality with respect to overprovisioned accuracy.

Just as the case for similarity detection, some processing is needed to characterize features of interest for the approximation. This phase is expensive (training), but can enable significant resource savings when the feature approximation model is stable enough to reuse the characterization repeatedly for subsequent data approximation (with occasional re-training). How often we use the approximation, once trained, may determine the breakeven point for RPD.

11.5.1.4   Proactive data gradation

Each of the above techniques provides a cue for prioritizing (grading) data based on how relevant or informative it is likely to be. Distilled data is more informative than raw data and could be placed in a fast storage tier, especially if it involves very common computations or significant dimensionality reduction. Similar data is less informative and can be graded lower (moved to a slower storage tier). Approximation selects a representative (informative) subset, which could be assigned a higher grade than the rest of the data. Resource proportional data gradation can be performed proactively when data is ingested by shifting some computation steps to occur in anticipation of the analytics computation it would be used for (e.g., as predicted from the memory of past workflow lineages). This could minimize the chances of needless data movement across storage tiers.

11.6   Streaming IoT Analytics Case Study

Step 1 in RPD (Chapter 4) provisions software components for peak functionality with a wide feature set, so it is extremely flexible. However, several data centric solutions tend to be implemented as specialized stacks for efficiency reasons, sometimes to the point for being optimized for a certain type of data or system configuration. Could RPD help us to evolve more general purpose stacks without losing that efficiency?

Consider the example of streaming analytics for IoT[13] use cases as our third case study. Continuously generated sensor data is processed at edge computing sites to guide local insights and action. Some data flows between the edges and the core (or cloud) for consolidated deep analysis and learning to guide global insights and action. The data could be both high volume and high velocity (e.g., especially when video and high precision sensors are involved). Further, response time and processing requirements for these applications could be

---

[13]Internet of Things.

fairly demanding (when fast and accurate insights are essential for reliable decisions and timely action). Latency and cost considerations at the edge (e.g., constraints on compute, storage, and WAN bandwidth) have driven early implementations of many edge computing solutions to be built as a specialized infrastructure stack packaged with IoT frameworks and application components for streaming analytics and data flows connecting edges to core sites and the cloud.

However, overspecialization can become limiting for edge platforms which may potentially need to evolve to run diverse workloads and varied analytics pipelines. The diversity of edge environments and specialized stacks can complicate integration with emerging ecosystems even as a variety of domain specific data reduction, filtering, and aggregation schemes give way to sophisticated edge to core function distribution and distributed learning techniques. Further edge infrastructure need not be limited to analytics use cases and could also be used to host traditional business applications close to the edge. This has encouraged the adoption of converged IT[14] and OT[15] infrastructure at the edge [94]. Standard operational features such as unified manageability, storage, and WAN optimization provided by software defined storage and hyper-converged infrastructure are also valuable for edge environments [348].

However hosting special purpose stacks over general purpose systems results in resource wastage due to cross-layer function duplication. This makes it difficult to match the efficiency of the specialized solutions, especially with respect to latency of action at the edge and data transfer costs. RPD principles can be used to co-design interfaces that unify this feature overlap.

For example, given the prevalence of repetitive patterns in normal sensor data, semantic similarity based de-correlation can improve data resource proportionality as discussed in Section 11.5.1.2. De-duplicating the effort expended on semantically similar data could save edge processing and storage costs and also significantly reduce edge to core data WAN data transfers. On the other hand, storage layers often include both data compression and a built in de-duplication mechanism that detects data blocks with identical content and stores them as clones, resulting in improved storage and data transfer efficiency. Despite the obvious overlap between these two features (de-duplication based on exact content vs. similar content) there is also an impedance mismatch which hinders direct reuse. In most IoT scenarios, the opportunity for exact block de-duplication is low and hence vanilla storage de-duplication only results in a runtime overhead. Meanwhile, semantic similarity within data streams is typically easier to detect and utilize at a higher layer, especially because cloning similar data is a lossy transformation which may be acceptable in some use cases but not others. Could these features be unified using RPD?

---

[14]Information technology.

[15]Operational technology.

Another example is the reliance on streaming analytics frameworks (such as Apache Kafka, Storm) to perform application defined selective streaming of data from edge to core. As data streams are backed by persistent storage (e.g., Kafka maintains streams as log files), this feature has a partial overlap with (asynchronous) edge to core storage replication to maintain remote copies or backups (e.g., for data protection or disaster recovery). These two different mechanisms for edge to core data transfer have arisen partly because of the impedance mismatch between management of streaming data vs. data at rest and partly because the application framework must be portable to storage systems that may not support replication. Could streaming framework and storage layer interfaces be RP co-designed to unify these features in deployment environments where this overlap exists?

StoreEdge RippleStream[57] explores how storage replication and de-duplication could be generalized to support the above features for edge to core streams in a resource proportional manner. Using RPD in this context improves the versatility of IoT stacks using general purpose infrastructure efficiently for edge computing, instead of requiring a specialized stack for edge analytics. An application layer streaming framework is used as is between local producers and consumers at each edge. Stored output streams intended for deep analysis at the core across are replicated from all edges by the storage layer (at application consistent intervals) to a second instance of the streaming framework at the core which serves local consumers at the core. The streaming framework broker at both ends may need to be carefully co-designed with minimalist storage replication feature extension interfaces that ensure application consistent transfers at the source and seamless notification of stream arrival at target. Figure 11.3 shows examples of edge to core stacks before and after co-design.

This approach provides a common cross-layer core mechanism for data replication from edge to core, with feature extensions to provide variations required for streaming vs. asynchronous backups. Unifying de-duplication and similarity detection mechanisms as part of a common RP data substrate with controlled specialization enables both storage savings and WAN transfer optimization. De-duplication may be expressed as three features: (i) computing a similarity measure when a new block arrives (SIM), (ii) maintaining a similarity index (INDEX), and (iii) cloning a block from a similar block (CLONE). From an RPD perspective, CLONE is the core feature (performs the actual de-duplication). SIM refines this core feature, proving multiple variations (similarity plugins co-designed between the application and storage layer), some of which would typically use INDEX to compare the similarity measure with previously stored blocks. Thus both SIM and INDEX are feature increments, which are typically used together, but may be independent in some variations. The default variation of SIM detects exact duplicates by supplying content hashes to INDEX. Ripplestream's similarity detection plugin customizes storage de-duplication to use semantic similarity measures as a basis for its index where

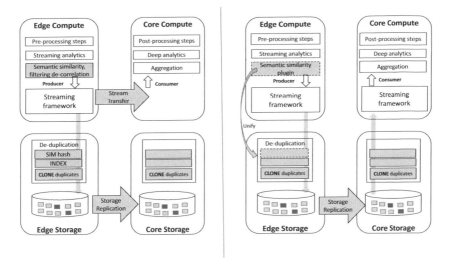

FIGURE 11.3: Edge to core analytics: Compute and storage stack layers with overlapping functionality (left). Cross-layer co-designed stack, e.g., RippleStream (right). Although only the data flow from edge to core is depicted above, data can also flow in the opposite direction (e.g., model updates can be pushed from the core to all edges).

appropriate instead of exact content hashes, while keeping the rest of the de-duplication infrastructure (`CLONE` and `INDEX`) as is.

Sometimes it is useful to retain the similar data for future reference, while assigning it lower importance and attention and still presenting the de-duplicated representation by default. This and other types of data gradation at the edge (e.g., placing data from correlated sources in a lower tier, having a separate full data view) in an RPD aware storage can help quickly eliminate processing and transfer of less relevant data in most typical situations, without losing the information in rare situations when the finer differences need to be analyzed.

11.7   Summary

In addition to evolution in program features and technology trade-offs, which have been described in previous chapters, data intensive applications also exhibit an additional dimension of RPD: data proportionality or input data relevance. Much of our discussion in this chapter has focused on this particular dimension.

Notice that both code features and technology factors can amplify the impact of non-resource proportional data and vice-versa. A common theme running across the case studies described in this chapter is the idea of separating the control plane of virtual memory, storage tiering, and replication from the data plane. This is a foundational RPD strategy that could be introduced in emerging systems to address a wide range of scenarios.

Traditionally, virtual memory management and storage services have offered only coarse-grained ways for software to specialize the basic mechanisms for addressing, caching, paging, protection, and advanced services such as de-duplication. As a result, many existing and most emerging applications rarely even attempt to exploit virtual memory features. This forces memory to be carefully sized upfront to avoid paging and managed directly by the application using custom buffer pools, caches, and heaps.

In RPD, we advocate moving towards a software-defined virtual memory approach, where custom virtual memories can be created using control plane knobs for different usage models that specialize the locality pattern (shape, size, and rules) assumed for paging decisions. Common usage models include JVM-like managed runtimes, relational databases, in-memory key-value stores, graph databases, with both regular (fixed stride) and irregular data access. For example, the shape and size of ideal paging units and their locality patterns are quite varied for managed objects (which are garbage collected), regular blocks, associative lookups, graph traversals, persistent structures, and irregular accesses. To identify such broad categories with different consistent structures and distinctive access patterns, both application specific rules and stack distance analysis in terms of application aligned locality may be used.

Further, we advocate the use of a common RPD aware storage layer, where instead of exact de-duplication and compression, similarity based de-correlation, insight reuse, approximate computing, and compressive sensing (such as impression store[431] for top K queries) could be supported as customizable software defined storage views and policies for different classes of use cases.

Part IV

The Road Ahead

Chapter 12

Adapting the Systems Software Stack to a Radically Non-Uniform Memory System

A number of technology change driven RP considerations were discussed and quantified in Chapters 8 and 9. In many cases, technology change is driving more and more memory non-uniformity into systems. To date, most software has dealt with memory non-uniformity in just a few ways.

- NUMA – Non-uniform memory access caused by large memory systems. This includes memory access across processors in a multi-processor server, and memory access across server components in a super-computer. Management practices and to some degree OS and application software are structured to recognize NUMA boundaries and align application software components with them.

- Page fault to IO – When a memory access cannot be completed quickly for reasons such as page swapping or remote access, exception interrupts such as page faults are used to convert memory access into an IO protocol that can involve a storage or local area network. This is generally triggered by CPU MMU supported memory virtualization functionality.

- RDMA – An application, knowing that a memory access is not local, converts it into a Remote Direct Memory Access (RDMA). Unlike NUMA or page fault driven approaches, RDMA is not directly supported by CPU MMU based virtualization so the distinction is visible in application source code and/or libraries.

- IO – An application, knowing that it is accessing data that is stored using block structured access, uses IO to access the device that is storing the data.

While the first two of these can be completely hidden from application data access using virtual memory techniques, this lack of awareness can lead to RP problems related to performance due to locality management shortfalls. Since the latter of these are not hidden from applications and they do not offer the cache coherency of the CPU's virtual memory system, applications must be explicitly designed for them from the start using classical clustering and/or storage access techniques including disk and/or network protocol stacks and data consistency control such as locking.

The PM driven dual stack scenario described in Chapter 8 and the expanded memory interconnects described in Chapter 9 conflate these four approaches, increasing the tendency of all to appear as memory accesses with radically different latencies. This is the ultimate implication of memory centric system architecture.

This chapter drills down into the problems of RP and memory non-uniformity by breaking it into three subproblems.

- Allocation – How can system resource proportionality be improved through RP memory allocation?

- Automation – How can RP memory allocation be made less manual?

- Blocking – How can thread or process blocking be constrained to increase RP?

Throughout this chapter we develop and explore an approach to RP with radically non-uniform memory that involves specialization of system work flow management in the context of memory pooling with Classes of Service that are defined for a given application.

12.1   Resource Proportionality of Memory Resource Allocation

In this section we will explore a response to non-uniform memory allocation in the form of memory pooling. We will make note of the RP issues related to pooled resource allocation overhead involved in placement decision making and extraneous data movement.

12.1.1   The evolution of memory pooling

The most obvious place to look for tools to manage non-uniform memory allocation is memory pooling. A significant amount of work has already gone into this area, and it is ongoing. CPU cache and memory hierarchies have created physical pools throughout most of their history. Over time the motivations and approaches to memory pooling have evolved beyond the basics of caching to manage the following issues.

- Cross-Processor Latency – Multi-socket processor systems create classical NUMA behavior. It is common to form memory pools that represent memory attached to particular processors. These are then used to increase affinity between application threads and the memory they access.

- Fragmentation – Pools are created within a homogeneous memory system to represent multiple allocation sizes. Application allocation library functions (i.e., malloc) attempt to choose a pool with a fixed size allocation unit that matches each allocation request. This eliminates the garbage collection processes that compensate for fragmentation caused by variable sized allocations. It does, however, introduce other types of fragmentation in applications with many different allocation sizes by rounding up to the nearest pool. It can also create underutilization if pool size requirements are unknown or unstable.

- Object creation overhead – Appalled by the time it takes to create objects in modern programming languages, programmers have resorted to maintaining pools of free objects having common types or classes. This circumvents the native language overhead of allocation and it allows pre-initialization of objects or groups of inter-related objects. Although it avoids fragmentation issues it must still deal with pool utilization. That said, object pools can capture significant application specific information that is otherwise hidden in addition to RPD aspects covered in Section 5.2.1.

- QoS – Applications may place specific requirements on particular data structures regarding aspects such as persistence, performance tier, data protection, redundancy, and security. The process of matching memory (or storage) consumers to providers also involves a type of pooling.

There has been considerable research published on memory pooling, some of which offers further insight into the resource proportionality of non-uniformity. For example, in Whirlpool [273], dynamic caching policies are constructed around memory pools, each of which represents a major application data structure. Static and automated approaches are compared to emphasize the point that locality shifts during execution so static approaches are not as RP as those that combine statically defined application relevant pools with dynamic cache management policies. In Whirlpool, the purpose of the dynamic policies is to re-configure placement of statically defined pools in physical memory. Periodic re-configuration is driven by monitoring of which cores are hitting each pool.

To quantify memory pooling RP we use equation 4.1 from Section 4.4 which describes resource amplification α as follows.

$$\alpha(S_j) = \frac{R^{actual(S_j)}}{R^{specialized(S_j)}}$$

Where S_j represents a deployment scenario, $R^{specialized(S_j)}$ represents the resources consumed in a specialized deployment of the scenario, and $R^{actual(S_j)}$ represents the resources used in in the actual deployment measured. The time to execute a PBBS MIS benchmark [353] using LRU cache management has been measured at 1.4 times that of Whirlpool [273]. If we equate execution time to processor resource consumption (a reasonable assumption if there is no blocking), we get that same number (1.4) for the resource amplification of LRU with conventional static allocation compared with the specialized Whirlpool implementation where cache management is the sole feature being evaluated.

Whirlpool also uses the notion of virtual caches, which organize similar performance physical memory resources together, as a second level of indirection. Its purpose is to keep application defined pool assignments more independent of the current physical resources during cache reconfiguration processing. In addition, an alternative semi-static approach, WhirlTool [273], can identify application level pools by grouping allocations based on where they appear in the application. Additional profiling and machine learning style clustering is used to train the pools to further optimize caching, hence the reference to a semi-static approach. The resource amplification of WhirlTool is about 1.2 indicating that it did better than totally static pool designations but not as well as the combination of static and dynamic.

Performance issues can arise having to do with the ability of GPU's to efficiently access CPU memory [20]. GPU's contain faster memory resources than CPU's, leading to trade-offs between data movement in and out of GPU memory and GPU processing efficiency. While results vary widely across benchmarks, 50% to 300% performance improvement has been demonstrated when all data that is needed by the GPU is copied to GPU memory on demand rather than leaving it in CPU memory [20]. One approach to further reducing the amount of migration is to wait until some number of GPU accesses to a page in CPU memory have occurred before migrating the page. Characterization of the waiting approach [20] shows that it is rarely more RP than copy on demand, although a few use cases showed amplification reduction of up to 33%. In other cases, waiting caused up to double the resource amplification compared to copy on demand.

Another approach is to add prefetching and bandwidth aware migration which seeks a mix of CPU and GPU memory occupancy weighted by the bandwidth of each [20]. These variations yielded between -33% (good) and $+80\%$ (bad) resource amplification. While the range of results was fairly consistent, various combinations of migration triggers benefit different benchmarks. One size does not fit all.

Another recent development in memory allocation is Mallacc [205], a memory allocation accelerator that operates in the context of hierarchical allocation pools that are already found in some current operating systems. Mallacc provides a hardware implementation of memory pool selection, memory block retrieval, and memory usage sampling that is accelerated by a very fast, very small cache. Mallacc is shown to reduce malloc time by 25-50% in benchmarks that spend 1-4% of their time in malloc. While this only yields up to a 2% improvement in resource amplification overall, acceleration could be increasingly important as allocation decisions become more complex due to the compounding of locality, fragmentation, and QoS factors into allocation decisions.

Object creation overhead has been addressed using a well known object pool design pattern in which objects are set aside for reused rather than freed [220, 66]. Although it has varied in popularity over time it is recently getting renewed attention [108] in the form of experiments showing high volume small object allocation performance improvement of up to a factor of 2, especially when 4 or more threads are involved. The movement towards transactional PM data structures [339, 391] may create additional demand for the object pool design pattern with larger objects as well.

In Chapter 8 we modeled atomicity and redundancy features in the context of a PM software stack. In Chapter 9 we described a strategy for making functionality placement decisions for similar features. For a successful implementation of dynamic function placement in these and other use cases it will be important to integrate function placement with data requirements. Ideally this would be accomplished in ways that enable static and dynamic policy elements to interact as they do in Whirlpool [273] but in an expanded scope in terms of features, physical hardware, and software layering. SNIA Swordfish [360] defines a management paradigm that could be leveraged to create Classes of Service that define additional functions needed by the data elements that subscribe to them. Resource pools of storage and/or memory are defined to provide the services described by a Class of Service. Data elements receive the services when they are placed in those pools. Allocation size and dynamic locality management could be factored into the Swordfish model using subpools that behave like those described above.

Finally, some applications such as analytic toolkits are driven by well defined workflows. Workflows are designed by data scientists to repeatedly perform desired analysis of various datasets over time. SeeSaw [206], for example, uses strategies for improving storage efficiency by reasoning about storage centric features in the context of an analytics workflow. Similar concepts appear in AnalyseThis [357]. The beauty of these strategies is that the workflow is known in advance. It recurs explicitly as it is repeatedly applied over time to various evolving datasets built from similar types. There is more analysis of this approach in Chapter 11. Many of the allocation strategies described above could use defined workflows to gain insight into future data access behavior.

12.1.2   Allocation system model

Figure 12.1 illustrates a model of a memory allocation system that combines several of the approaches described in Section 12.1.1. The centerpiece is a hierarchical memory pool construct with two distinct purposes associated with different layers. One or more upper layers of the hierarchy define pools that align with allocation sizes, Classes of Service, and/or partially initialized data structures. Space from these pool layers is accessed through a virtualization layer (Virt 1) that provides addresses to applications. Lower hierarchy layers (Virt 2) exist to allow placement flexibility to group physical memory and enable transparent migration as demand changes.

The access path from code to memory shows the two virtualization layers that are associated with upper and lower pool hierarchy layers. The light gray circle represents the need for the allocation process initiated by the application to assign all of the pool layers and virtualization translations that apply to the data structure being allocated. This allows pool management approaches such as Whirlpool [273]. Optimized implementations may be able to merge translation of Virt 1 and Virt 2. Hardware acceleration can be applied to the allocation performance path as in Mallacc [205].

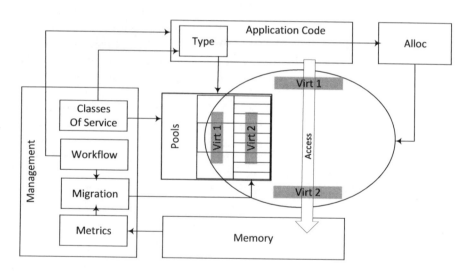

FIGURE 12.1: Allocation system reference model.

Application code defines a set of types (structs or classes) that are assigned to particular Virt 1 pools during allocation based on information that is associated with the type by the application. This includes Class of Service, allocation size, and pre-initialization information. Allocation processes related to Virt 1 may require some pre-processing to associate Classes of Service with pools and to pre-initialize pre-allocated resources. This pre-processing is not repeated when resources are reused. The upper pool hierarchy enables approaches such as the object pool pattern [220] and persistent memory data structures [339] to be integrated with the Virt 1 pools.

The management block includes several functions that control allocation policies and real time dynamics. Classes of Service are defined in advance to specify memory tiers and groups of related functionality such as redundancy, atomicity, encryption, compression, and so forth. Physical memory that is capable of providing a Class of Service can be manually or automatically assigned to upper layer pools. Specific subsets of the assigned memory will be chosen later during allocation and migration. The definition, pooling, and type association of Classes of Service enable approaches such as Swordfish [360] to be incorporated.

The workflow block defines how the services that comprise an application are combined and/or sequenced to accomplish a task. This may include composition of microservices (see Section 7.2.1.1) or analytics streams as discussed in Chapter 11 and in SeeSaw [206]. It may also identify functional blocks that can be placed within or below a memory interconnect as described in Section 9.5. Workflow information controls the application and it may provide information to enable timely migration of data. Other functionality described in SeeSaw and in Chapter 11 can be derived from workflows and layered into this allocation system.

The migration block relocates allocation blocks in memory based on application dynamics such as working set and thread placement in a multi-processor system. It can also adapt to resource changes. This is especially relevant when long lived types are repeatedly allocated to threads on different processors or when different parts of long lived data structures are manipulated from different server nodes. Migration is informed by a set of real time metrics generally aimed at tracking access correlation between cores and memory needed, as in approaches such as Whirlpool [273].

Figure 12.1 does not show all of the infrastructure necessary to support the various allocation approaches that can be integrated. For example, tracking of physical configuration and memory type is required. Large pieces of infrastructure may be introduced to simplify human interaction with the management system and to automate placement decisions. This infrastructure can be avoided in the access path if the pool hierarchy and metrics can capture sophisticated decisions made in the background using policies that direct machine learning driven automation. This dichotomy is consistent with the separation of data and control planes as analyzed in Section 7.1.2.1.

The reference model of Figure 12.1 forms a framework that allows ongoing PM innovations to inter-operate similarly in concept to external brokering described in Section 7.1.1. Selective inclusion of functionality and replacement of both policy and algorithm can be accommodated with a mix of static and dynamic behavior. These can be anchored in data type and work flow specifications that were already needed by applications. This is the kind of infrastructure blueprint necessary to enable diverse contributions from across the industry to solve complex problems while improving RP in critical areas at the same time.

12.2   Comparison of Guided vs. Automated Allocation Policies

One pervasive theme throughout Section 12.1 is that of static vs. dynamic behavior. Humans and programs make many direct and indirect data placement decisions including equipment purchases, how to configure software, what caches to use, and how often to change the rules. Figure 12.2 illustrates the relationship between level of automation, data placement decision overhead, and data placement decision quality.

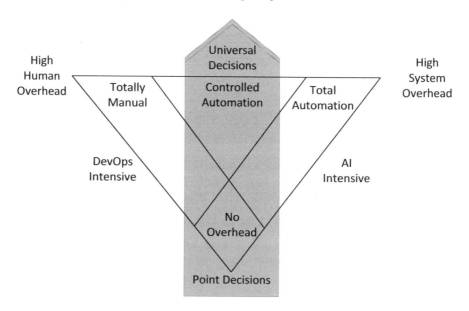

FIGURE 12.2: Placement decision quality.

A low quality decision from an RP point of view is one that only achieves good RP for a small number of scenarios. In some systems such as small embedded applications, low quality decisions satisfy requirements because the system never experiences any workload or resource dynamics relevant to data placement. In other systems, the numerous factors explored in this book may demand frequent changes in data placement decisions to the point where system stability becomes an issue. The spectrum of decision quality is represented by the central gray arrow moving from "point" to "universal" decisions.

The left side of the outer triangle represents overhead in the form of human intervention ranging from physical (re)configuration to setting or tweaking parameters that control automated placement behavior. Intervention involves system administration and devops personnel. The right side represents system overhead for making placement decisions in the background or in the data path. System overhead includes memory, interconnect, and processor resources used to gather statistics, choose free space, analyze demand patterns, change placement policies, and migrate data. Artificial intelligence such as machine learning algorithms may be needed as automation sophistication increases.

Low quality placement decisions can be made once and never changed, hence they require no human or computational overhead. Higher quality decision making requires more resources depending on system and scenario dynamics such as shifting applications and bottlenecks. The inner triangle represents the holy grail of controlled automation wherein the right static information is provided to enable efficient and effective dynamic decisions to be made in all relevant scenarios.

Metrics to evaluate allocation policy RP should take into account the resource consumption of the user application and that of the allocation decision making necessary to achieve it. This is difficult to quantify because it is hard to separate the RP of application primary features from the RP of placement decision making that surround them both within and around the application. Primary features and placement decisions interact in application specific ways that need to be either expressed or learned, hopefully without much human intervention.

When human intervention is required it is crucial to match the skills of the administrator or developer to the intervention context, tools, and process. Also, the work of people in different roles is leveraged across different system populations. For example, the work of administrators and devops personnel applies to a single data center or organization, while the work of application developers applies to everyone who uses the application. Application developers generally have a deeper understanding of how the application interacts with systems, but they are unaware of the breadth of scenarios involved in all deployments. Administrators and devops must acquire application specific knowledge through training and sharing information but they see the results of their work from the front line of data center operations. In all cases, experience is a huge factor.

These are the reasons why RP metrics do not take human factors into account. Accounting for both humans and systems leads to a competitive total cost of ownership (TCO) situation. If the RP of the application and system do not yield an affordable TCO for consumers, they will seek alternatives. RP driven competitive dynamics may occur in the microcosm of an IT small shop or at the magnitude of a cloud provider.

Even without quantifying TCO, Figure 12.2 requires a closer look at how an efficient relationship between human intervention and automation can be achieved. We will examine this from two perspectives: Class of Service driven allocation and automated cache allocation.

12.2.1   Class of Service driven allocation

Over the last 10 years, several storage management tool-kits have been developed to manage storage with attributes or capabilities that can be used to automatically match storage media to applications [188, 359]. The more recent SNIA Swordfish [360] standard is particularly relevant to Class of Service driven allocation for the following reasons.

- Swordfish defines Classes of Service with a detailed structure including many of the attributes that are common to today's data storage systems

- Swordfish defines relationships between Classes of Service, Storage Pools, and Sources of Capacity that facilitate static policy definitions driving dynamic provisioning

- Swordfish storage pools already support memory as a source of capacity

- Swordfish definitions are extensible to include hooks for work flow and data type associations

Figure 12.3 is a simplified diagram of allocation flow based on Classes of Service that aligns with parts of Figure 12.1. The leftmost column includes elements of the SNIA Swordfish model. To the right, the figure conceptually represents application and system elements. The solid arrows represent one way to use the SNIA and application elements during allocation. The dotted connections represent the association of data structures with the allocated capacity, i.e., the result of allocation.

Pre-conditions for allocation include the following.

1. Initialize Classes of Service to include QoS information such as performance, redundancy, data protection, and security. Classes of Service can also be decorated with information about how, when, and where applications will access data including connections to work flow.

2. Provision Swordfish pools that match Classes of Service by creating volumes to represent physical or virtualized capacity. Pools may have a hierarchical structure as described in Figure 12.1.

3. Associate Classes of Service with application specific types using application specific tools and/or shared keywords.

4. Create lower level pools to represent allocation units for particular data types, co-located with locations where the application will run.

The following steps occur when the application's allocation library function (e.g., malloc) is called.

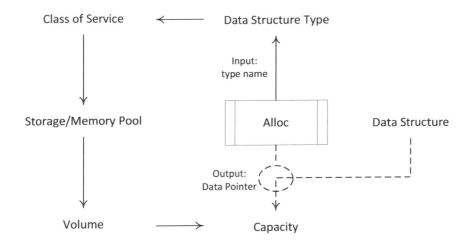

FIGURE 12.3: Allocation flow.

1. If there is a lower level pool pre-designated for the type and preferred location of the data, allocate directly from it.

2. Otherwise, use information associated with the data type being allocated to identify a Class of Service.

3. Use references to storage pools in the Class of Service to select one or more storage pools where appropriate free space resides.

4. Allocate capacity designated by volumes within the selected storage pool(s) and place it in the application's address space (if not already there)

Since it would be unwieldy to pause for human intervention during an allocation library call such as malloc, this part of the allocation process is always entirely programmatic. While the whole process described here may seem heavyweight, most allocations can be made very efficient by maintaining references to low level pools with data structure types. The elements of Figure 12.3 illustrate one way to organize relatively static information for policies that guide more dynamic behavior. Generally the use and manipulation of lower level pools in the hierarchy are more automated than the top level. Automation varies when it comes to processing or other activities that may be required prior to allocation. For example, Class of Service definitions tend to be manual but ML style training is automated.

Section 12.1.1 gave several examples of application RP driven by placement decision quality. These included the RP of run time allocation but not the RP of policy decision overhead that is not in the application's allocation path. These overheads do not increase the latency of allocation but they do compete with the application for system resources. Two overheads in particular deserve further attention.

- Memory resources for data structures that support allocation policy decisions

- Processing such as machine learning that combine static information with system and application behavior observations to make efficient real time allocation decisions

Let's consider a numerical example to see how memory overhead for allocation management might accumulate. Suppose we have a system with 4 Tby of memory and an average allocation size unit of 4K. Suppose each allocation unit requires 24 bytes for pointers to track allocation and pool associations, and 8 bytes for metrics. In this case the memory overhead for allocation unit management is under .8% of the memory capacity.

Suppose we have a 4 level pool hierarchy in a tree with a fan-out of 8x at each level. The pool hierarchy tree could have up to 4K leaf pools but suppose only 1K of them are populated. Suppose there are 4 pointers per leaf pool to anchor allocation free lists, and 16 per inner pool for hierarchy navigation. Suppose each pool has 8 attributes where each is a 128 byte key-value pair. The entire pool hierarchy would require under 7 Mby or .0002% of the total memory capacity. Let's further suppose there are 32 Classes of Service with 64 attributes and 64 pointers each to track policy parameters and relationships between classes and pools. The capacity for pool definitions would come to less than 300K bytes which is minuscule compared to the overhead to manage allocation units.

In this example there are many more allocation units than pools or classes, so they dominate memory resource consumption for allocation management even though the overhead for each allocation unit is small. It would be unwise to associate key-value style attributes with allocation units as overhead would explode. That is why attributes are associated with large groups of resources using pools and Classes of Service. Those attributes are then indirectly associated with allocation units based on the pool they are in.

To get a sense of the computational resources that might be involved in training a machine learning system management application, we can look at training time measurements of neural networks used for future traffic prediction [290]. Datasets consisted of network traffic samples taken every 5 minutes for periods in the range of 6-12 hours. Several types of neural networks were tested on a modest size (Intel Core i5-2430M) processor. Training times ranged from about 86 seconds to 140 minutes depending on neural network sophistication and training set size. The prediction error that resulted did not always correlate well with training time. Overall the best prediction value occurred with training times between 1 and 2 minutes.

If machine learning that requires training is used, the training time should be correlated with the value of the decision. This should include the following factors.

- RP improvement – What amplification or bloat reduction does the decision enable when it is working well?

- Decision robustness – How often is the decision accurate based on surrounding application and system conditions?

- Decision change rate – Once the decision is activated, how long is it likely to remain accurate until a different decision is needed?

- Retraining – How often does the decision need to be retrained? Is the retraining incremental? How much load does retraining place on the system?

Low value, short lived decisions should be lower priority than high value, long lived decisions. The value of short lived decisions may be increased if they can be enabled and disabled quickly with minimal warm-up. This is most often typical of real time placement decisions such as caching.

12.2.2 Automated caching and phase change

Even though automated caching is a well studied field, new algorithms continue to emerge. Considering the wealth of algorithms available in the context of the hierarchical storage pool architecture of Figure 12.1 it seems that the trick is deciding which ones to use over time in a changing application and system environment. Information related to workload change may take several forms.

- Human provided information such as time based scheduling or static data structure placement constraints

- Application or OS related information such as data grouping and work flow or task scheduling descriptions

- System metrics such as resource utilization and traffic flow

- Results of machine learning (ML) style training for phase change detection

Considerable R&D has gone into sophisticated caching policies, some of which require parametric tuning or ML style training. Examples at these extremes include look-ahead parameter settings to optimize sequential performance and access prediction engines driven by workload history. For any real system other than a point solution at the bottom of the triangle shown in Figure 12.2, a single parameter setting or training result is not sufficient. For example, Belief Cache [326] is a caching engine that uses Bayesian logic to predict which data should be prefetched into a cache based on access history. This approach requires workload specific training that can demand considerable processing resources, making re-training decisions key to prediction accuracy and computational resource consumption.

In this type of situation it is likely that if a given training result falls out of favor there is a reasonable chance that it will become relevant again in the future. Figure 12.4 illustrates a framework intended to accommodate patterns of behavior like this. This framework is an overlay on the reference model shown in Figure 12.1. All of the elements and relationships in the reference model are relevant to Figure 12.4 even though some are omitted for ease of interpretation.

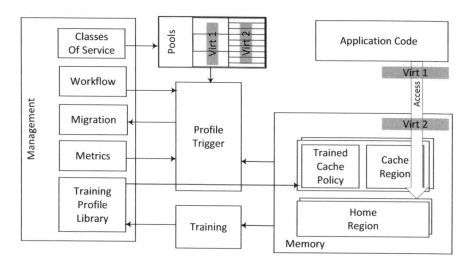

FIGURE 12.4: Training profile library.

As in Figure 12.1 applications are shown accessing memory through two layers of virtualization that reflect two parts of the memory pool hierarchy. Figure 12.4 adds some detail associated with the memory system including some memory regions that cache data based on a cache policy and others that represent the more permanent home locations of data. There are many caching policies that do not require training or parametric tuning. Figure 12.4 is relevant to more sophisticated policies or policy layers as described in Sections 12.1 and 12.2 in which caching and other data placement policies are controlled by static information, be it parametric or derived from ML style training.

In the management block there is a new "Training Profile Library" to preserve training information in the form of profiles that can be used repeatedly over time. Training is shown to the right of the library with arrows indicating that behavior of application access to the memory system is captured and used to drive ML engines such as neural networks. Often the information needed for ML is more detailed than the metrics already captured in the management block. Although it is not shown here, training profiles may also comprise rules or tuning parameters for automating allocation as described in Section 12.2 captured in the upper layers of the storage pool hierarchy.

The key to driving memory RP higher is the profile trigger function which can use storage pool hierarchy, metrics, memory behavior, and work flow information to detect situations where a given profile already in the library would be advantageous. Profile trigger rules may be captured as part of each profile. One could also envision the use of ML to train triggers resulting in additional profiles that only apply to trigger detection. When a profile is triggered it is installed as a trained cache policy for a part of the memory system that is defined by the pool hierarchy to be associated with the relevant application data.

In some cases a change in policy may involve data migration. This implies expansion of profile library scope to include data placement that is broader than caching. Although Figure 12.4 does not show all of the information flows required for scope expansion, the profile library and trigger elements serve to illustrate concepts that can be applied.

While the application of ML to data placement is gaining traction in the industry, the notion of choosing training profiles is relatively new so the RP benefit of this approach is difficult to quantify. Still, it should not be surprising that singular or limited data placement tuning may yield narrowly applicable decisions. If on the other hand, libraries of profiles containing tuning or training settings are created, systems may be able to adapt to change by automatically choosing policies with the best RP for many combinations of application and system behaviors.

12.3 Resource Proportionality of Waiting, Polling, and Context Switching

Radical non-uniformity may also affect the way applications wait for memory accesses to complete. This can have a large impact on RP. In today's processor architectures, once an application starts a memory access it has no choice but to wait for it to complete. This is reflected in the Processing Efficiency (PE) models of Chapter 9. In Chapter 8, on the other hand we considered both PM and disk accesses which brought out the question of polling vs. context switching. Using the model from Chapter 9 we can compare some of the RP impacts of all three approaches.

In order to do this we must hypothesize that one could trigger any of the three approaches using a memory access. Normally, memory accesses just wait for completion. Theoretically a processor could be designed to start a memory read but not complete it during the same instruction. Instructions could then be provided to poll for a read completion state afterwards, or to immediately context switch and resume when read completion occurs. This is less far fetched with respect to writes in that write pipelines already do this;

however if persistence is to be assured by flushing data to PM, then the same hypothetical completion approaches could be applied. Such a hypothetical feature would also need to avoid triggering polling or context switching on CPU cache accesses. With this hypothetical processor in mind, equation 9.4 can be used to explore an approximation of this trade-off as follows.

- wait – This completion approach is the one modeled throughout Chapter 9 so the calculations for it are used from there.

- poll – In this completion approach we assume that as soon as the processor detects a cache miss it completes the instruction and allows subsequent instructions to poll for completion of the cache miss including writes to persistence. As a result this approach does not benefit from stall probability P_S in equation 9.4 so it is set to 1. This is because the processor goes into polling for the duration of the memory access which is tantamount to stalling from the point of view of the processor core.

- switch – For this completion method we assume that as soon as the processor runs into the cache miss it blocks the thread during the instruction that caused the memory access. To model this in equation 9.4 the memory access time is set to a nominal context switch overhead time. In this case 2 μS was used as it enabled the primary phenomena of interest to be observed, but actual values are very implementation specific. The stall probabilities were retained for each workload as they appear in Table 9.2.

Analysis is performed for the same Enterprise, Big Data, and HPC workloads as in Section 9.4 with the following exceptions.

- MI characteristics are not varied.

- both read and write access times are varied from .2 μS to 3 μS to observe RP effects at various memory access times.

- as noted above, media access time for the context switch approach was fixed at the context switch time for all trials.

Figure 12.5 compares waiting vs. context switching for all three workloads. As expected, the PE of waiting at typical memory speeds is much better than context switching. Also as expected, the cross-over between waiting and context switching occurs precisely when wait time equals context switch time even though the PE of context switching varies by workload. This result is obvious based on the modeling approach.

More interestingly, Figure 12.6 compares polling with context switching for the same workloads. The scale of the vertical axis is expanded because waiting has a much higher PE than polling for low memory latencies. In fact for a given workload the PE of waiting is always better than polling. A second interesting observation is that the cross-over between polling and

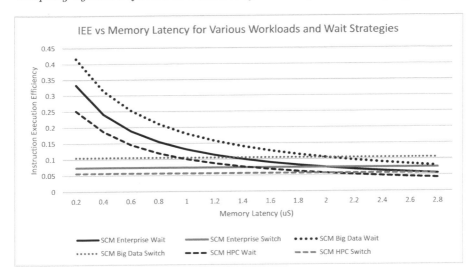

FIGURE 12.5: Comparison of PE for waiting vs. context switching for three workloads. The three dark curves show PE for waiting which is more efficient in all workloads for small memory latency. The less efficient group of three straight lines shows constant time context switches in each workload.

context switching occurs well below the context switch latency due to the influence of the stall probability. If the native stall probability of the workload were used for polling, the result would be the same as waiting; however, this would be a misleading experiment. Once a stall is detected it is too late for the processor instruction flow to go into polling so unless the stall probability is ignored the difference between polling and waiting is moot. Still, for all but the HPC workload there is a memory latency range where polling has better PE than context switching. Polling is always a huge disadvantage for HPC because of its low stall probability as shown in Table 9.2.

We can see from Figures 12.5 and 12.6 that this decision can have a very large impact on PM RP. The poll vs. context switch decision also applies to block storage devices such as disk drives. It has been suggested that disk IO's be guided into synchronous and asynchronous streams depending on various temporal and spatial criteria [427]. This is a similar optimization to the polling vs. context switching decision. An even simpler criteria would be a prediction of how long an IO might take.

In a memory centric architecture it might be reasonable to treat everything as memory. In that context, some memory interconnect protocols [110] distinguish between read/write memory access, which may stall, and put/get access which can be either synchronous or asynchronous. All addresses in both read/write and put/get are memory addresses; however, some may be known in advance to reference slower devices such as SSD's or disks. With this type of MI, the choices modeled in this section can be made during normal system

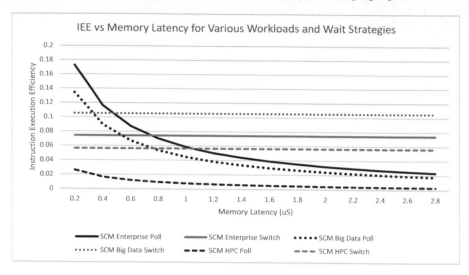

FIGURE 12.6: Comparison of PE for polling vs. context switching for three workloads. The three dark curved lines show PE for polling. The three straight lines show the same context switch times as in Figure 12.5.

operation in the data path, guided by prior knowledge that could be integrated into the hierarchical storage pool management described in Section 12.1.2.

12.4   Managing Non-uniformity Using Work Flow Scheduling and Classes of Service

Throughout Chapters 8, 9, and 12 we have been exploring examples of the RP implications of memory and memory interconnect technologies and the importance of both function and data placement in the resulting emerging systems. The examples described are far from exhaustive. They were chosen to illustrate a combination of fundamental and representative RP challenges culminating in the description of an approach to RP best represented by Figure 12.1.

A recurring theme throughout this book has been RP awareness through the careful dissemination of new information that allows applications and systems to detect and respond to RP improvement opportunities. Chapter 6, for example, describes how information on the usage frequency of functionality can be leveraged to tune software deployments. As another example in Chapters 8 and 9, information about application and system interaction is used to tune data and functionality placement. In each case it is crucial to organize such information and respond to it in a way that can resolve a wide range of technology and functionality choices.

In this chapter we broadened the common notion of allocation pools into a general framework for organizing data and function placement information. Relatively static information like data type, grouping, and technology affinities are captured in upper layers while more dynamic information like processing and caching affinities are captured below. The key to making this work is to drive placement decisions by matching Classes of Service, and to anticipate activity patterns by sharing work flow information with applications and OS's.

Figure 12.7 illustrates this idea as a set of information flows that converge on a data and function placement policy. Application needs and technology capabilities are brought together in the form of Class of Service definitions. Application task scheduling through the use of OS and middleware frameworks contributes to a description of impending or recurring work flow. OS based system monitoring and technology capabilities combine to form a description of ongoing system behavior which, along with Classes of Service and work flow knowledge, may be distilled by rule based or machine learning frameworks for consumption by policies.

The ultimate test of an RP solution is total cost of ownership (TCO) for a system that is able to do a given amount of work. This can be decomposed along the following lines.

- Physical resources: equipment, power, cooling, and real estate. Improving RP reduces all of these, or allows more work to be performed by the same amount of physical resource.

- Human costs: IT staff including administrators, devops, support, procurement, finance, and more. The framework described here for managing RP reduces administration through automated tuning. It also captures information that can be leveraged for easier and more precise devops, procurement, and support.

- Reliability, Availability, and Security costs: These non-functional requirements carry liability in the event of omission or failure to function properly. Loss of business due to downtime or breach have costs above and beyond the human and physical costs necessary to support them. RP improvement, if applied during recovery, can reduce some of these costs as well. RP aspects of security are covered in Section 7.3.1

By managing RP using frameworks such as those illustrated by Figures 12.1 and 12.4, IT can prioritize optimization so that the greatest benefit applies to the most important applications. This practice of matching need to delivery is at the core of good business.

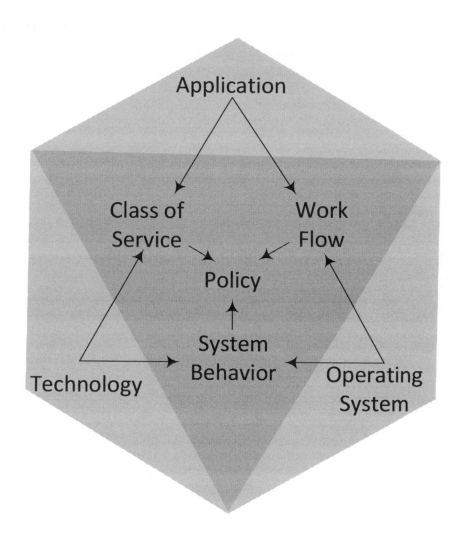

FIGURE 12.7: Allocation policy information flow.

Chapter 13

Bridging the Gap from What We Know Today to Open Challenges and Research Topics

13.1   Introduction

As discussed in the previous chapters, a systemwide focus is needed to design RP systems. In the future, as more systems are attempted to be integrated such as

- a cyber physical system based on drones that could be integrated with electronic commerce, agriculture, or resource management systems, or

- a video-based instructional system that integrates multiple devices (such as mobile phones, tablets, and other "IoT" devices), image processing using GPUs and AI for tracking and intervening in learning outcomes while being sensitive to privacy considerations, a thorough systems analysis is needed.

Some interesting issues in the designs for the future:

- How to design systems that attempt to reduce interactions as a first principle? Chapter 5 discusses some of the approaches from a linguistic "features" perspective. We now discuss this issue from a "systems" perspective here: designs that minimize state spill[75] but that can also introduce some overhead.

- How to engineer the right type systems for the systems at hand? Precise types provide highly specific information about how some value will be used; this has the potential to reduce bloat in general.

- In large scale systems, error management is of paramount importance.[1] Similarly, intermittent computing[347] can be a critical aspect for making cloud servers efficient. How to make error handling RP is therefore of considerable importance.

- How to incorporate approximate computing across the software stack? Given that the digital world is only a part of the whole system and, depending on the resolution needed in the system, a design *ab initio* or a redesign of the system (keeping in mind the accuracies needed across the layers in the system) can be more resource efficient.

We will discuss these issues in turn but in a different order.

13.2   Approximate Computing

Due to the rise of the Internet and explosion in the number of connected devices, cloud computing (including the newer enhancements with edge and fog computing) as well as Internet of Things (IoT) have become well established paradigms fairly quickly. The first is a significant attempt at resource proportional computing while the latter tries to reduce the computing at the edge and move major computations to the cloud. However, one downside is that the datacenters that provide the infrastructure to the cloud have high energy consumption[2] which is expected to double every four years.[3] Due to

---

[1]For an interesting perspective on this aspect and its role in Google's rise, see https://www.newyorker.com/magazine/2018/12/10/the-friendship-that-made-google-huge.

[2]As of 2018, its use is now higher than 2 percent of the world's electricity and emits around the same amount of CO_2 as the airline industry.

[3]The higher accuracy, flexibility, or even just plain feasibility that the digital domain provides often comes with a higher cost than previous solutions or "non-solutions." For example, considering the storage aspect only, it is said that about 25TB of SSD storage is collected (or needs to be replaced) everyday (even after compression by a factor of 2000) in an autonomous car driving context[13].

this scale, a carefully thought out RP systems design is important. A newer RP model, "Approximate Computing," can provide a part of the solution to this conundrum of increasing demands on energy and responsible computing. As approximate computing "exploits the gap between the level of accuracy required by the applications/users and that provided by the computing system, for achieving diverse optimizations"[269], its relevance in the use of resources proportional to the computing task is central. For a redesign across the stack, approximate computing may be needed and approximations treated as a first class concern. Furthermore, some hardware platforms that provide sensing with adaptive approximation have also become available recently for evaluation and use[370].

Many applications are resilient to approximations due to various factors such as sufficiency of approximate solutions and human imperceptibility to some data when presented with a higher resolution than necessary (image/audio/video, for example). Approximate computing exploits this and makes a trade-off between accuracy and efficiency; this can be viewed as another facet of RPD. For example, almost 50x energy savings can be achieved by tolerating a loss of 5% accuracy in a k-means clustering algorithm[105]. Similarly, up to 26x energy savings can be achieved in inverse kinematic application by using neural approximations[167].

Approximate computing can be performed, at the level of software for example, by skipping some iterations of loops (especially in software for numerical analysis), using memoization (or, using "stale" values), skipping memory accesses[333], reducing branch divergence in SIMD (single instructional multiple data) code, and using techniques such as chaotic relaxation. Another major approach is the use of neural networks in approximate computing. Neural networks, for example, can be used to mimic a portion of imperative code but which is much simpler ("approximate"). First, a portion of code is marked to be approximated by a programmer[136] or even automatically[338]. A neural network is then created and trained by observing the behavior of imperative code by logging the inputs/outputs by the compiler. These input output pairs represent both the training and testing examples for the selected neural network. Once the neural network provides enough efficiency and accuracy, the calls to marked portions of code are replaced with calls to these neural networks. These tasks may further be accelerated by modifications to ISA (instruction set architecture) as in [136] or can be completely software based as in [255]. Note that approaches such as JIT and dynamic binary instrumentation are related: these are roughly at the instruction level, whereas the neural processing unit (NPU) is at the level of a hot region that can be approximated with a clear I/O relation. Instead of just using "digital" NPUs, to reduce power consumption or latency,[4] an analog NPU can also be considered; one study[366] finds that analog neural acceleration provides a speedup

---

[4]Microsoft has reported use of 152 layers for image classification but energy considerations limit use of such large DNNs even if accurate. LightNNs[122] have been developed that replace multiplications with shifts or a constrained number of shifts and adds.

of 2.3x and energy savings of 3x with quality loss less than 10% across diverse applications except one benchmark. One may also view "bloat" as an unintentional outcome of not-sufficiently (or, non-adversarially) trained or evaluated software models that work well in some contexts but behave poorly when previously unseen (or, adversarial) contexts are presented.

In addition to the "simpler" aspect of deciding the level of approximations from a computational point of view, there is a need for handling uncertainty also and both are needed together sometimes. Newer applications use noisy data from low-cost sensors or (the not always available) GPS, or use probabilistic models (often learned through machine learning, data mining, or big data), and also may need to work with uncleansed human data. Current programming models (say, C++11/14/17) do not provide good notations or adequate support for representing such uncertain data; when such values, each with its own error distribution or noise, are used in computations, the error propagation is not tracked and can lead to unexpected results (such as higher false positives and negatives).

Consider tracking cars on a freeway. It is clear that as long as there is no abnormal situation, the car has to be within the lanes. Using that as the prior, approximate and possibly conflicting data from sensors can be refined or evaluated using the prior ("car inside the lane") and the Bayes rule. Hence, sensor readings can be filtered for outliers, or even ignored for small durations. Next, accuracy of computations needed (for example, number of iterations or the convergence condition) can be tailored to the extent the system is affected or is "controllable" using the higher order error terms.

Probabilistic computing is another model[162] that works with distributions directly where probabilistic inference is posed as the problem of computing an explicit representation of the probability distribution that is implicitly specified by a probabilistic program. We do not discuss this, nor do we concern ourselves with another and different model of probabilistic computing[294] or stochastic computing[25] that both are typically at the circuit level.

Approximate Computing (AC) can be at various levels:

- Processor Level. Since all computation involving real numbers need not be done in standard IEEE FP but some possibly at lower precision, Tensor Processing Units (TPUs)[202] allow 8b and 16b precision that can also be used if that is sufficient.

- Linguistic level such as enerJ, FlexJava, Uncertain<T>[341, 298, 77], and TensorFlow[379]. While enerJ and FlexJava provide support for AC by inferring and propagating approximate types (inspired by information flow models, the parallel being that just like information flow labels form a lattice, the approximations can do likewise, and many issues in both similar), Uncertain<T> deals with uncertainty primarily. Both enerJ and FlexJava are based on a small set of language extensions to Java that enable developers to show non-interference between approximate and critical parts of a program. Since the problem of determining if a given set of operations is approximable in a given annotated FlexJava program is undecidable, Flexjava uses conservative static analysis to infer the set of operations and data that are safe to approximate from programmer annotations. RPD in the context of AC therefore cannot be automated except in a heuristic manner.

In frameworks like TensorFlow, it is possible to provide variable precision as needed using optimizations that use approximate operation definitions to result in an optimized model, along with specialized kernel implementations that use a mix of fixed- and floating-point arithmetic. Expensive computations are executed in lower precision for speed, but the most sensitive ones with higher precision, thus typically resulting in low losses in accuracy but with a significant speed-up over pure floating-point execution. If there is no matching "hybrid" kernel for some operations, or where necessary, the framework reconverts the parameters to the higher floating point precision for execution[379]. Similarly, polyhedral optimization of TensorFlow Computation Graphs[312] has been attempted on the structure of the DNNs.

Uncertain<T> provides new programming language abstractions for uncertain data. It has a new generic data type Uncertain<T> along with operations on such uncertain data based on their probability distributions; these operations propagate operand distributions through computations and conditionals.

- Algorithmic level. The area of approximate algorithms is now well established and we do not discuss this any further[271].

- Systems or Application Level. We discuss two examples. To investigate the impact of performing individual subsystem-level approximations on system-level energy consumption and application quality in the context of an overall computing system, some insights are discussed in [322] that can be useful in the design and construction of full-system approximation techniques. They observe that the full-system quality vs. energy (Q-E) trade-off is impacted by approximating different subsystems at different granularities. If the sampling rate of the sensor is changed, it affects the full-system Q-E trade-off substantially since the resultant energy savings (as well as quality) affect the entire system. However, they report that computation and memory approximations affect Q-E much less.

To store multimedia content, approximations are used in [198] thus: first, any given format is transcoded into an approximation-aware format which is again transcoded into an approximation-aware encryption that finally gets transcoded for approximation-aware decryption. Some approximation approaches use techniques across the hardware/software stack; a good example[428] is the development of spatial query operators that exploit GPUs to take advantage of their native support for rendering while using "bounded raster join," an efficient approximate approach that eliminates the need for costly point-in-polygon tests but provides close to accurate results in real-time.

However, security is an important concern. Although there have been studies about resilience of applications to approximation, these models have not been studied in adversarial contexts. In adversarial machine learning settings, it has been shown that carefully crafted inputs, or wild patterns[70], can be provided during testing or training phase so that the output of the system is perturbed from normal. The neural networks must be trained against enough adversarial examples to make them robust against such wild patterns. Even though the field of adversarial machine learning has been around for a decade, similar studies are needed for making approximate computing more robust.

13.3   Stateless Designs Revisited, or Reducing "State Spill"

In software engineering, the notion of coupling has been used to discuss the degree of interdependence between software modules. Operating systems, especially the kernel, have been widely known to be highly coupled in spite of using many advanced linguistic notions (e.g., notions similar to objects and classes in device drivers, schedulers, filesystems) and strict layering models (such as between virtual memory subsystems and filesystems, especially in Unix-based systems). Recently, reducing the coupling, also called "state spill"[75], is being investigated systematically. This approach is aligned with some aspects of RPD but not so in other aspects. We discuss this approach below.

We first give a quick summary of some of the major trends/issues in the design of an OS:

- Generality: Ability to handle heterogeneity of devices, evolution over time, support, and be supported by multiple languages. Types and type hierarchies can help in handling the heterogeneity.

- Performance: "Zero" overhead cost of any feature, or at most a very low proportional overhead cost. If specific type information is available, it may be possible to support the principle of "pay for what you use."

- Security: Support information flow models. The type system can help to statically check the flows at the level of assignments or higher.

At the hardware level, virtual machines have been used since the '60s as a way of running arbitrary OS on arbitrary hardware. We do not discuss this further except to note that non-virtualizable architectures made this approach unviable (due to expensive emulation) and a RP approach (by JIT binary rewriting) in the late '90's made VMs usable leading to another RP approach, that of cloud computing.

The solutions at the OS level have been monolithic kernels in the past, microkernels 2-3 decades back, and recently Unikernels[248, 433, 221] that reduce the generality of an OS kernel by specializing it for a specific application. In the OSv approach[221] that is targeted for cloud applications running inside a VM, some of the layers are merged (with application, runtime/JVM, and kernel merged), simplified models are assumed (for example, single application running in a single AS), and the resulting systems are further simplified (for example, no spin locks anywhere thus avoiding lockholder preemption but now needing threads to effect interrupt processing, and, needing threads also for balancer threads to schedule fairly across multiple CPUs[221]). The Mirage unikernel[248], on the other hand, goes further: the Mirage compiler takes the application source, configuration files, the hardware architecture, and using the whole system optimization approach, produces the application code that runs on a hypervisor directly using the Mirage runtime. However, there is no support for fork and other dynamic aspects; this has been relaxed in more recent designs such as KylinX[433].

The solutions at the programming languages level have been "language-based isolation" of subsystems since at least from the '80's. Some examples are

- SPIN[53] is an extensible microkernel that has been used, for example, to implement Digital Unix through dynamically loaded modules that in turn implement interfaces (defined by "Modula-3 INTERFACE") that represent domains. All kernel extensions are written in a "Modula-3 safe subset with metalanguage constructs and type safe casting system"; it uses a special run-time extension compiler.

- Singularity[235] is a high dependability microkernel that has components that run in the same address space (process) as it contains software-isolated processes (SIPs). The kernel, device drivers, and application software are all written in managed code (Sing#). Instead of hardware memory protection, internal security uses type safety.

- Java Information Flow (JIF) Language[129] extends Java with information labels and uses static type inferencing. However, dynamic information flow analysis (for example, a fine-grained dynamic bit-tracking analysis to measure the information revealed during a particular execution) may be needed in many contexts.

- Rust is a newer systems programming language with newer type constructs (for example, linear types). We discuss this further below.

However, in spite of many advances in the design of OS, newer requirements have again cast doubts on the designs so far. For example, the newer goals (or, "features") for OS are[75]

- Process migration. But this is negated by residual dependencies on the system from where migration takes place.

- Fault isolation/tolerance and software virtualization. But this is made difficult by large number of states that are shared across ("fate sharing"), and hence makes isolation difficult along with a need for a complex multiplexing logic across the redundancy sets and the virtualization levels.

- Live update and hot-swapping. But this is complicated by the inability to modify an individual entity in isolation as the state transfer functions are non-trivial.

- Maintainability. But this is negated by interactions that remain despite modularization.

- Security. But this is negated by loss of control over propagated data unless information flow is modeled across applications and kernels.

If an application interacts with a system service, a state spill can occur if the service stores application-relevant states in its memory. Microkernel designs do not as such reduce state spill[75] even if vast majority of OS functionality is moved into userspace servers with only a very small kernel core. For example, in Minix 3, userspace servers are often indirection layers that proxy between applications and microkernel (e.g., convert POSIX API calls into MINIX-specific (microkernel) functions). This results in state spill. For example, MINIX 3's userspace scheduler SCHED works as an indirection layer between user processes and microkernel's context switch mechanism to control the system's scheduling policy but SCHED does not have the context switch privileges as it can only choose the next process for scheduling[75]. Therefore, it just delegates the function to the kernel's context switching mechanism (*sys_schedctl*). The latter system call copies relevant process parameters into the kernel's list of *struct procs*, a good example of state spill, before actually initiating the context switch. The spilled states include the target process's scheduling flags, endpoint, parent endpoint, priority, timeslice quantum, CPU affinity mask, and schedule-enqueue bit. Similarly, state spill impedes live update and hot-swapping of microkernel servers; hence live update in MINIX 3 is non-trivial[156].

State spill depends on the choice of the unit chosen for analysis. It often is a module (for example, an entity can be a group of related functions and the data they modify) or a class (for example, when analyzing Android system services, a Java class). State spill often results from caching or the attempt to compute locally. The common design patterns identified are indirection layers (discussed already), multiplexers, dispatchers, and inter-entity collaboration[75]. Multiplexers arise temporally or spatially due to resource sharing (like schedulers, window managers, and high-level drivers). Dispatchers arise to register callbacks from clients such as device event callbacks or upcalls. The last variety arises due to multiple user space servers needing to coordinate themselves to deliver a service such as NFS and using storage and network user space servers.

There are many design patterns to avoid state spill; the stateless design of NFS servers is well known just like the early HTTP protocol without cookies. Broadly they can be termed as RESTful principles[75]:

- Client provided resources. All state is kept in the clients in the form of opaque handles; a handle is also sent to the server on each interaction after the first.

- Stateless communication. The servers create opaque handles and send it to the clients.

- Hardening of entity state. To reduce the amount of state in the opaque handle, the state is divided into a hard state (present in the handle) and a soft state that can be rebuilt as necessary.

- Modularity without interdependence. Encapsulating state into a handle makes the server and per-client interactions simpler. Otherwise, the state of one client can spill onto the others especially in a security context (e.g., attacks on tabs if implemented as threads in a web browser process).

- Separation of multiplexing from indirection. In the multiplexing case, some common state may need to be kept across the multiplexed entities. Hence, the indirection case should be distinguished separately.

Using the above, it is possible to design a spill-free OS and this has been reported in Theseus[76]. To isolate modules, Theseus also uses language-based isolation using the Rust language[9]. Rust gives memory safety guarantees for data and fault isolation between modules for the safe part of the code. The unsafe part of the code that Rust cannot guarantee memory safety (e.g., port I/O and MMIO, register accesses, interrupt configuration) is clearly demarcated, and Theseus can use MMU-based isolation without having to depend on the Rust language alone for safety. Theseus uses affine types[5] to effect zero-copy communication even between isolated entities, along with caching and shared mappings.[6] Co-design of the language (or, its careful selection) for writing the OS and the OS itself is an important design aspect. Safe languages are helpful but low level manipulations require proper support.

13.3.1 "Spill free designs" and failures

The above discussion is fine as long as failures are not in the picture; an additional wrinkle is needed to handle failures. Consider a server-client

---

[5]Linear type systems allow references but not aliases. A reference goes out of scope after appearing on the RHS of an assignment, thus only one reference to any object exists and that can be used only once. An affine type is a version of a linear type that permits non-use (i.e., discard) of a resource, corresponding to affine logic. An affine resource can only be used once, while a linear one must be used once as in Rust.

[6]Singularity also exploits linear types for zero-copy transfer.

protocol; a good one here is NFSv3. Some of the operations are idempotent (example: a client asks for listing of the files in a directory) but some may not be (example: a client requests delete of an existing file.)[7] If the server gets a non-idempotent request and the server completes its processing, its response (along with the opaque state handle) is expected to be received by the client. If due to a network error (temporary partition, for example), the response is not received and the server then crashes, the request will be retried by the client who will get an incorrect response; this is the case unless persistent storage is present in the server (or, by some other means) to capture the incomplete protocol exchange for the non-idempotent operation.

Hence, in spite of a stateless server design, logging of requests[384] is needed to handle non-idempotent requests in the context of failures. Or, the server should send a unique integer that changes on every crash as part of the protocol (as in NFSv3). Some well-designed protocols also have this property as part of their design; a good example is TCP/IP[215]. A random sequence number is chosen at the beginning of a TCP session so that packet insertion, whether accidental or malicious, is not possible unless this random sequence number can be guessed. However, on a crash, an earlier random sequence number must not be repeated to avoid accidental packet insertion due to an earlier packet arriving through a slower path. This requires persistent storage to keep track of earlier ones (a costly solution), or (as in TCP/IP) the system cannot be rebooted (reused) after a crash for as much as 2 minutes typically[8] until after all the earlier protocol exchanges and their retransmissions have been drained out of the system using a time-to-live or other mechanism. Since the TCP/IP-like solution is likely to be non-attractive (from an availability perspective), random number management on a large distributed system becomes quite critical.

More generally, it has been known that consensus is impossible in a strictly asynchronous system with process failures[143]. It also has been shown that without persistence, no data link layer can work correctly (this includes well known protocols such as HDLC, etc.)[140]; this result work has been extended to study the fault span of crash failures[387] when there is no persistent memory at nodes. Based on whether packets can be dropped, delivered FIFO, or reordered, protocols can be driven to both node states and link states that are arbitrary mixtures of states of runs, or for only the node states or for a acyclic component of the network[387]. Such results have to be taken into account when we design protocols for stateless server management.

---

[7]The first time the server should respond with success; if repeated next, as the file no longer exists, the server should respond with failure: "No such file."

[8]This is widely violated! The Maximum Segment Lifetime (MSL) value, typically set to 1 minute, is used to determine the TIME_WAIT interval (2*MSL) after a connection close or crash, which is therefore typically 2 minutes.

13.3.2 "Spill free designs" and "serverless" computing/ microservices

The above discussion of spill free designs naturally ties into current developments in scalable deployment of applications; such designs make these developments even more attractive.

As discussed earlier (Section 10.1.3), "serverless" computing decouples application logic from resource management. However, the store needed in these designs can become a bottleneck just as memory can become one[276]; we need to optimize across "chains" or flows that form the sequence of calls. Additionally, a serverless function constitutes a natural unit of information flow tracking but we need a termination-sensitive non-interference (TSNI) model.[9] Otherwise, in a concurrent system, the termination channel can leak a secret in time linear in the size of the secret.[10]

Under a safe assumption that all sensitive data obtained during the evaluation of a function propagate to all its outputs, one can track the global flow of information in the system by monitoring inputs and outputs of all functions in the system. If we encapsulate each unmodified serverless function in a sandbox (as in Trapeze[29]), the sandbox can intercept all interactions between the function and rest of the world, including other functions, shared data stores, and external communication channels, and redirect them to a security shim to track information flow and enforce a global security policy.

13.3.3 Pointer management

In the context of spilled state, pointers with state information are often passed up and down some software stack. Due to this, the lifetime management of these objects can become problematic. Some recent languages like Rust[9] allow the user to annotate the lifetime of objects that are being passed as references. Additionally, a pointer can be marked as weak so that the traditional reference counting does not apply to it; an object pointed to only by one or more weak pointers can be freed. In essence, ownership can be disassociated with the ability to point; this is similar to the ownership of objects across address spaces that needs to be specially handled.

Pointers also may have to be made persistent across address spaces. Depending on the design, the lifetimes of such persistent pointers may be indefinite (e.g., NFS handles) which is, however, difficult to achieve over long periods as we need a persistent lookup table. Furthermore, such pointers may or may not be opaque; both designs, for example, are seen in web addresses

---

[9]In this model, termination or slow progress of functions can reveal sensitive information to curious bystanders. An easy way to understand this concept is to think of a highly competitive and equally competent staff in an office; depending on who is progressing or not in their career, in many situations, one can guess who may be running the office.

[10]The termination channel in serverless systems may not be low bandwidth; in one study, it has been measured to be 170b/sec[29] in AWS Lambda. While this seems low, note that one can elastically scale the computation and therefore the termination channel.

that encode pointers whether opaque or not.[11] Depending on the usage patterns, maintaining these pointers may have non-RP costs.

Furthermore, pointers may be implicit instead of being explicit. For example, two objects may not seem to be related but, using camera information, for example, it may be inferred that they are adjacent and an explicit pointer therefore instantiated to register the nearness. Or, some relationships might be virtual and made physical only on demand. For example, there might be some relationships across certain entities. For some analyses, only part of that may be needed and only those instantiated based on some criterion.

The pointers may also optionally encode distance in some metric space. "Unoptimized" pointers may encode relationships based on structure whereas "optimized" ones may be based on use value temporally or spatially, with the discovery of such relationships based on traces using either simple statistical models or deep machine learning models. The metric can be used to decide what virtual pointers to instantiate; one such is nearness which can be across multiple levels in the system. In some situations, especially in the context of security, distant ones are to be preferred to avoid interference.

Given the above discussion, it is clear that pointer management is an important issue for resource proportionality. We will now present two aspects of these issues, one from a systems perspective and another from a learning perspective.

13.3.3.1   Memory management, pointer management, and read-copy-update

Memory management is a critical function of an OS with memory allocation being a specific aspect. Though this aspect has been studied for the last 6 decades or more, newer problems keep arising in different forms. For example, is it possible to reduce memory power consumption in a mobile from a memory management perspective? As only an OS knows about where the processes are, it seems natural that an OS can move processes around to pack them into a fewer number of memory banks and ensure that as many other memory banks as possible are unoccupied. The hardware can then sense inactivity in the unoccupied banks and can therefore shut them off or go into low power mode, thus ensuring resource proportionality w.r.t. energy. The research problem then is how to ensure this by careful memory allocation or how to minimize the movement to get the maximum benefit, or both; there is also the loss of parallelism as fewer banks are being used.

Approaches such as workload modeling (for example, memory access patterns of applications) and lifetimes of objects are important parts of a solution

---

[11] However, some design choices across address spaces can become problematic. For example, CORBA's IORs (interoperable object references) are opaque entities but this can cause difficulties: there is now a need to use a naming service as clients cannot create object references without the help of an external service and this can also become an additional failure point. Also, opaque references require RPCs to compare object identity reliably and this overhead can become high.

to this problem. However, moving kernel pages to create contiguous free areas is unexpectedly not feasible in some cases in many kernels; kernel memory allocations in Linux kernel (and many others too) are not movable. This is due to some deep conviction amongst kernel developers that having fully movable (or even re-mappable) kernel pages adds considerable difficulties/inefficiencies in a core part of an OS due to this generality of treating kernel ("manager") pages and user pages alike; there is also work that has shown that if this insight is also employed for user pages *where* useful, it can lead to better performance (about double) at the cost of some complexity[52]. Note that having just *one* kernel page (4KB) in a memory bank of say 256MB can prevent that bank from going to sleep as we cannot move the kernel page. Recent research[296] has shown that keeping the memory bank structure in mind when allocating kernel pages can help in keeping all the unmovable kernel pages only in a few banks and thus enable the possibility of shutting down a substantial number of other banks; such a system could in principle reduce the power consumption of mobiles. A properly designed but simple algorithm during memory allocation may leave only a *small* opportunity for further optimizations such as those based on ML. A similar issue shows up when rowhammer attacks are to be avoided[79]; reverse engineering physical memory layout can help avoid allocating memories for different security domains close to one another at the physical level.

This unmovable kernel page aspect also has ramifications in systems that have variable pages (4KB and 2MB for example). Due to the same problem of "polluting" a large contiguous memory (2MB) with a 4KB kernel (unmovable) page, finding and allocating large pages also can become increasingly unlikely with time and systems suffer from excessive page faults due to the inability to keep mapping information small as only small pages can be used[297]. Note that this problem affects long running systems more; the unmovable aspect could therefore be implicated as one factor in the general observation that a system keeps slowing down with time and needs to be rebooted now and then.

To effect resource proportionality with respect to reader-heavy synchronization, and hence a scalable model for such synchronization, the read-copy-upate (RCU) mechanism is used extensively in the Linux kernel. When a new writer arrives, it gets its own copy while the previous readers still use the old version. When newer readers arrive, they see the new updated value; there is thus a weaker model of consistency. The old value is freed when its corresponding readers all finish; thus memory availability now can depend on synchronization activity. One can thus be inefficient, suffer jitter, or even run out of memory if the regular memory allocation is not integrated with RCU[313]; we can therefore have significant memory bloat in a systems sense. It turns out that there are other subsystems such as I/O that also impact memory allocation in a major way just as RCU does. Going by all the above, there is a need for a more general approach for memory allocation in the Linux kernel which is not only informed about efficiency issues but also availability.

When allocating huge pages, it can be done lazily or eagerly; both being suboptimal. If done lazily (i.e., aggregate small pages into a huge page using compaction), TLB faults can be high till the page is promoted but memory is not wasted (no bloat). If done eagerly, a good part of the huge page may not be used (high memory bloat) but with lower number of TLB faults. A recent design therefore integrates compaction, defragmentation, huge pages, and the RCU mechanism[297]. RCU also has serious implications for virtual machines (used heavily in cloud computing), namely excess memory consumption[315]. Furthermore, there are many other issues from a resource proportionality point of view: RCU on HPC systems[314], reducing the cost of update in RCU itself, redesign of some parts of an OS kernel to accommodate the application sensitivity to the need for huge pages[295], etc.

13.3.3.2   Learned indices

If managing kernel and user memory is difficult for the above reasons, one newer idea is the use of deep learning to learn the mappings[227, 252]. One can view the keys of a monotonic mapping (such as a sorted B+ tree) as defining a CDF that gives the cumulative probability of finding a value or lower ($P(X <=$value)). Learning can be used to discover this CDF followed by inference for any query. From a resource proportionality perspective, if disk accesses are more costly than inference from a learned model, then the deep learning model (inspite of the extra training cost) may be useful. Using recursive mixture models[227], it has been shown that deep learning models can be competitive or better (upto 70%) than disk-based B+ trees on modern hardware with good GPUs. However, for learning TLB mappings, it has been not effective[252] as the inference budget is much smaller. Further research is needed to see how the inference cost can be reduced.

13.3.3.3   Systems design vs. ML-based methods

From the above, one general observation is that systems design tends to be RP through approaches such as piggybacking,[12] caching/memoization, and fast/slow path optimization. For example, RCU is an attempt at resource proportionality with read heavy workloads. But this can cause memory bloat if other kernel mechanisms/structures (such as slab caches) are not co-designed with it[313]. When multiple RP considerations come into play, it can be difficult to predict the outcome; issues like security usually present difficulties for resource proportional designs. If detailed data is available, machine learning models can be used and may provide better insights especially when multiple RP considerations have to be taken into account at the same time but, in core parts of the system, it may not yet be practical due to the cost of training or monitoring. Even here, what data is to be collected for building the ML/DL

---

[12]RCU, for example, piggybacks on context switches to signal that deferred objects for freeing can now be freed; similarly, many kernels use system call return to do "housekeeping" work.

models is not clear as the number of interesting variables or data points is huge for any modern microprocessor, compiler, or operating system.

13.4   Type Analysis

As types provide precise information on how to use an object, an important issue for resource proportionality is therefore to specify/discover and use types that are tailored to the problem at hand. It will also be helpful if they are general enough to be used in related contexts using some kind of parametrization such as in C++ and Java. While the C++ design's goal has been "pay for what you use," its need to support C has created complexities, in spite of its careful design including that for libraries[371].

If the system that is being modeled can be described by a well-typed static type system, the type information can be used to effect many optimizations including eliminating some simpler types of bloat; RPD is helped by strong type systems. Considering Figure 1.3, we can map the sides of the triangle as those corresponding to designs with overgeneral types (framework based software), type remapping (code generation approaches), and dynamic types (used by OS kernels in a highly disciplined way); what is missing is implicit types discovered by data analysis (such as that of dependent types that encode value constraints). In the context of concurrency and transactions, abstract datatypes (which do emphasize encapsulation) have been shown to have higher performance[184] through the use of datatype semantics (such as commutativity), reduction of bookkeeping and limiting false conflicts (as it is not necessary to model the system at the level of unstructured memory but at the higher object level), and enabling efficient concurrency control. However, many complex systems need dynamicity in their type systems and hence such optimizations may not be possible all through.

A recent example designed from ground up is the language Julia, a language designed for high performance and increasingly used for scientific computation[54]. It has a type system[55] that is dynamic (only values have types, not variables), nominative (name equivalence, not structural equivalence; or, hierarchical relationships between types that have to be explicitly declared, rather than implied by compatible structure[203]), and parametric (generic types can have parameters; when a generic function is applied, the most specific definition that matches the run-time argument types is called). Some of its extensive use of types to optimize code are optional type annotations, multiple dispatch using types to select implementations, code specialization, code generation using macro processing to avoid runtime overhead, and libraries that use type information for efficiency. It also avoids the permissive type systems where, for example, types can be mutable, or where the type of a value can change during execution (as in CLOS) as many compiler optimizations are not possible. Similarly, it does not have dependent types[13] as efficient type inferencing either becomes undecidable or difficult even after some restrictions; it does support intersection and union types as they can be compiled well. Automatic type inference is extensively used as it is well established and highly effective[114]. The types of program expressions and variables are inferred by forward dataflow analysis; essentially, this analysis propagates eventually type information from each statement to all other statements reachable by control flow.

In general while automatic type inference can be effective, allowing the user to provide type annotations may be useful (especially across address spaces). In a recent system[179], a deep learning model is used that can analyze types that naturally occur in certain contexts and relations in a corpus, and also can provide type suggestions. These suggestions can often be verified by the type checker, even if the system could not infer the type initially. Similarly, in a large infrastructure for machine learning at Facebook[178], many of its tools pass type information across one another when run on deployed systems that are likely to be heterogeneous. Recently there has also been work to incorporate polyhedral shapes at the intermediate representation level for machine learning workloads, so that all the downstream processing is aware of the types of the objects being worked on[6].

Embedded systems have very tight constraints and type safe languages can help in realizing them as many checks can be carried out at compile time; for example, we can check the safety of a zero-copy transfer using linear types. For this reason, some recent languages like Rust restrict the semantics of assignment using linear types. Each value in Rust has a unique owner, the variable it is bound to. When the owner goes out of scope, it can be freed. However, this by itself is too restrictive. Hence it also allows limited borrowing so that additional variables can also reference the value (with the restriction that there is no other borrow already and borrowers' lifetime cannot last beyond the loaners). Across procedures, lifetimes of input parameters and output values may also need to be annotated (for example, that a specific input parameter outlives the lifetime of the returned value) so that a compiler can check the safety of using a reference to the returned value. In this design, double-free and use-after-free are not possible. However, it has been found that some further restrictions are needed in very resource constrained systems[237]. For example, embedded OSes are mostly event-driven and the code must be written to manually pass state between callers and callbacks using techniques such as "stack ripping"[237, 19] that copies and passes values from the stack needed in the callbacks. This complexity is not needed if a single threading model is assumed, or it can be guaranteed that borrows are never shared between threads, which can be done with static analysis[237]. Since languages such as

---

[13]Such as the type of a pair of integers where the first value is greater or equal to the second value. For greater efficiencies, where possible, dependent types may also need to be supported. One way is to discover the precise types using deep learning techniques[179].

Rust do not have such restrictions (it is multi-threaded), the complexity of a design can be reduced with design constraints that can be checked at compile time.

13.4.1 Information flow

Information flow analysis can be carried out using types, for a variety of reasons such as security, approximations, homomorphic computation (see below), and for runtime optimization across the layers. For example, reclassification and declassification of security labels may be necessary in systems designed for security. Similarly, in approximate computing, some values may be needed to be loosened or tightened[298] or endorsed[341].

Making information flow analysis RP is an important issue as many analyses are based on it. Compile time analysis augmented with runtime fine-grained bit-level tracking is necessary in the general case[256, 160]. Furthermore, a spill-free design is likely to be RP as the server does not keep the state but returns an opaque representation to the client; hence client to client information flow is already minimized to the extent possible. Using linear types or affine types allows compile time checks for information flow across variables.

13.4.2 Homomorphic computation

In a cloud setting, if all the data is encrypted for privacy, any computation that needs a value has to suffer the cost of its decryption. To avoid this cost, one approach is to attempt *homomorphic* computation on the encrypted data itself where feasible (i.e., if $a + b = c$, for example, is to be computed, is there an additive encryption scheme where $encr(a) + encr(b) = encr(c)?$). Since it has been shown that such encryption is very costly if it has to work with multiple operators such as add and multiply, one approach is to check if the particular application needs only a specific operation in a specific area of the computation. If so, each such segment is encrypted with the appropriate scheme for the operator and then transformed into a different scheme, if necessary, outside that segment. As the encryption schemes can be put in a partial order, a type system is defined in the analysis where variables have qualified types that take on a qualifier (such as the encryption scheme used or none). Type inference is then used to check for type soundness, i.e., an encryption scheme is used to perform only supported operations and operations are only applied to operands that use the same encryption scheme[380, 262]. Using such inference, it is now possible in many useful cases to provide performance on encrypted data that is comparable to computations on the original unencrypted data. This is an example of RP cloud computing that uses types to provide privacy also.

13.4.3 Intermittent computation

Let us next consider intermittency which is an important feature as computing becomes embedded in society at both the individual and organizational level with the expectation of "any where, any time, any device" computing. Due to the lack of access to energy sources in many human spaces and batteries with limited capacity, investigating intermittent computing is natural. Intermittent computing can also arise in IaaS (infrastructure-as-a-service) cloud systems where some vendors rent VM resources organized as a spot market. However, if the demand increases suddenly and the vendor increases the costs of renting the resources, clients may lose the VMs in this market. While this model emphasizes market allocation efficiency (and hence possibly resource proportionality in a deep way, as it is not just at the client level but at the server level), intermittency of computing resources becomes an important issue. In such an environment, risk and cost aware server management designs are required that maximize availability[347]; this can be achieved using a derivative cloud platform. In the derivatives market, resources from the main market are offered to customers with different pricing models and availability guarantees that are not possible in the main market itself. This is also important in the context of stranded power that we discuss below.

13.5 When Is Resource Proportionality Not Applicable?

While we have discussed some issues with RPD (Chapter 10.1.5), there is a deeper issue of what resource proportionality is intrinsically worth and with respect to what issues of concern. Instead of resource proportionality considered so far, the larger principle could be conservation of resources critical for system equilibrium or for the design of a self-sustaining "universe"[14]; here both the system or the "universe" can be circumscribed as appropriate. If resource proportionality is considered only in the context of, say, energy, it might be counterproductive when the larger issue is something different, such as that of carbon neutrality or that of sustainability of the biosphere.

Consider a new issue that has become important: the problem of stranded power when unreliable renewable energy sources become a dominant part

---

[14]In the context of Bitcoin, even though the ability to put a transaction on the blockchain is proportional to the work required for solving the puzzle in the asymptotic case, it is extremely energy consuming and therefore other proposals have been mooted. In the economic sphere, debates whether continuous growth (which is the goal of almost all governments) or stability is desirable center around the possibility of proportional decrease in harmful impacts of growth in many areas including emission of greenhouse gases or even the consumption of goods such as plastics. Increasing amount of greenhouse warming gases (which has usually historically correlated with increasing economic activity) is clearly untenable unless sublinear or even greater decrease of production of such gases is possible.

of power in the context of data center design[281]. Since large amounts of energy cannot be stored (energy production should temporally equal energy consumption), excess energy production by renewables may have to be sold sometimes at negative cost if such production cannot be curtailed. If it cannot be sold, it has to be "burnt" in some way as it cannot be stored (as high capacity batteries may not be available); there is now possibly a carbon cost. By reducing renewable power available ("curtailing power"), we can prevent carbon addition but possibly at the cost of both fossil energy and also more carbon at a later time if some timeliness guarantees have also to be met. Since this excess energy is of variable quantity, intermittent computing at the data center level to absorb the stranded power is an attractive choice to reduce carbon footprint overall. Since curtailing power at short notice is not easy, colocated additional computing equipment may be required to provide the resources for intermittent computation to absorb the excess power. This may prove to be necessary if the stranded power is much more than the curtailed power (which is true as of 2019[281] in some power sectors in the US, Germany, and China.). Here large batch jobs (for example, those needed for training in AI applications) can be scheduled on spare computing units when non-curtailable power is present at the cost of higher capital costs.

We may see this as "power bloat" in the system: additional power/energy capacity that is harmful just as with bloated computation. By adding software bloat judiciously, we have seen in Section 3.3 that peak power can be reduced; by suitable rescheduling of jobs or their work intensity, it may be possible to also mitigate carbon production. Some of the challenges in this context are[281] stochastic cloud resource management (as capacity is variable, either the grid power drives the cloud or the stranded power drives the grid), distributed protocols for highly unavailable resources (especially correlated availability and reliability models for long periods), making cloud workload supply-following (especially exploiting geographic supply-following load), coupling power grid and multiple very large scale clouds, and power markets and grid management with stochastic quantities.

In the future, computation (in the form of workloads) and resources[15] (cloud and non-cloud) can all become stochastic; it is already so with caches and the like. To do a good job of using such a system in an intermittent context, some predictive models are necessary. How are the resources for running these models in a timely way guaranteed? The issue is also "what the resource proportionality is now about" as there is no one "center"; distributed stochastic control is one attractive design option. It can be at a market or game-theoretic sense across all the players in the game but there is the "price of anarchy"[225] to pay. Furthermore, if market-based incentives are in play, possibilities of oscillations also cannot be ruled out.

---

[15]Note that this can also occur at the network level, for example, with backscatter communication in passive systems[244].

Having alerted the reader to the suitableness of the concept of resource proportionality itself in the context of a carbon economy very briefly, we note that the RP notion is very much relevant in many significant areas.

13.6   Conclusions

In the future, given the highly disparate and demanding environments where computing systems may be deployed, there is a need to rethink many of the current design assumptions. We have discussed approximate computing as one important new area and also briefly touched upon homomorphic computation and intermittent computing as other important areas. Research in specifying, checking, engineering, or discovering newer type systems is a continuing effort that involves programming language and compiler designers. Furthermore, data-driven semi-automated modelers may also be necessary in the future.

Chapter 14

Conclusions

Although industry buzzwords, models, and techniques ebb and flow, designing software for emerging systems will remain important as long as demand for computing continues to propel the evolution of new hardware and software technologies. We have explored this challenge through the lens of resource proportional design (RPD), a simple enduring principle that helps enable software and system designers adapt to three key trends:

- the leap in flexibility needed to meet the expanding demands of a digital universe with its explosion of data and functionality, and the associated price of escalating complexity and resource usage.

- the leap in productivity enabled by software advancements such as framework based software, cloud native deployment and machine learning, and the associated price of loss of visibility, interpretability, and low level control.

- the leap in computational, storage, and network capability enabled by hardware and systems advancement, and the associated price of increased non-uniformity and diversity of resources to harness.

RPD is motivated by the observation that only a fraction of the large space of possibilities collectively induced by these trends is relevant at a time in a given situation. Hence designing components which consume resources in proportion to this relevant fraction can result in significant savings. As we have noted throughout the book, the key challenge, however, is that different fractions of the possibilities are relevant in different situations as technology and application needs change. A traditional separation of concerns approach maximizes cohesion within modules (by reducing indirections) and minimizes coupling/needless interaction across modules (by adding indirections). Such module boundaries may be too rigid when combinations are more fluid/variable and can change at fine-grained levels.

313

RPD simultaneously aims to reduce the burden (generality tax) of possibilities (concerns) that are unutilized while optimizing (for efficient reuse) the utilization of the combination of possibilities (concerns) that are relevant in a given scenario. This approach scales better with change, when we cannot predict future needs upfront. Software that is designed to be RP from ground up can thus allow new capabilities to use the deep insights distilled across generations but not be burdened by past boundaries. The RP principle can be separately applied across the dimensions of code (Chapter 5, 6), technology (Chapter 8, 9), and data (Chapter 11). Table 14.1 summarizes some of the high level strategies for each dimension.

TABLE 14.1:   Resource Proportional Practices for Code, Technology, and Data

Resource proportional code (framework features)	Resource proportional technology use (bottleneck trade-offs)	Resource proportional data use (input relevance)
Minimize feature interactions (without limiting efficient reuse)	Extract embedded bottleneck assumptions	De-prioritize correlated inputs
Reduce recurring overheads	Amortize recurring trade-offs	Reduce redundant data and compute (Insight reuse, Similarity detection, Approximation)
Activate high overhead feature on demand	Activate control plane dials to adapt trade-offs	Scale computation effort with difficulty (Scalable effort classifiers, Relaxed synchronization)
Program constructs and runtime with RP-awareness	RP resource shifter	RP aware storage with proactive data gradation Software defined virtual memory

14.1   Applicability to Large Scale Solution Stacks, Rising Complexity, and Software Evolution

How does RPD apply to large software systems which are typically composed of many subsystems and contain numerous flows that traverse a deeply layered stack? Chapters 7, 12, and 13 expanded the RP discussion beyond a local perspective to encompass complex global architectural RP considerations which involve balancing multiple non-functional trade-offs (a few of which are summarized in Table 14.2). For example:

- Control and data plane separation enables RP optimizations where the control plane wires up a desired combination of capabilities which is then utilized efficiently in the data plane. This approach works well when the subset of concerns utilized is stable, not overly dynamic or too fine-grained and changes are explicitly activated. Each subsystem has its own control and data plane and these are hierarchically combined for a systemwide view. The data planes are largely non-mediated to minimize overheads, except for the ability to provide some form of dynamic feedback to the control plane through state based triggers using (potentially hardware accelerated) resource proportional counters.

- De-duplicating or unifying functionality across layers spanning loosely coupled subsystems is a common RP optimization opportunity (as illustrated in Section 11.6). A large solution typically reuses framework components which are designed to be platform agnostic and cannot make assumptions about a specific underlying platform or system layer. Making it easier to detect similar/overlapping features across layers is an open RPD challenge.[1] Thereafter, it is possible to devise mechanisms to unify them resource proportionally in a way that the layers can still be used independently as well (e.g., using common helper libraries).

- Approximate computing enables RP optimization to be implemented by the system or a runtime framework for operations where resource usage can be traded for accuracy when acceptability criteria can be verified.

- The implications of RP on security need to be carefully considered. On the one hand, RP principles can be used when designing security features such as information flow tracking, protecting secrets, side channel neutralization, and reducing attack surface due to unused code. On the other hand an RP implementation introduces a greater risk that resource usage could become a side channel for information leakage. An anti-RPD design for providing DoS resistance or a noisy RPD design for preventing inference may be safer in certain situations.

Notice that RPD principles continue to be applicable as programming languages, models, and software deployment frameworks evolve to support new environments with higher degrees of productivity (e.g., the advent of function as a service for cloud native applications, the replacement of hard coded rules with neural network models trained to perform data driven decisions). The choice of languages and software models is often independently determined based on various factors such as the expressivity requirements, operational conditions, and developer preferences.[2]

---

[1] This could be attempted via semiautomatic concern analysis or meta annotations during component construction.

[2] E.g., the trend towards polyglot solutions, where components written in multiple programming languages may be composed into a single application or service.

TABLE 14.2:   Applying RPD to Large Complex Software Systems: A Rich Design Space with RP Trade-offs

Architecture challenge	RPD consideration
Deeply layered stack	Non-strict layering allows cross-layer optimization (e.g., replacing mid-layers with helper libraries), inefficiencies at lower layers of the stack have a multiplicative effect.
RP across large number of subsystems	Hierarchy/mesh of subsystem control planes (stateful) and data planes (stateless) allows a systemwide view. Data planes have accelerated RP counters for state based triggers and may even use analog models. Union of multiple designs may be achieved via internal or external brokering.
Large number of flows or paths	Composing RP optimized virtual chains of critical path flows. If many near-critical paths exist then system has poor interpretability.
RP Information flows	Minimize rechecking by introducing just enough for sufficient redundancy to catch non-authorized flows.
RP neutralization of covert flows	Resource usage as a side channel. Learning to neutralize side channels only where necessary.
State spill problem	Extensive caching can cause high state spill, where migration incurs runtime overhead of scanning for consistency.
RP Failure handling and fault tolerance	Variable cost at higher layers vs. fixed HA cost when handled at lower layers. Lazy/deferred computation models. Blame proportional logging.
RP Accuracy Trade-off	Approximate computing and analog computing models.

We observe a recurring debate between collapsing and splitting layers as hardware, software, and data processing needs evolve[32]. At one extreme, stacks could be built with fine-grained separation and RP optimized mechanisms could then be provided for composition at runtime (collapsing interaction overheads dynamically, using caching, layer folding, function as-a-service fusion[48]). At the other extreme, stacks could be maximally composed with coarse grained modules and then provided with RP optimized mechanisms for fine-grained separation at runtime (e.g., library specialization techniques, feature oriented programming, slicing using type information, or CAPA techniques to detect interactions and exclude what isn't needed). In both cases predicting what would be needed and adapting accordingly (whether by dynamically collapsing/dynamically splitting layer interactions) becomes the next challenge. This could either be based on the environment and technology capabilities (which can be detected automatically and adjusted, either directly or through a control plane or an RP aware runtime) or the context of the caller (which requires explicit information to be passed in, or set by a control plane).

14.2 Resilience to Future Changes as Hardware Technology Advancements Are Raising the Bar

It is difficult to adapt to major technology shifts such as the emergence of PM before the feasibility and usefulness of said technologies are known. One should analyze this impact speculatively to some degree; however the deep analysis necessary to construct RPD solutions for a new technology is time consuming enough that it must be focused on the most likely and beneficial candidates.

That said, once an important technology transition such as PM has been identified, RP becomes a key metric in deciding how and when various applications make the transition as described in Chapter 8. The following high level guidelines apply.

- Identify software and protocol patterns that are disrupted by the new technology. In the case of the PM transition, examples include atomicity, marshaling, and context switching, among others, that may disappear or be recast during the disruption. Use RP as a metric to analyze the net benefit of the new technology to various types of applications as a result of such design pattern changes.

- Create design patterns such as the dual stack scenario that allow application implementations and technology deployments to inter-operate across the disruption. Ideally, old and new applications can use both old and new technologies with different RP results.

- Account for changes in system bottlenecks caused by technology transitions, ideally in real time. For example while it is clear that the provisioning of PM must take performance and capacity into account, the impact does not end there. Assumptions built into applications and changing system conditions can have an unexpected impact on bottlenecks. One response to this is to feed resource utilization information back into real time application library decisions that favor different resource bottleneck scenarios.

- Use RPD to identify and optimize permutations of functionality that are impacted by the disruption. For example, in the case of PM, the interplay between redundancy and atomicity within a dual stack scenario creates optimization opportunities.

The evolution of memory interconnects provides another example of a technology transition with RP implications. In Chapter 9 we observed that this transition generally demands less application modification to achieve RPD than the PM transition alone; however it impacts the distribution of data and functionality between CPU's and accelerators in a memory centric architecture. Analysis of this transition adds the following additional response guidelines.

- If possible, cast the measurement and analysis of RP as the ability to do work (i.e., instruction execution rate) in a systemwide context. For example, memory and memory interconnects impact instruction execution rate in different ways for various applications.

- The placement of various functions within a memory centric architecture can be facilitated using RPD and adjusted in real time when appropriate.

The ability to access remote PM directly through a memory interconnect instead of traditional network protocols also enables applications to shift from scale out (shared nothing) to scale up (shared something) models, e.g., optimizing shuffles in Apache Spark to leverage a memory centric architecture improves its resource proportionality. Finally, as non-uniformity of memory access time increases it becomes more and more important to use a combination of control plane and data plane elements to enable systems to adapt without inhibiting real time performance. Chapter 12 elaborated on a use case of this recurring theme. There we suggest an architecture for real time caching, pooling, and tiering in the data plane that is guided by configuration and machine learning in the control plane.

14.3   RP Impact on Productivity and Other Metrics

Computer engineering involves both technical and social aspects. Social aspects of computer engineering range from macroscopic matters of ethics, rights, and environmental stewardship to the microscopic interaction between developers involved in large projects and evolving software ecosystems.

While this book has focused primarily on resource usage corresponding to utility in terms of software capabilities, the concept of resource proportionality can be viewed from different lenses for value and cost. For example, improving the accuracy of a machine learning model by even what seems to be a small percentage may reduce prediction errors by an order of magnitude as a result of which it becomes practical to rely on it, which amounts to a leap in value[22]. Spending extra resources on the model inference or training may not be an issue (i.e., a high value of k-RPD in terms of runtime resources is acceptable), compared to the cost and effort required to distill clean data or the cost of wrong predictions. In Chapter 13 we discussed another example where run-time resources (such as "stranded power") are abundant or perishable and not worth optimizing. The operational overhead cost (human effort) may be a more important factor in this situation and AI intensive operations may be preferable despite the expensive computation cost involved.

What about the overarching economic trade-off between time to market of software and sustained advantage that comes from ensuring longer term operational quality of life of the software? The impact of RPD on software development productivity is difficult to model objectively because programmers have complex trade-offs to manage, some of which are not well articulated. One way to reason about this is to note that productivity is impacted by the cognitive burden of a number of different things a programmer has to keep in mind or specify when implementing code, i.e., programmer's attention budget. RP designs attempt to (i) unify similar/overlapping features and data and (ii) minimize the overhead of fine-grained interactions both when they are needed and when they aren't. As a result, using RPD components can reduce programmer burden (and hence improve productivity) by simplifying the number of interactions and features to worry about. However, implementing an RP feature requires some conscious effort to refactor code in this manner or expose enough information so that interactions can be discovered and tracked automatically using static and dynamic analysis or run-time mechanisms. This has a productivity cost, but is limited only to the interactions induced by the feature being implemented and can largely be automated using the CAPA tool and an RP-aware runtime.

RPD also forces one to consciously evaluate what the design space is with respect to some important concern. Given that there are multiple dimensions on which to base RP analysis in any complex system, the ability to articulate it and make allowances for bottlenecks may make designs better overall. Furthermore such clear articulation helps in building data driven models, especially in the beginning stages with machine learning models (when data is scarce); later, after system (or prototype) realization, one can build deep learning models (when data is abundant) that can be used to rearchitect the system.

RPD can be adopted without mandating any specific linguistic abstraction or language (e.g., C++ vs. Java vs. Scala) and thus does not limit productivity improvements from the use of newer software models. How a chosen linguistic abstraction is realized in practice may have implications for RPD. For example, a linguistic representation with types offers structure and ease of interpretability for RPD analysis. A library based realization on the other hand offers zero overhead when unused, but needs copy elimination/concurrency optimization across interfaces. Another extreme is a software 2.0 approach (as proposed by Andrej Karpathy) which generates code from datasets (pre-labeled or available from previous runs or test cases) using neural network models and differentiable programming.[3] This offers automation at the cost of explainability and the risk of adversarial attacks.

How can RPD be made economical for real businesses? By working towards RPD we have more benefit than just resource efficiency. For example, complexity of software tends to get out of hand unless we have the tools to contain it. The complexity of operating environments has driven a shift from purchasing software towards more subscription based models; this mirrors the transition from internally brokered models to external ones: essentially, the introduction of market mechanisms. Perhaps the latter is geared towards a longer term sustainable solution.

A deeper productivity argument for RPD can be made in the context of large software development projects[402]. The traditional waterfall method of software development is fitful; many features are consolidated for inclusion, some are finally added as part of a release but without any feedback regarding usefulness until the next release, which can be many months or years away. One Microsoft study found that only "one-third of well-designed features delivered value to the customer"[223]. The DevOps method, with its tighter integration of development and operations, is more likely to be responsive to the feedback from the operations on what is being used and therefore is likely to focus only on features needed, resulting in a proportional expenditure of efforts.

14.4   Future Systems Design

Traditionally, system software enables application software and hardware advancements to progress in a de-coupled fashion. This decoupling has facilitated tremendous development productivity increases through generations of hardware advancements as application developers assume that the systems software layer would take care of the hard issues of optimizing resource usage and managing key trade-offs. However, as we have seen through examples described in this book, if the software isn't resource proportional, it is hard to make up for lost efficiency without sacrificing flexibility or productivity.

We have explored a few RP awareness mechanisms that could be incorporated in operating systems, storage systems, runtime engines, profilers, and optimizers to enable RP execution of software stacks if they are designed in a way that is friendly for resource proportional analysis and optimization. For example, an RP-aware runtime can optimize interaction overheads during program execution (when annotations are available either for fine-grained virtual separation [210] or optimized composition of virtual chains) and an RP-aware storage system can improve data resource proportionality through similarity

---

[3]Leveraging the duality between code and data, per Erik Meijer.

detection, approximation, and insight reuse (when workflow interception or similarity detection plugins are available).

A question that continues to intrigue us is how systems (hardware and software) could be co-designed to enable whole systems to become more resource proportional.

There are many axes along which we can discuss these issues (some of which we have considered in the book); these axes intertwine with each other and hence it is difficult to reason about them in real complex systems. Notions of control and data plane, their dynamic and static aspects, what is predictable and what is not what is secure and what is not, are all related to what is RP and what is not. On top of this, there is also layering or DAG structures, with each layer or node having the same sets of issues along with newer aspects such as information that flows from top to bottom (such as QoS, realtime requirements) and information that flows from bottom to top layers (e.g., performance indices). Furthermore there could also be information flows laterally in some of the layers.

Finally, as the system evolves, to handle unanticipated aspects, there could be inband signalling where control information is embedded in the data streams; with a new design, such information could be separated out into the control plane. If new designs are difficult (due to heavy investment in the older models), statistical techniques (such as ML, DL) can again recapture the inband signalling aspects. The new designs therefore rest not only on software foundations (such as language design) and hardware advancements, they can also use statistical insights obtained by effective realtime inferences on traces of data generated by the system during its normal operation. There is thus a cyclic aspect of introducing newer features or capabilities ("becoming") that over time become well established and require redesign ("being"). RPD provides a helpful perspective during such redesign: Are we handling the common cases well? Are we adding something really useful?

We look forward to the journey of weaving RP across code, technology, and data into the being and the becoming of future systems to realize flexible software that is also green.

Glossary

analytics: Covers a broad set of techniques to process data in order to understand and discover meaningful insights or guide decisions, ranging from structured database queries to sophisticated machine learning and deep learning or artificial intelligence models. Analytics could be descriptive (interpreting what happened), diagnostic (understanding why it happened), predictive (anticipating what could happen), or prescriptive (recommending what should be done).

anti-pattern: Common practices, mistakes, or approaches known to lead inefficient software, performance, and scaling issues, i.e., recurring patterns that represent poor engineering choices to avoid when designing software (the word "anti-pattern" can be viewed as the opposite of "design patterns" that lead to well-engineered software).

aspect mining: Approaches for discovering and locating cross-cutting concerns (aspects) in a program.

aspect oriented programming (AOP): AOP is a programming paradigm that allows the separation of cross-cutting concerns in a program such as transactions or logging which are spread across many modules and hence difficult to separate using traditional module representations such as methods, classes, components. AOP expresses these concerns (also called aspects) in a special form that may be subsequently woven into the base program as needed. For example, aspects may be described as code snippets to be embedded at specified join points in the program (Section 5.3.1).

atomicity: The "all or nothing" property of a data update which guarantees that the result of any subsequent access will be precisely the data value that existed before the update or the updated value but never any other value. This property is often important to avoid data corruption. Atomicity requires careful design of hardware and software, especially when data is shared or when failures such as power loss arise.

bandwidth: The rate at which data can travel (maximum) or is traveling (measured) through a component or interface measured in an amount of data (e.g., bytes) per second.

Big Data: Computing involving search and analysis within large data sets where less data internal structure and semantics is known in advance than is the case with Enterprise on line transaction processing.

blocking: The suspension of instruction stream (thread or process) execution so as to wait for an event such as an IO completion or a lock acquisition. Other instruction streams can run after a context switch. When the event occurs the blocked instruction stream picks up where it left off.

bottleneck: A resource, device, or component that limits the performance or capability of a system or program.

cache: A temporary location for data intended to accelerate access by increasing the proximity of data to processing.

cache coherent: A property of multiple caches containing the same data wherein all consumers of the data perceive the same data image and update order at the same time in spite of temporary differences in physical cache contents. In CPUs this is achieved using a cache coherency protocol.

cache line: The unit of data transfer between CPUs and memory. Cache lines are generally larger than processor registers in order to improve memory performance by distributing bytes across DRAM chips (interleave) and by pre-fetching adjacent memory locations.

Class of Service: A description of behavior that defines a group of services or resources. Generally the services or resources described by a class of service provide the same function but with different non-functional requirements such as performance, recoverability, and security.

co-design: Designing multiple components of a system in a synergistic way to provide a flexible and effective means to optimize characteristics of the final solution. For example, hardware-software co-design involves designing the hardware components and the software layers that would utilize that hardware side-by-side so that the software gets the maximum benefit of advanced hardware capabilities.

concept assignment: Recognizing and mapping concepts in a program to their corresponding implementation in code.

concerns: Features, design idioms, or other conceptual considerations that can impact the implementation of a program.

consistency: An application requirement that all of its data adhere to design rules to enable the application to make sense of the data. Consistency rules inhibit structural or logical contradictions caused by data structure use or manipulation. This ensures the overall integrity of whole sets of data but does not preclude higher order logic errors or contradictions that may originate outside the application.

container: A self-contained environment to run an application packaged with all its dependencies and dynamically bound to its persistent data, in effective isolation from other application containers hosted on same operating system. The container image is composed as reusable layers built using Docker. Resource isolation is achieved using lightweight virtualization mechanisms such as control groups, namespaces, overlay filesystems (Section 2.3.1).

context switch: In a processor, the act of temporarily suspending one instruction execution stream (a thread or process) and resuming execution of another. During a context switch, processor state is saved and restored. Context switches are usually used to allow the processor to continue with some activities while others are blocked, waiting for IO.

Cumulative Distribution Function (CDF): A CDF of a random variable x is the probability that x will take a value less than or equal to some x.

Deep learning (DL): A term used to refer to a broad family of machine learning approaches enabled by the renaissance of artificial neural networks with deep hierarchies; the word "deep" refers to multiple layers in a neural network that automatically learn features representing different levels of abstraction as the model is trained with vast amounts of data.

DES/AES: Encryption and decryption standards (1970's and 2010's).

DevOps: A confluence of development and operations in a datacenter that engenders an agile relationship between application development and IT operations. Development in this case often involves the customization and integration of largely pre-existing application components to meet the needs of a specific business.

Direct Memory Access (DMA): The rapid copying of bulk data using electronics designed for that purpose. Data flows in and/or out of memory regions under the control of state machines for memory address sequencing and data transmission.

Double Data Rate (DDR): The current pervasive interconnect between CPU's and memory. Over time, numbered generations of DDR technology have been rolled out. DDR5 is likely to be introduced by 2020.

Dual Inline Memory Module (DIMM): The current pervasive packaging form factor for main memory (DRAM). The "Dual" refers to the placement of memory chips on both sides of the memory module. "Inline" refers to the placement of multiple memory chips in a single row on each side of the memory module. DIMMs interface to CPU's using DDR.

Dynamic Random Access Memory (DRAM): The technology currently used as main memory in most computer systems. DRAM is volatile because it operates by storing a charge in a memory cell which decays quickly (e.g., in seconds) and must be dynamically refreshed frequently to retain contents.

edge computing: Shift analytics computation close to the source of data and physical world action at remote edges instead of having to transport huge volumes of data to the main data centers (core) or cloud in order to perform the computation. Typically some data is still transferred to the core for consolidated analysis across all edges for long term learning.

endurance: The expected number of times a memory cell can be updated before becoming too degraded to reliably store data. This metric is often translated into a total write rate of the entire component capacity (write passes per day) given the space management algorithms within the component. One function of those algorithms is often wear leveling, which refers to the equal distribution of writes across all of the memory cells in the component.

enterprise computing: Computing associated with business operations, often involving online transaction processing databases.

escape analysis: A compiler analysis method for determining the scope of pointers in a managed language such as Java (where there is no explicit free() operation and the JVM is expected to garbage collect the object when it is no longer referenced). For example, the analysis determines whether a pointer reference "escapes" the scope of a method to its caller or other threads and hence cannot be allocated on stack inside that method.

Ethereum: A blockchain-based distributed computing platform with smart contract (scripting) functionality.

Ethernet: The pervasive underlying technology of the internet. Over time the term has been generalized from its original definition to refer to many physical and low level link protocols used to access the internet.

feature: a unit of functionality of a software system.

fence: An action to ensure that no effective result of any action (within a defined scope) that preceded a defined point in a data stream (i.e., a fence instruction) can be delayed until after that point in the data stream. "Effective" in this case means "perceivable by a data consumer." In the context of PM, data consumption may occur after a failure or reset. "Scope" is generally defined by identifying a complete or ongoing data stream (sequence of transmission and/or update) or a memory address range. In processors, fence instructions are generally used to ensure that the results of a series of flush instructions are in place before continuing with the next instruction.

flash: The current pervasive non-rotating block structured storage technology.

flush: The act of moving data from one level of a caching and memory hierarchy to a lower level. Flush is generally used to ensure persistence (flush to PM) or to free up space in an upper hierarchy level. In processors, flush instructions trigger flush actions but do not wait for them to complete.

flush on failure: Persistence solutions in which data is not flushed until after a power failure is detected. This requires a guarantee of power hold time sufficient to complete all accumulated flushes.

Garbage Collector (GC): An automatic memory manager that safely releases memory allocated to objects when those objects are no longer referenced, without requiring programmers to explicitly deallocate them.

General Purpose Graphics Processing Unit (GPGPU): A graphics processing unit that has been repurposed and enhanced to offload other processing besides graphics, e.g., vector processing. A GPGPU may not have a video interface.

Graphics Processing Unit (GPU): Electronics to accelerate graphics processing (primarily shading of polygons displayed) by offloading it from CPU's. GPU's are placed between CPU's and video displays. The current pervasive interface between GPU's and CPU's is PCIe. Various video interfaces are used to connect GPU's to displays including VGA, HDMI, and DVI.

Hadoop: An open source mapreduce framework[116].

Hard Disk Drive (HDD): A block structured storage device that uses magnetic recording on rotating media.

High Availability (HA): The property of a system that guarantees continuous operation (uptime) in spite of limited failures of compute, network, and/or storage components such as servers, switches, and disk drives. Various fault tolerance techniques involving redundant components are employed to compensate for failure of specific component permutations.

High Bandwidth Memory (HBM): A high-performance RAM interface for 3D-stacked DRAM.

High Performance Computing (HPC): Scientific computing generally associated with super-computers.

Host Bus Adapter (HBA): A PCIe Card that connects to a storage network such as FibreChannel or Serial Attached SCSI.

hybrid cloud: A computing environment that utilizes a combination of on-premises private cloud(s) and public cloud(s) including the ability to orchestrate, move, or split data and workloads across them.

InfiniBand (IB): An alternative network to Ethernet that specializes in RDMA.

Input/Output (IO): A processor interaction with a peripheral device that causes and subsequently responds to the completion of an action by the peripheral device. The peripheral device may be a server internal component such as a PCIe card or a disk drive, or it may be an external device such as a disk array. Related terms include DMA, interrupt, and context switch.

Interrupt: A processor internal action that diverts the flow of instructions in order to respond to an event. The event generally indicates a potentially urgent processor internal or external situation such as an IO completion, a timer expiration, or a hardware fault.

iWarp: An RDMA implementation for Ethernet.

KMeans: An unsupervised machine learning algorithm for partitioning a set of data observations into K clusters based on feature similarity so as to minimize the within cluster variance. Each cluster is represented by its centroid (mean).

L2/L3: Level 2 (data link layer) and 3 (IP layer) in an Internet protocol Stack. Also, a subset of processor internal caches which are generally numbered from level 1 to level 3. Level 1 is the fastest, smallest cache while higher numbers are slower and larger.

latency: The elapsed time required to access data, communicate data, and/or perform a computation. In addition to the nature of the access, communication, or computation and the underlying technologies involved, latency is sensitive to context of measurement including where, when, and how it occurred.

load instruction (LD): A processor instruction that causes a small amount (up to a cache line) of data to be fetched from a processor cache or main memory location into a processor register.

Local Area Network (LAN): A network that spans a room, building, or campus, generally connected to the internet.

Low Level Virtual Machine (LLVM): A modular and reusable compiler and toolchain framework designed to support analysis and transformation of programs.

marshaling: The reformatting of data structures surrounding the transmission of data between components within or across systems.

media: The underlying physical technology used to store or transmit data.

memoization: An optimization that caches past computation, e.g., to save computation costs by reusing the result of a costly function when the function is called again with the same inputs.

memory centric: An approach to computer architecture and computation which positions memory, rather than processors, as the focal point of the system. This typically involves paths that allow multiple components including adapters and accelerators to directly access memory without forcing data to flow through processor memory management units.

Memory Interconnect (MI): A generalized term for the interface between CPU's and memory currently dominated by DDR. MI technology alternatives to DDR are emerging that better serve PM and play a larger role in computer systems, overlapping with PCIe and networks.

Memory Management Unit (MMU): Part of a processor system that orchestrates memory access, usually on behalf of multiple core processing units.

Memory Mapped (MM): A condition of a file or region of memory wherein the contents can be manipulated by applications using virtual addresses. Without memory mapping the contents of files can only be accessed using read and write commands. The use of the term "memory mapping" in this book should not be confused with the older concept of "memory mapped IO" wherein memory locations, instead of IO instructions, are used to control peripherals.

microservice: A microservice's architectural style is very typical of cloud native applications, where an application is composed of a collection of small loosely coupled fine-grained distributed services (microservices) that can be scaled independently and are connected using lightweight protocols.

mmap: An OS or file system call that causes part of a file to become memory mapped.

move instruction (MOV): A processor instruction that causes a series of data items (i.e., cache lines) to be copied through the processor's instruction execution electronics from one memory location to another. Move instructions function like a series of store instructions with the same caveats as store. Move instructions are less efficient than DMA as they engage more of the processor's instruction execution electronics.

msync: An OS or file system call that uses flush and fence to ensure that data reaches PM.

Network Interface Card (NIC): A PCIe card that connects to a network.

non-cacheable: Data that is prohibited from caching due to a processor or other system configuration setting, usually describing a memory address range.

Non-Uniform Memory Access (NUMA): The condition, in a system, where some memory accesses take longer than others. When a processor accesses local DRAM, accesses generally have relatively uniform latency. Access to non-local memory attached to other processors generally takes longer even though the location of the memory is hidden from the application. This difference in access latency causes the NUMA condition.

Non-Volatile Memory (NVM): A broad category of storage and memory technologies that includes persistent memory but is often used to also refer to SSD's and HDD's as in "NVMe."

Non-Volatile Memory Express (NVMe): A storage protocol layered on PCIe that uses IO to access SSD's and HDD's.

Non-Volatile RAM: See Persistent Memory.

opaque pointer: A pointer whose bits have some structure as defined by the creator but cannot be parsed by the user.

overprovisioning: The deployment of excess resources or features that are infrequently used.

page fault: In virtual memory systems, a page fault is an event indicating an attempt to access data that is not currently in memory. Page faults are expected after memory contents are moved to storage in order to reuse physical memory for ongoing accesses when the virtual address space does not fit in physical memory. The typical response to a page fault is to execute one or more SSD or HDD IO's to get the desired data in memory.

Peripheral Component Interconnect Express (PCIe): The current pervasive standard for connecting processors to interface cards including network adapters, storage, storage adapters, GPU's, and other peripherals. Earlier versions omitted the "express."

Persistent Memory (PM): A category of technologies that provide non-volatile data retention (persistence) at near memory access times. Battery backed RAM is a type of persistent memory. Emergent persistent memory technologies include Phase Change Memory (PCM), Spin Torque Transfer (STT), and Resistive RAM (ReRAM).

polling: The repeated reading of a status or state indicator or memory location in order to detect a change.

power hold time: The interval between the detection of the loss of a system's external power and halting of the system due to internal power exhaustion. Internal power may be stored using capacitance and/or batteries. Power hold time depends on the rate of power consumption, which, in turn, depends on how much work is demanded of whatever part of the system remains operational after external power loss. Generally any power generation capability is viewed as external.

Quality of Service (QoS): Characteristics of a hardware or software function other than the function itself that relate to the fitness of the function for particular uses. QoS characteristics may relate to performance, robustness, safety, or any other aspect (other than cost) that may make the function implementation more suitable for some uses than others. Storage or memory capacity and network connectivity are generally excluded from QoS because they are quintessential to storage, memory, or network function.

RDMA Over Converged Ethernet (ROCE): A protocol for RDMA over Ethernet that layers parts of the IB protocol stack over parts of the Ethernet protocol stack. ROCE only uses the lowest layers of Ethernet protocol so that wires can be shared, whereas ROCE2 also uses the TCP/IP protocol.

Recovery Point Objective (RPO): In the context of high availability, a metric indicating the amount of work that may be lost in the event of a recoverable failure that involves the loss of recently updated data. RPO=0 implies no work loss other than transactions or updates that were not complete before the failure (see atomicity). Other values generally indicate a time interval or a number of transactions' worth of work that may be lost in the course of restoring data consistency after a system fails.

Remote Direct Memory Access (RDMA): A type of DMA where data is copied through a network. Infiniband, iWarp, and ROCE protocols all enable RDMA.

scale out: A system architecture pattern wherein work is distributed across a large number of loosely coupled (i.e., networked) low power processing nodes as an alternative to tightly coupled (i.e., cache coherent) processors in a smaller number of high power nodes.

serialization-deserialization: Serialization converts the state of an object into a formatted sequence of bytes which can be stored or transmitted over a network. De-serialization is the reverse process of reconstructing the object from that byte sequence.

single point of failure: Any part (however large or small) of a system component that, by failing, can single-handedly stop the system from operating. Avoiding single points of failure is viewed as a standard of HA excellence for mission critical systems which requires redundancy to be separated across groups of components that are very unlikely to fail at the same time.

smart contracts: Scripting functionality for Ethereum.

Software Defined Network (SDN): An architecture for designing, building, and managing networks that separates the network's control and forwarding data planes.

Solid State Disk (SSD): A block structured storage device that uses flash or other non-rotating storage technology.

Storage Class Memory (SCM): A type of persistent memory.

storage pool: A grouping of storage or memory resources that can be drawn upon to create virtual storage or memory resources. In SNIA swordfish, storage pools represent groups of resources that can provide a given class of service. In Swordfish, storage or memory resources are registered as sources of capacity for storage pools to enable their use.

store instruction (ST): A processor instruction that causes a small amount (up to a cache line) of data to be sent from a processor register to main memory. The data may take time to reach memory after the instruction. Depending on the configuration and implementation of the processor, data may reach memory in a different order than that implied by a sequence of store instructions.

Stream Control Transmission Protocol (SCTP): A network protocol that operates at the transport layer and is message-oriented like UDP but ensures reliable, in-sequence transport of messages with congestion control like TCP. It additionally provides multi-homing and redundant paths to increase resilience and reliability.

Swordfish: A SNIA standard [360] which extends the RedFish system management standard to include storage services with classes of service and storage pools.

TensorFlow: An open source machine learning platform especially suited for deep learning, often used to build neural networks for a variety of applications.

Term Frequency - Inverse Document Frequency (TF-IDF): An information retrieval metric used in text mining that reflects how well a word (term) characterizes or matches a given document relative to an entire corpus of documents. If a term occurs very frequently in one document and very rarely in other documents, then it is a strong match.

tiering: Moving data between different storage tiers (e.g., classes of media with different cost vs. performance trade-offs) to maximize overall efficiency by keeping data in its most appropriate storage tier based on policies or other optimization considerations.

Total Cost of Ownership (TCO): A cost metric that takes into account all of the capital and operating expenses involved in deploying and running an application to provide a business service. This includes not only the cost of software, equipment, power, and real estate but also the human workers who deploy, operate, and administer systems and applications.

transaction: A set of data changes that work together to transition an arbitrarily complex set of data from one state to another while never exposing intolerable inconsistency regardless of data sharing or power failure. This normally implies atomicity of the transition even if a number of disjoint data structures are involved. Transactions are often used when applications cannot tolerate any data inconsistency.

transfer learning: A machine learning approach where a model that was trained on one task is reused as a starting point for efficiently learning a model for another task.

Transmission Control Protocol/Internet Protocol (TCP/IP): The predominant data streaming and routing protocol for the internet.

Wide Area Network (WAN): A network that connects LANs across cities, countries, or continents, generally forming the internet.

write amplification: Overhead, usually in the form of extra writes over and above those demanded directly by an application to deliver its user data to memory or storage media. Features such as wear leveling, atomicity, and higher levels of recoverability may cause write amplification.

write through: Data that is immediately flushed from a cache after update due to a configuration setting, instruction implementation, or system state condition. The connotation is different from non-cacheable in that data does not necessarily bypass the cache and triggers for write-through may be more dynamic.

AI: Artificial Intelligence

AOP: Aspect Oriented Programming

AS: Address Space

ASIC: Application Specific Integrated Circuit

BER: Bit Error Rate

CDF: Cumulative Distribution Function

CI/CD: Continuous Integration/Continuous Deployment

CMOS: Complementary Metal-Oxide-Semiconductor

CNN: Convolutional neural networks

COW: Copy-on-Write

CPI: Cycles per Instruction

DAG: Directed Acyclic Graph

DDR: Double Data Rate

DIMM: Dual Inline Memory Module

DL: Deep Learning

DMA: Direct Memory Access

DNN: Deep neural networks

DRAM: Dynamic Random Access Memory

DSL: Domain Specific Language

DVFS: Dynamic Voltage and Frequency Scaling

EBR: Epoch Based Reclamation

ESS: Evolutionarily Stable Strategy

ETL: Extract Transform Load

GPU: Graphics Processing Unit

GPGPU: General Purpose Graphics Processing Unit

HA: High Availability

HBA: Host Bus Adapter

HDD: Hard Disk Drive

HDFS: Hadoop Distributed File System

HPC: High Performance Computing

IB: InfiniBand

IDE: Integrated Development Environment

IO: Input/Output

IOR: Interoperable Object Reference

IoT: Internet of Things

ISA: Industry Standard Architecture

JIF: Java Information Flow

JIT optimizer: Just-in-time optimizer

JRE: Java Runtime Environment

JVM: Java Virtual Machine

LD: load instruction

LAN: Local area Network

LLVM: Low Level Virtual Machine

LR: Logistic Regression

MDD: Model Driven Development

MI: Memory Interconnect

ML: Machine Learning

MM: Memory Mapped

MMU: Memory Management Unit

MOV: move instruction

NFV: Network Function Virtualization

NIC: Network Interface Card

NLP: Natural Language Processing

NPU: Neural Processing Unit

NUMA: Non-Uniform Memory Access

NVM: Non-Volatile Memory

NVMe: Non-Volatile Memory Express

NVRAM: Non-Volatile RAM

OS: Operating System

PCIe: Peripheral Component Interconnect Express

PIM: Processing In Memory

PM: Persistent Memory

QoS: Quality of Service

RAN: Radio Access Network

RDD: Resilient Distributed Datasets

RDMA: Remote Direct Memory Access

RNN: Recurrent neural networks

ROCE: RDMA Over Converged Ethernet

RPM: Remote Persistent Memory

RTTI: Run Time Type Information

SCM: Storage Class Memory

SCTP: Stream Control Transmission Protocol

SDN: Software Defined Network

SGX: Software Guard Extensions

SIP: Software Isolated Process

SMP: Symmetric Multiprocessing

SOAP: Simple Object Access Protocol

SSD: Solid State Disk

ST: store instruction

SVM: Support Vector Machine

TCAM: Ternary Content Addressable Memory

TCO: Total Cost of Ownership

TCP/IP: Transmission Control Protocol/Internet Protocol

TF-IDF: Term Frequency - Inverse Document Frequency

TLB: Translation Look-aside Buffer (a virtual to physical address translation cache)

TPU: Tensor Processing Unit

VLIW: Very Long Instruction Word

VM: Based on context, Virtual Machine or Virtual Memory

WAN: Wide Area Network

WCC: Weakly Connected Components

XML: eXtensible Markup Language

Bibliography

[1] Classical alternative to quantum recommendation algorithm. https://www.quantamagazine.org/teenager-finds-classical-alternative-to-quantum-recommendation-algorithm-20180731/.

[2] Deep learning made easier with transfer learning. https://blog.fastforwardlabs.com/2018/09/17/deep-learning-is-easy-an-introduction-to-transfer-learning.html.

[3] Ethereum. https://www.ethereum.org/.

[4] Kubernetes. `http://kubernetes.io`.

[5] Kubernetes. https://kubernetes.io/.

[6] Mlir: a new intermediate representation and compiler framework. https://github.com/tensorflow/mlir/.

[7] Nvm express. https://nvmexpress.org/.

[8] Rdma consortium. http://www.rdmaconsortium.org/home.

[9] Rust language. https://www.rust-lang.org/.

[10] Serverless computing. https://aws.amazon.com/serverless/.

[11] Snow cooling experiment in Japan melts away hvac costs. https://www.datacenterdynamics.com/news/snow-cooling-experiment-in-japan-melts-away-hvac-costs.

[12] Soot: a Java Optimization Framework.

[13] Training ai for self-driving vehicles: the challenge of scale. https://devblogs.nvidia.com/training-self-driving-vehicles-challenge-scale/.

[14] Ucla-visual. https://visual.ee.ucla.edu.

[15] Via architecture. http://www.viaarch.org.

[16] Why Microsoft thinks underwater data centers may cost less. https://www.datacenterknowledge.com/microsoft/why-microsoft-thinks-underwater-data-centers-may-cost-less.

[17] Zeromq. http://zeromq.org/.

[18] Zipkin. https://zipkin.io/.

[19] Atul Adya, Jon Howell, Marvin Theimer, William J. Bolosky, and John R. Douceur. Cooperative task management without manual stack management. In *Proceedings of the General Track of the Annual Conference on USENIX Annual Technical Conference*, ATEC '02, pages 289–302, Berkeley, CA, USA, 2002. USENIX Association.

[20] Neha Agarwal, David Nellans, Mike O'Connor, Stephen W. Keckler, and Thomas Wenisch. Unlocking bandwidth for gpus in cc-numa systems. *2015 IEEE 21st International Symposium on High Performance Computer Architecture, HPCA 2015*, pages 354–365, 03 2015.

[21] Sameer Agarwal, Barzan Mozafari, Aurojit Panda, Henry Milner, Samuel Madden, and Ion Stoica. Blinkdb: Queries with bounded errors and bounded response times on very large data. In *Proceedings of EuroSys '13*, 2013.

[22] Ajay Agrawal, Joshua Gans, and Avi Goldfarb. *Prediction Machines: The Simple Economics of Artificial Intelligence.* Harvard Business Review Press, USA, 2018.

[23] Khajonpong Akkarajitsakul, Ekram Hossain, and Dusit Niyato. Coalition-based cooperative packet delivery under uncertainty: A dynamic bayesian coalitional game. *IEEE Trans. Mob. Comput.*, 12(2):371–385, 2013.

[24] B. Aktemur and S. Kamin. A comparative study of techniques to write customizable libraries. In *Proceedings of the ACM Symposium on Applied Computing, SAC*, 2009.

[25] Armin Alaghi, Weikang Qian, and John P. Hayes. The promise and challenge of stochastic computing. *IEEE Trans. on CAD of Integrated Circuits and Systems*, 37(8):1515–1531, 2018.

[26] Hanna Alam, Tianhao Zhang, Mattan Erez, and Yoav Etsion. Do-it-yourself virtual memory translation. *SIGARCH Comput. Archit. News*, 45(2):457–468, June 2017.

[27] Dan Alistarh, William Leiserson, Alexander Matveev, and Nir Shavit. Threadscan: Automatic and scalable memory reclamation. *ACM Trans. Parallel Comput.*, 4(4):18:1–18:18, May 2018.

[28] Alluxio. Open source memory speed virtual distributed storage. http://www.alluxio.org/.

[29] Kalev Alpernas, Cormac Flanagan, Sadjad Fouladi, Leonid Ryzhyk, Mooly Sagiv, Thomas Schmitz, and Keith Winstein. Secure serverless computing using dynamic information flow control. *Proc. ACM Program. Lang.*, 2(OOPSLA):118:1–118:26, October 2018.

[30] Erik Altman, Matthew Arnold, Stephen Fink, and Nick Mitchell. Performance analysis of idle programs. In *Proceedings of the ACM International Conference on Object Oriented Programming Systems Languages and Applications*, OOPSLA '10, pages 739–753, New York, NY, USA, 2010. ACM.

[31] Amazon. What are microservices. `https://aws.amazon.com/microservices/`.

[32] Kyle Anderson. Layers: A historic and futuristic look at the abstractions we build and destroy. https://www.socallinuxexpo.org/scale/17x/presentations/layers-historic-and-futuristic-look-abstractions-we-build-and-destroy, 2019.

[33] Thomas E. Anderson, Brian N. Bershad, Edward D. Lazowska, and Henry M. Levy. Scheduler activations: Effective kernel support for the user-level management of parallelism. In *Proceedings of the Thirteenth ACM Symposium on Operating Systems Principles*, SOSP '91, pages 95–109, New York, NY, USA, 1991. ACM.

[34] Sven Apel and Christian Kästner. An overview of feature-oriented software development. In *Journal of Object Technology*, 2009.

[35] Robert S. Arnold. *Software Change Impact Analysis*. IEEE Computer Society Press, Los Alamitos, CA, USA, 1996.

[36] Sanjeev Arora, Carsten Lund, Rajeev Motwani, Madhu Sudan, and Mario Szegedy. Proof verification and the hardness of approximation problems. *J. ACM*, 45(3):501–555, May 1998.

[37] K. Arpith. IO pattern aware methods to improve the performance and lifetime of NAND SSD. Masters Thesis, Indian Institute of Science, Bangalore, 2018.

[38] Kenneth Arrow. A difficulty in the concept of social welfare. *Journal of Political Economy*, 58, 1950.

[39] Frank Arute et al., Quantum supremacy using a programmable superconducting processor. *Nature*, Vol. 574, 24 Oct 2019.

[40] Krste Asanovic. Firebox: A hardware building block for 2020 warehouse-scale computers. `https://www.usenix.org/node/179918`.

[41] C. Ashokkumar, Bholanath Roy, M. Bhargav, Sri Venkatesh, and Bernard L. Menezes. S-box implementation of aes is not side channel resistant. https://eprit.iacr.org/2018/1002.pdf.

[42] Manos Athanassoulis, Michael S. Kester, Lukas M. Maas, Radu Stoica, Stratos Idreos, Anastasia Ailamaki, and Mark Callaghan. Designing access methods: The rum conjecture. *EDBT 2016*, 2016.

[43] G. Ayers, J. H. Ahn, C. Kozyrakis, and P. Ranganathan. Memory hierarchy for web search. In *2018 IEEE International Symposium on High Performance Computer Architecture (HPCA)*, pages 643–656, Feb 2018.

[44] Adam Back. Hashcash - a denial of service counter-measure. Technical report, 2002.

[45] Woongki Baek and Trishul M. Chilimbi. Green: a framework for supporting energy-conscious programming using controlled approximation. In *Proceedings of the 2010 ACM SIGPLAN Conference on Programming Language Design and Implementation*, PLDI '10, pages 198–209, New York, NY, USA, 2010. ACM.

[46] Pierre F. Baldi, Cristina V. Lopes, Erik J. Linstead, and Sushil K. Bajracharya. A theory of aspects as latent topics. In *Proceedings of the ACM SIGPLAN Conference on Object-Oriented Programming, Systems, Languages and Applications, OOPSLA*, 2008.

[47] Ioana Baldini, Paul Castro, Kerry Chang, Perry Cheng, Stephen Fink, Vatche Ishakian, Nick Mitchell, Vinod Muthusamy, Rodric Rabbah, Aleksander Slominski, and Philippe Suter. *Serverless Computing: Current Trends and Open Problems*, pages 1–20. 01 2017.

[48] Ioana Baldini, Perry Cheng, Stephen J. Fink, Nick Mitchell, Vinod Muthusamy, Rodric Rabbah, Philippe Suter, and Olivier Tardieu. The serverless trilemma: Function composition for serverless computing. In *Proceedings of the 2017 ACM SIGPLAN International Symposium on New Ideas, New Paradigms, and Reflections on Programming and Software*, Onward! 2017, pages 89–103, New York, NY, USA, 2017. ACM.

[49] L.A. Barroso, U. Hölzle, and P. Ranganathan. *The Datacenter as a Computer: Designing Warehouse-Scale Machines, Third Edition*. Synthesis Lectures on Computer Architecture. Morgan & Claypool Publishers, 2018.

[50] Luiz Barroso, Mike Marty, David Patterson, and Parthasarathy Ranganathan. Attack of the killer microseconds. *Commun. ACM*, 60(4):48–54, March 2017.

[51] Luiz André Barroso and Urs Hölzle. The case for energy-proportional computing. *Computer*, 40(12):33–37, 2007.

[52] Arkaprava Basu, Jayneel Gandhi, Jichuan Chang, Mark D. Hill, and Michael M. Swift. Efficient virtual memory for big memory servers. In *Proceedings of the 40th Annual International Symposium on Computer Architecture*, ISCA '13, pages 237–248, New York, NY, USA, 2013. ACM.

[53] B. N. Bershad, S. Savage, P. Pardyak, E. G. Sirer, M. E. Fiuczynski, D. Becker, C. Chambers, and S. Eggers. Extensibility safety and performance in the spin operating system. In *Proceedings of the Fifteenth ACM Symposium on Operating Systems Principles*, SOSP '95, pages 267–283, New York, NY, USA, 1995. ACM.

[54] Jeff Bezanson, Alan Edelman, Stefan Karpinski, and Viral B. Shah. Julia: A fresh approach to numerical computing. *CoRR*, abs/1411.1607, 2014.

[55] Jeff Bezanson, Stefan Karpinski, Viral B. Shah, and Alan Edelman. Julia: A fast dynamic language for technical computing. *CoRR*, abs/1209.5145, 2012.

[56] Kumud Bhandari, Dhruva Chakrabarti, and Hans R. Boehm. Implications of cpu caching on byte-addressable non-volatile memory programming. *HP Labs Technical Reports*, 2012.

[57] Madhumita Bharde, Annmary K Justine, Suparna Bhattacharya, and Dileep Deepa Shree. Store-edge ripplestream: Versatile infrastructure for iot data transfer. In *USENIX Workshop on Hot Topics in Edge Computing (HotEdge 18)*, Boston, MA, 2018. USENIX Association.

[58] Manish Bhardwaj, Rex Min, and Anantha P. Chandrakasan. Quantifying and enhancing power awareness of vlsi systems. *IEEE Transactions on VLSI Systems*, 9(6):757–772, 2001.

[59] Koustubha Bhat, Erik van der Kouwe, Herbert Bos, and Cristiano Giuffrida. Probeguard: Mitigating probing attacks through reactive program transformations. In *Proceedings of the Twenty-Fourth International Conference on Architectural Support for Programming Languages and Operating Systems*, ASPLOS '19, pages 545–558, New York, NY, USA, 2019. ACM.

[60] Eshan Bhatia, Gino Chacon, Seth Pugsley, Elvira Teran, Paul V. Gratz, and Daniel A. Jiménez. Perceptron-based prefetch filtering. In *Proceedings of the 46th International Symposium on Computer Architecture*, ISCA '19, pages 1–13, New York, NY, USA, 2019. ACM.

[61] S. Bhattacharya, K. Gopinath, K. Rajamani, and M. Gupta. Software bloat and wasted joules: Is modularity a hurdle to green software? *Computer*, 44(9):97–101, Sep. 2011.

[62] Suparna Bhattacharya. Linux and the art of minimalist development. In *FOSS.IN*, 2006.

[63] Suparna Bhattacharya. *A Systems Perspective of Software Runtime Bloat – Origin, Mitigation and Power-Performance Implications*. PhD thesis, IISc, 2013.

[64] Suparna Bhattacharya and K. Gopinath. Virtually Cool Ternary Content Addressable Memory. *Proceedings of the 13th USENIX Conference on Hot Topics in Operating Systems, HotOS*, 2011.

[65] Suparna Bhattacharya, Kanchi Gopinath, and Mangala Gowri Nanda. Combining concern input with program analysis for bloat detection. In *Proceedings of the 2013 ACM SIGPLAN International Conference on Object Oriented Programming Systems Languages and Applications*, OOPSLA '13, pages 745–764, New York, NY, USA, 2013. ACM.

[66] Suparna Bhattacharya, Mangala Gowri Nanda, K. Gopinath, and Manish Gupta. Reuse recycle to debloat software. In *European Conference on Object-Oriented Programming, ECOOP*, 2011.

[67] Suparna Bhattacharya, Karthick Rajamani, K. Gopinath, and Manish Gupta. The interplay of software bloat, hardware energy proportionality and system bottlenecks. In *Proceedings of the 4th Workshop on Power-Aware Computing and Systems, HotPower'11*, 2011.

[68] Suparna Bhattacharya, Karthick Rajamani, K. Gopinath, and Manish Gupta. Does lean imply green? A study of the power-performance implications of java runtime bloat. In *Proceedings of the 12th ACM SIGMETRICS/PERFORMANCE Joint International Conference on Measurement and Modeling of Computer Systems, SIGMETRICS/Performance*, pages 259–270, June 2012.

[69] Ted J. Biggerstaff, Bharat G. Mitbander, and Dallas Webster. The concept assignment problem in program understanding. In *ICSE '93*, pages 482–498, Los Alamitos, CA, USA, 1993. IEEE Computer Society Press.

[70] Battista Biggio and Fabio Roli. Wild patterns: Ten years after the rise of adversarial machine learning. *Pattern Recognition*, 2018.

[71] Eli Biham and Adi Shamir. Differential cryptanalysis of des-like cryptosystems. In *Proceedings of the 10th Annual International Cryptology Conference on Advances in Cryptology*, CRYPTO '90, pages 2–21, Berlin, Heidelberg, 1991. Springer-Verlag.

[72] Kashif Bilal, Samee Khan, Limin Zhang, Hongxiang Li, Khizar Hayat, Sajjad Madani, Nasro Min Allah, Lizhe Wang, Dan Chen, Majid Iqbal Khan, Cheng-Zhong Xu, and Albert Zomaya. Quantitative comparisons of the state-of-the-art data center architectures. *Concurrency and Computation: Practice and Experience*, 25, 08 2013.

[73] D. Binkley, G. Gold, M. Harman, Z. Li, and K. Mahdavi. An empirical study of the relationship between the concepts expressed in source code and dependence. In *The Journal of Systems and Software*, volume 81, 2008.

[74] Manuel Blum. On effective procedures for speeding up algorithms. *J. ACM*, 18(2):290–305, April 1971.

[75] Kevin Boos, Emilio Del Vecchio, and Lin Zhong. A characterization of state spill in modern operating systems. In *Proceedings of the Twelfth European Conference on Computer Systems*, EuroSys '17, pages 389–404, New York, NY, USA, 2017. ACM.

[76] Kevin Boos and Lin Zhong. Theseus: A state spill-free operating system. In *Proceedings of the 9th Workshop on Programming Languages and Operating Systems*, PLOS'17, pages 29–35, New York, NY, USA, 2017. ACM.

[77] James Bornholt, Todd Mytkowicz, and Kathryn S. McKinley. Uncertain<*T*>: A first-order type for uncertain data. In *Proceedings of the 19th International Conference on Architectural Support for Programming Languages and Operating Systems*, ASPLOS '14, pages 51–66, New York, NY, USA, 2014. ACM.

[78] Amirali Boroumand, Saugata Ghose, Youngsok Kim, Rachata Ausavarungnirun, Eric Shiu, Rahul Thakur, Daehyun Kim, Aki Kuusela, Allan Knies, Parthasarathy Ranganathan, and Onur Mutlu. Google workloads for consumer devices: Mitigating data movement bottlenecks. In *Proceedings of the Twenty-Third International Conference on Architectural Support for Programming Languages and Operating Systems*, ASPLOS '18, pages 316–331, New York, NY, USA, 2018. ACM.

[79] Ferdinand Brasser, Lucas Davi, David Gens, Christopher Liebchen, and Ahmad-Reza Sadeghi. Can't touch this: Software-only mitigation against rowhammer attacks targeting kernel memory. In *26th USENIX Security Symposium (USENIX Security 17)*, pages 117–130, Vancouver, BC, 2017. USENIX Association.

[80] Eric Brewer, Lawrence Ying, Lawrence Greenfield, Robert Cypher, and Theodore Ts'o. Disks for data centers. *White paper for FAST 2016*, February 2016.

[81] Michael D. Brown and Santosh Pande. Is less really more? Why reducing code reuse gadget counts via software debloating doesn't necessarily lead to better security. *CoRR*, abs/1902.10880, 2019.

[82] Neil Brown. Linux kernel design patterns - part 2. *Linux Weekly News*, 2009.

[83] Neil Brown. Linux kernel design patterns - part 3. *Linux Weekly News*, 2009.

[84] Neil Brown. Object-oriented design patterns in the kernel, part 2. *Linux Weekly News*, 2011.

[85] Trevor Alexander Brown. Reclaiming memory for lock-free data structures: There has to be a better way. In *Proceedings of the 2015 ACM Symposium on Principles of Distributed Computing*, PODC '15, pages 261–270, New York, NY, USA, 2015. ACM.

[86] R. E. Bryant. Data-intensive scalable computing for scientific applications. *Computing in Science Engineering*, 13(6):25–33, Nov 2011.

[87] Yingyi Bu, Vinayak Borkar, Guoqing Xu, and Michael J. Carey. A bloat-aware design for big data applications. In *Proceedings of the 2013 International Symposium on Memory Management*, ISMM '13, pages 119–130, New York, NY, USA, 2013. ACM.

[88] Mark Buckler, Suren Jayasuriya, and Adrian Sampson. Reconfiguring the imaging pipeline for computer vision. *CoRR*, abs/1705.04352, 2017.

[89] Raymond P. L. Buse and Westley Weimer. The road not taken: Estimating path execution frequency statically. In *Proceedings of the 31st International Conference on Software Engineering*, ICSE '09, pages 144–154, Washington, DC, USA, 2009. IEEE Computer Society.

[90] Fabia Bustamante, Greg Eisenhauer, and Karsten Schwan. Native data representation: An efficient wire format for high-performance distributed computing. *IEEE Transactions on Parallel and Distributed Systems*, 13(12):1234–1246, 2002.

[91] Dries Buytaert, Kristof Beyls, and Koen De Bosschere. Hinting refactorings to reduce object creation in java. In *ACES*, pages 73–76, 1 2005.

[92] J. P. Buzen. Fundamental laws of computer system performance. In *Proceedings of the 1976 ACM SIGMETRICS Conference on Computer Performance Modeling Measurement and Evaluation, SIGMETRICS'76*, 1976.

[93] Benedikt Bünz, Jonathan Bootle, Dan Boneh, Andrew Poelstra, Pieter Wuille, and Greg Maxwell. Bulletproofs: Short proofs for confidential transactions and more. Cryptology ePrint Archive, Report 2017/1066, 2017. https://eprint.iacr.org/2017/1066/.

[94] Olu Campbell. The convergence of it and ot. https://dzone.com/articles/the-convergence-of-it-and-ot.

[95] Julie Chang, Vincent Sitzmann, Xiong Dun, Wolfgang Heidrich, and Gordon Wetzstein. Hybrid optical-electronic convolutional neural networks with optimized diffractive optics for image classification. *Scientific Reports, Nature Online*, 2018.

[96] Meng-Fan Chang, Pi-Feng Chiu, and Shyh-Shyuan Sheu. Circuit design challenges in embedded memory and resistive ram (rram) for mobile soc

and 3d-ic. In *Proceedings of the 16th Asia and South Pacific Design Automation Conference*, ASPDAC '11, pages 197–203, Piscataway, NJ, USA, 2011. IEEE Press.

[97] Swarat Chaudhuri, Sumit Gulwani, Roberto Lublinerman, and Sara Navidpour. Proving programs robust. In *Proceedings of the 19th ACM SIGSOFT Symposium and the 13th European Conference on Foundations of Software Engineering*, ESEC/FSE '11, pages 102–112, New York, NY, USA, 2011. ACM.

[98] Guoxing Chen, Sanchuan Chen, Yuan Xiao, Yinqian Zhang, Zhiqiang Lin, and Ten H. Lai. Sgxpectre attacks: Leaking enclave secrets via speculative execution. *CoRR*, abs/1802.09085, 2018.

[99] Qi Alfred Chen, Zhiyun Qian, and Z. Morley Mao. Peeking into your app without actually seeing it: UI state inference and novel android attacks. In *Proceedings of the 23rd USENIX Conference on Security Symposium*, SEC'14, pages 1037–1052, Berkeley, CA, USA, 2014. USENIX Association.

[100] Shuang Chen, Christina Delimitrou, and José F. Martínez. Parties: QOS-aware resource partitioning for multiple interactive services. In *Proceedings of the Twenty-Fourth International Conference on Architectural Support for Programming Languages and Operating Systems*, ASPLOS '19, pages 107–120, New York, NY, USA, 2019. ACM.

[101] Yurong Chen, Tian Lan, and Guru Venkataramani. Damgate: Dynamic adaptive multi-feature gating in program binaries. In *Proceedings of the 2017 Workshop on Forming an Ecosystem Around Software Transformation*, FEAST '17, pages 23–29, New York, NY, USA, 2017. ACM.

[102] Zhifei Chen, Bihuan Chen, Lu Xiao, Xiao Wang, Lin Chen, Yang Liu, and Baowen Xu. Speedoo: Prioritizing performance optimization opportunities. In *Proceedings of the 40th International Conference on Software Engineering*, ICSE '18, pages 811–821, New York, NY, USA, 2018. ACM.

[103] David Cheriton, Amin Firoozshahian, Alex Solomatnikov, John P. Stevenson, and Omid Azizi. Hicamp: Architectural support for efficient concurrency-safe shared structured data access. *SIGPLAN Not.*, 47(4):287–300, March 2012.

[104] M. Chiang, S.H. Low, A.R. Calderbank, and J.C. Doyle. Layering as optimization decomposition: A mathematical theory of network architectures. *Proceedings of the IEEE, 95, 1*, Jan 2007.

[105] Vinay Kumar Chippa, Debabrata Mohapatra, Kaushik Roy, Srimat T. Chakradhar, and Anand Raghunathan. Scalable effort hardware design.

*IEEE Transactions on Very Large Scale Integration (VLSI) Systems,
Volume: 22, Issue: 9, Sept. 2014.*

[106] Adriana E. Chis, Nick Mitchell, Edith Schonberg, Gary Sevitsky, Patrick
O'Sullivan, Trevor Parsons, and John Murphy. Patterns of memory
inefficiency. In *Proceedings of the 25th European Conference on Object-
Oriented Programming*, ECOOP'11, pages 383–407, Berlin, Heidelberg,
2011. Springer-Verlag.

[107] Jungsik Choi, Jiwon Kim, and Hwansoo Han. Efficient memory mapped
file i/o for in-memory file systems. In *9th USENIX Workshop on Hot
Topics in Storage and File Systems (HotStorage 17)*, Santa Clara, CA,
2017. USENIX Association.

[108] Ioannis Christou and Sofoklis Efremidis. To pool or not to pool? Revis-
iting an old pattern. 2018. https://arxiv.org/abs/1801.03763.

[109] Russell Clapp, Martin Dimitrov, Karthik Kumar, Vish Viswanathan,
and Thomas Willhalm. Quantifying the performance impact of memory
latency and bandwidth for big data workloads. In *2015 IEEE International Symposium on Workload Characterization*, 2015.

[110] GenZ Consortium. Gen-z core specification 1.0. 2018, https://
genzconsortium.org/specification/core-specification-1-0/.

[111] Jonathan Corbet. Supporting filesystems in persistent memory. *Linux
Weekly News*, 2014.

[112] Diego Costa and Artur Andrzejak. Collectionswitch: A framework for
efficient and dynamic collection selection. In *Proceedings of the 2018
International Symposium on Code Generation and Optimization*, CGO
2018, pages 16–26, New York, NY, USA, 2018. ACM.

[113] William J. Dally, James Balfour, David Black-Shaffer, James Chen,
R. Curtis Harting, Vishal Parikh, Jongsoo Park, and David Sheffield.
Efficient embedded computing. *IEEE Computer*, 41:27–32, 2008.

[114] Luis Damas and Robin Milner. Principal type-schemes for functional
programs. In *Proceedings of the 9th ACM SIGPLAN-SIGACT Symposium on Principles of Programming Languages*, POPL '82, pages 207–
212, New York, NY, USA, 1982. ACM.

[115] Ivan Damgård, Chaya Ganesh, and Claudio Orlandi. Proofs of replicated
storage without timing assumptions. Cryptology ePrint Archive, Report
2018/654, 2018. https://eprint.iacr.org/2018/654.

[116] Jeffrey Dean and Sanjay Ghemawat. Mapreduce: Simplified data processing on large clusters. In *Proceedings of the 6th Conference on Symposium on Operating Systems Design & Implementation - Volume 6*,
OSDI'04, pages 10–10, Berkeley, CA, USA, 2004. USENIX Association.

[117] Ulan Degenbaev, Jochen Eisinger, Manfred Ernst, Ross McIlroy, and Hannes Payer. Idle time garbage collection scheduling. In *Proceedings of the 37th ACM SIGPLAN Conference on Programming Language Design and Implementation*, PLDI '16, pages 570–583, New York, NY, USA, 2016. ACM.

[118] Brian DeHammer. Optimizing docker images. `https://www.ctl.io/developers/blog/post/optimizing-docker-images/`.

[119] Luca Della Toffola, Michael Pradel, and Thomas R. Gross. Performance problems you can fix: A dynamic analysis of memoization opportunities. *SIGPLAN Not.*, 50(10):607–622, October 2015.

[120] Dorothy E. Denning. Toward more secure software. *Commun. ACM*, 58(4):24–26, March 2015.

[121] Monica Dhok. *Automated Test Generation and Performance Improvement using Dynamic Program Analysis*. PhD thesis, IISc, 2018.

[122] Ruizhou Ding, Zeye Liu, R. D. (Shawn) Blanton, and Diana Marculescu. Lightening the load with highly accurate storage- and energy-efficient lightnns. *ACM Trans. Reconfigurable Technol. Syst.*, 11(3):17:1–17:24, December 2018.

[123] Bogdan Dit, Meghan Revelle, Malcom Gethers, and Denys Poshyvanyk. Feature location in source code: a taxonomy and survey. *Journal of Software: Evolution and Process*, 25(1):53–95, 2013.

[124] Cuong T. Do, Nguyen H. Tran, Choongseon Hong, Charles A. Kamhoua, Kevin A. Kwiat, Erik Blasch, Shaolei Ren, Niki Pissinou, and Sundaraja Sitharama Iyengar. Game theory for cyber security and privacy. *ACM Comput. Surv.*, 50(2):30:1–30:37, May 2017.

[125] Pedro Domingos. A few useful things to know about machine learning. *Commun. ACM*, 55(10):78–87, October 2012.

[126] Pedro Domingos. *The Master Algorithm: How the Quest for the Ultimate Learning Machine Will Remake Our World*. Basic Books, Inc., New York, NY, USA, 2018.

[127] Chris Dragga and Douglas J. Santry. GCTrees: Garbage collecting snapshots. *ACM Trans. Storage*, 12(1):4:1–4:32, January 2016.

[128] Bruno Dufour, Barbara G. Ryder, and Gary Sevitsky. A scalable technique for characterizing the usage of temporaries in framework-intensive java applications. In *SIGSOFT '08/FSE-16*, pages 59–70, 2008.

[129] Dominic Duggan and Frederick Bent. Explaining type inference. *Sci. Comput. Program.*, 27(1):37–83, July 1996.

[130] Cynthia Dwork and Aaron Roth. The algorithmic foundations of differential privacy. *Found. Trends Theor. Comput. Sci.*, 9, 3–4 (August 2014), 211–407.

[131] David E. Lowell and Peter M. Chen. Free transactions with rio vista. volume 31, pages 92–101, 12 1997.

[132] M. Eaddy, A. V. Aho, G. Antoniol, and Y. G. Guéhéneuc. Cerberus: Tracing requirements to source code using information retrieval, dynamic analysis, and program analysis. In *Proceedings of the IEEE International Conference on Program Comprehension, ICPC*, pages 53–62, 2008.

[133] Justin Ellingwood. Building optimized containers for kubernetes. https://www.digitalocean.com/community/tutorials/building-optimized-containers-for-kubernetes.

[134] Stanford Engineering. https://engineering.stanford.edu/magazine/artic le/new-ai-camera-recognizes-objects-faster-and-more-efficiently?utm_so urce=SoEnewsletter&utm_medium=email&utm_campaign=1aSep2018 &utm_content=recognize081718.

[135] Hadi Esmaeilzadeh, Adrian Sampson, Luis Ceze, and Doug Burger. Architecture support for disciplined approximate programming. In *Proceedings of the Seventeenth International Conference on Architectural Support for Programming Languages and Operating Systems*, ASPLOS '12, pages 301–312, New York, NY, USA, 2012. ACM.

[136] Hadi Esmaeilzadeh, Adrian Sampson, Luis Ceze, and Doug Burger. Neural acceleration for general-purpose approximate programs. In *Proceedings of the 2012 45th Annual IEEE/ACM International Symposium on Microarchitecture*, MICRO-45, pages 449–460, Washington, DC, USA, 2012. IEEE Computer Society.

[137] Shacham et al. Chameleon: Adaptive selection of collections. In *Programming Language Design and Implementation (PLDI)*.

[138] Facebook. Facebook outage 2010. https://www.facebook.com/notes/facebook-engineering/more-details-on-todays-outage/431441338919/.

[139] Paolo Faraboschi, Kimberly Keeton, Tim Marsland, and Dejan Milojicic. Beyond processor-centric operating systems. In *15th Workshop on Hot Topics in Operating Systems (HotOS XV)*, Kartause Ittingen, Switzerland, 2015. USENIX Association.

[140] Alan Fekete, Nancy Lynch, Yishay Mansour, and John Spinelli. The impossibility of implementing reliable communication in the face of crashes. *J. ACM*, 40(5):1087–1107, November 1993.

[141] Antonio Filieri, Henry Hoffmann, and Martina Maggio. Automated design of self-adaptive software with control-theoretical formal guarantees. In *Proceedings of the 36th International Conference on Software Engineering*, ICSE 2014, pages 299–310, New York, NY, USA, 2014. ACM.

[142] Antonio Filieri, Martina Maggio, Konstantinos Angelopoulos, Nicolás D'ippolito, Ilias Gerostathopoulos, Andreas Berndt Hempel, Henry Hoffmann, Pooyan Jamshidi, Evangelia Kalyvianaki, Cristian Klein, Filip Krikava, Sasa Misailovic, Alessandro V. Papadopoulos, Suprio Ray, Amir M. Sharifloo, Stepan Shevtsov, Mateusz Ujma, and Thomas Vogel. Control strategies for self-adaptive software systems. *ACM Trans. Auton. Adapt. Syst.*, 11(4):24:1–24:31, February 2017.

[143] Michael J. Fischer, Nancy A. Lynch, and Michael S. Paterson. Impossibility of distributed consensus with one faulty process. *J. ACM*, 32(2):374–382, April 1985.

[144] Cormac Flanagan and Stephen N. Freund. Fasttrack: Efficient and precise dynamic race detection. *SIGPLAN Not.*, 44(6):121–133, June 2009.

[145] Sally Floyd and Van Jacobson. Random early detection gateways for congestion avoidance. *IEEE/ACM Trans. Netw.*, 1(4):397–413, August 1993.

[146] Apache Software Foundation. Hdfs erasure coding. `https://hadoop.apache.org/docs/r3.0.0/hadoop-project-dist/hadoop-hdfs/HDFSErasureCoding.html`.

[147] Alan Freedman. Definition of Software Bloat, Author's note.

[148] João Gama, Indrė Žliobaitė, Albert Bifet, Mykola Pechenizkiy, and Abdelhamid Bouchachia. A survey on concept drift adaptation. *ACM Comput. Surv.*, 46(4):44:1–44:37, March 2014.

[149] Erich Gamma, Richard Helm, Ralph Johnson, and John Vlissides. *Design Patterns: Elements of Reusable Object-oriented Software*. Addison-Wesley Longman Publishing Co., Inc., Boston, MA, USA, 1995.

[150] A. S. Ganapathi. *Predicting and Optimizing System Utilization and Performance via Statistical Machine Learning*. PhD thesis, EECS Department, University of California, Berkeley, Dec – 2009.

[151] Hector Garcia-Molina and Kenneth Salem. Sagas. *SIGMOD Rec.*, 16(3):249–259, December 1987.

[152] Jayesh Gaur, Mainak Chaudhuri, Pradeep Ramachandran, and Sreenivas Subramoney. Near-optimal access partitioning for memory hierarchies with multiple heterogeneous bandwidth sources. *HPCA2017*.

[153] Malcom Gethers, Bogdan Dit, Huzefa Kagdi, and Denys Poshyvanyk. Integrated impact analysis for managing software changes. In *ICSE*, 2012.

[154] Jongmin Gim, Taeho Hwang, Youjip Won, and Krishna Kant. Quantifying the cost of context switch. *ACM Transactions on Storage*, 11(2), 2015.

[155] Giraph. Apache giraph: Open source project. http://giraph.apache.org/.

[156] Cristiano Giuffrida, Anton Kuijsten, and Andrew S. Tanenbaum. Safe and automatic live update for operating systems. In *Proceedings of the Eighteenth International Conference on Architectural Support for Programming Languages and Operating Systems*, ASPLOS '13, pages 279–292, New York, NY, USA, 2013. ACM.

[157] Inigo Goiri, Ricardo Bianchini, Santosh Nagarakatte, and Thu D. Nguyen. Approxhadoop: Bringing approximations to MapReduce frameworks. In *Proceedings of ASPLOS '15*, 2015.

[158] O. Goldreich. Towards a theory of software protection and simulation by oblivious rams. In *Proceedings of the Nineteenth Annual ACM Symposium on Theory of Computing*, STOC '87, pages 182–194, New York, NY, USA, 1987. ACM.

[159] Ian Goodfellow, Yoshua Bengio, and Aaron Courville. *Deep Learning*. The MIT Press, 2016.

[160] K. Gopinath. Static program analysis for security. *The Compiler Design Handbook: Optimizations and Machine Code Generation*, 2007.

[161] K. Gopinath, Nitin Muppalaneni, N. Suresh Kumar, and Pankaj Risbood. A 3-tier raid storage system with raid1, raid5 and compressed raid5 for linux. In *Proceedings of the Annual Conference on USENIX Annual Technical Conference*, ATEC '00, pages 30–30, Berkeley, CA, USA, 2000. USENIX Association.

[162] Andrew D. Gordon, Thomas A. Henzinger, Aditya V. Nori, and Sriram K. Rajamani. Probabilistic programming. In *Proceedings of the Future of Software Engineering*, FOSE 2014, pages 167–181, New York, NY, USA, 2014. ACM.

[163] Vivek Goyal, Eric W. Biederman, and Hariprasad Nellitheertha. Kdump, a kexec-based kernel crash dumping mechanism. In *Ottawa Linux Symposium*, 2005.

[164] Scott R. Graham, Girish Baliga, and P. R. Kumar. Abstractions, architecture, mechanisms, and a middleware for networked control. *IEEE Trans. Automat. Contr.*, 54(7):1490–1503, 2009.

[165] Jim Gray and Andreas Reuter. *Transaction Processing: Concepts and Techniques*. Morgan Kaufmann Publishers Inc., San Francisco, CA, USA, 1st edition, 1992.

[166] Brendan Gregg and Jim Mauro. *DTrace: Dynamic Tracing in Oracle Solaris, Mac OS X and FreeBSD*. Prentice Hall Press, Upper Saddle River, NJ, USA, 1st edition, 2011.

[167] Beayna Grigorian and Glenn Reinman. Accelerating divergent applications on simd architectures using neural networks. *ACM Trans. Archit. Code Optim.*, 12(1):2:1–2:23, March 2015.

[168] Sumit Gulwani, Krishna K. Mehra, and Trishul Chilimbi. Speed: Precise and efficient static estimation of program computational complexity. *SIGPLAN Not.*, 44(1):127–139, January 2009.

[169] Pradeep Kumar Gunda et al. Nectar: Automatic management of data and computation in datacenters. In *Proceedings of the 9th USENIX Conference on Operating Systems Design and Implementation*, OSDI'10, 2010.

[170] Peizhen Guo, Bo Hu, Rui Li, and Wenjun Hu. Foggycache: Cross-device approximate computation reuse. In *Proceedings of the 24th Annual International Conference on Mobile Computing and Networking*, Mobi-Com '18, pages 19–34, New York, NY, USA, 2018. ACM.

[171] Peizhen Guo and Wenjun Hu. Potluck: Cross-application approximate deduplication for computation-intensive mobile applications. In *Proceedings of the Twenty-Third International Conference on Architectural Support for Programming Languages and Operating Systems*, ASPLOS '18, pages 271–284, New York, NY, USA, 2018. ACM.

[172] Qing Guo, Xiaochen Guo, Yuxin Bai, and Engin İpek. A resistive tcam accelerator for data-intensive computing. In *Proceedings of the 44th Annual IEEE/ACM International Symposium on Microarchitecture*, MICRO-44 '11, pages 339–350, New York, NY, USA, 2011. ACM.

[173] Rehan Hameed, Wajahat Qadeer, Megan Wachs, Omid Azizi, Alex Solomatnikov, Benjamin C. Lee, Stephen Richardson, Christos Kozyrakis, and Mark Horowitz. Understanding sources of inefficiency in general-purpose chips. In *Proceedings of the 37th Annual International Symposium on Computer Architecture*, ISCA '10, pages 37–47, New York, NY, USA, 2010. ACM.

[174] Kevin W. Hamlen, Greg Morrisett, and Fred B. Schneider. Computability classes for enforcement mechanisms. *ACM Trans. Program. Lang. Syst.*, 28(1):175–205, January 2006.

[175] Jiawei Han, Micheline Kamber, and Jian Pei. *Data Mining: Concepts and Techniques*. Morgan Kaufmann Publishers Inc., San Francisco, CA, USA, 3rd edition, 2011.

[176] Shi Han, Yingnong Dang, Song Ge, Dongmei Zhang, and Tao Xie. Performance debugging in the large via mining millions of stack traces. In *Proceedings of the 34th International Conference on Software Engineering*, ICSE '12, pages 145–155, Piscataway, NJ, USA, 2012. IEEE Press.

[177] Tyler Harter, Brandon Salmon, Rose Liu, Andrea C. Arpaci-Dusseau, and Remzi H. Arpaci-Dusseau. Slacker: Fast distribution with lazy docker containers. In *14th USENIX Conference on File and Storage Technologies (FAST 16)*, pages 181–195, Santa Clara, CA, 2016. USENIX Association.

[178] Kim M. Hazelwood, Sarah Bird, David M. Brooks, Soumith Chintala, Utku Diril, Dmytro Dzhulgakov, Mohamed Fawzy, Bill Jia, Yangqing Jia, Aditya Kalro, James Law, Kevin Lee, Jason Lu, Pieter Noordhuis, Misha Smelyanskiy, Liang Xiong, and Xiaodong Wang. Applied machine learning at Facebook: A datacenter infrastructure perspective. In *IEEE International Symposium on High Performance Computer Architecture, HPCA 2018, Vienna, Austria, February 24-28, 2018*, pages 620–629, 2018.

[179] Vincent J. Hellendoorn, Christian Bird, Earl T. Barr, and Miltiadis Allamanis. Deep learning type inference. In *Proceedings of the 2018 26th ACM Joint Meeting on European Software Engineering Conference and Symposium on the Foundations of Software Engineering*, ESEC/FSE 2018, pages 152–162, New York, NY, USA, 2018. ACM.

[180] David P. Helmbold, Darrell D. E. Long, Tracey L. Sconyers, and Bruce Sherrod. Adaptive disk spin-down for mobile computers. *Mob. Netw. Appl.*, 5(4):285–297, December 2000.

[181] Michi Henning. The rise and fall of corba. *Queue*, 4(5):28–34, June 2006.

[182] Kihong Heo, Woosuk Lee, Pardis Pashakhanloo, and Mayur Naik. Effective program debloating via reinforcement learning. In *Proceedings of the 2018 ACM SIGSAC Conference on Computer and Communications Security*, CCS '18, pages 380–394, New York, NY, USA, 2018. ACM.

[183] Kihong Heo, Woosuk Lee, Pardis Pashakhanloo, and Mayur Naik. Effective program debloating via reinforcement learning. In *Proceedings of the 2018 ACM SIGSAC Conference on Computer and Communications Security*, CCS '18, pages 380–394, New York, NY, USA, 2018. ACM.

[184] Nathaniel Herman, Jeevana Priya Inala, Yihe Huang, Lillian Tsai, Eddie Kohler, Barbara Liskov, and Liuba Shrira. Type-aware transactions for faster concurrent code. In *Proceedings of the Eleventh European Conference on Computer Systems*, EuroSys '16, pages 31:1–31:16, New York, NY, USA, 2016. ACM.

[185] Mark Hill and Vijay Janapa Reddi. Gables: A roofline model for mobile socs. In *25th IEEE International Symposium on High Performance Computer Architecture, HPCA 2019, Washington, DC, USA, February 16-20, 2019*, pages 317–330, 2019.

[186] Mark D. Hill. On the meltdown& spectre design flaws. Talk, Feb 2018.

[187] Berthold K. P. Horn and Liang Wang. Wave equation of suppressed traffic flow instabilities. *IEEE Trans. Intelligent Transportation Systems*, 19(9):2955–2964, 2018.

[188] HP. HP storage provisioning manager (spm) user guide. 2010.

[189] Randall Hyde. The fallacy of premature optimization. *Ubiquity*, 2009(February), February 2009.

[190] iguazio. The future of serverless computing: Nuclio. `https://www.iguazio.com/nuclio-future-serverless-computing/`.

[191] Intel.

[192] IOTA. https://www.iota.org/.

[193] Michael Isard, Mihai Budiu, Yuan Yu, Andrew Birrell, and Dennis Fetterly. Dryad: Distributed data-parallel programs from sequential building blocks. In *Proceedings of the 2nd ACM SIGOPS/EuroSys European Conference on Computer Systems 2007*, EuroSys '07, pages 59–72, New York, NY, USA, 2007. ACM.

[194] Anand Padmanabha Iyer, Aurojit Panda, Shivaram Venkataraman, Mosharaf Chowdhury, Aditya Akella, Scott Shenker, and Ion Stoica. Bridging the gap: Towards approximate graph analytics. In *Proceedings of the 1st ACM SIGMOD Joint International Workshop on Graph Data Management Experiences & Systems (GRADES) and Network Data Analytics (NDA)*, GRADES-NDA '18, pages 10:1–10:5, New York, NY, USA, 2018. ACM.

[195] H. Zimmerman and J. D. Day. The osi reference model. *Proceedings of the IEEE*, 71(12):1334–1340, 1983.

[196] Raj Jain. Operational laws, ch 33. In *The Art of Computer Systems Performance Analysis, Wiley India Edition*, pages 555–567.

[197] Samuel Jero, Endadul Hoque, David Choffnes, Alan Mislove, and Cristina Nita-Rotaru. Automated attack discovery in tcp congestion control using a model-guided approach. In *Proceedings of the Applied Networking Research Workshop*, ANRW '18, pages 95–95, New York, NY, USA, 2018. ACM.

[198] Djordje Jevdjic, Karin Strauss, Luis Ceze, and Henrique S. Malvar. Approximate storage of compressed and encrypted videos. *SIGARCH Comput. Archit. News*, 45(1):361–373, April 2017.

[199] Somesh Jha, Tom Reps, Sibin Mohan, Rakesh Bobba, David Lie, and Eric Schulte. Techniques and tools for debloating containers. https://security.csl.toronto.edu/wp-content/uploads/2018/06/ONR_Debloating.pdf.

[200] Y. Jiang, D. Wu, and P. Liu. JRed: Program customization and bloatware mitigation based on static analysis. In *2016 IEEE 40th Annual Computer Software and Applications Conference (COMPSAC)*, volume 1, pages 12–21, June 2016.

[201] Abhilash Jindal, Y. Charlie Hu, Samuel Midkiff, and Prahlad Joshi. Unsafe time handling in smartphones. In *2016 USENIX Annual Technical Conference (USENIX ATC 16)*, pages 115–127, Denver, CO, 2016. USENIX Association.

[202] Norman P. Jouppi, Cliff Young, Nishant Patil, David Patterson, Gaurav Agrawal, Raminder Bajwa, Sarah Bates, Suresh Bhatia, Nan Boden, Al Borchers, Rick Boyle, Pierre-luc Cantin, Clifford Chao, Chris Clark, Jeremy Coriell, Mike Daley, Matt Dau, Jeffrey Dean, Ben Gelb, Tara Vazir Ghaemmaghami, Rajendra Gottipati, William Gulland, Robert Hagmann, C. Richard Ho, Doug Hogberg, John Hu, Robert Hundt, Dan Hurt, Julian Ibarz, Aaron Jaffey, Alek Jaworski, Alexander Kaplan, Harshit Khaitan, Daniel Killebrew, Andy Koch, Naveen Kumar, Steve Lacy, James Laudon, James Law, Diemthu Le, Chris Leary, Zhuyuan Liu, Kyle Lucke, Alan Lundin, Gordon MacKean, Adriana Maggiore, Maire Mahony, Kieran Miller, Rahul Nagarajan, Ravi Narayanaswami, Ray Ni, Kathy Nix, Thomas Norrie, Mark Omernick, Narayana Penukonda, Andy Phelps, Jonathan Ross, Matt Ross, Amir Salek, Emad Samadiani, Chris Severn, Gregory Sizikov, Matthew Snelham, Jed Souter, Dan Steinberg, Andy Swing, Mercedes Tan, Gregory Thorson, Bo Tian, Horia Toma, Erick Tuttle, Vijay Vasudevan, Richard Walter, Walter Wang, Eric Wilcox, and Doe Hyun Yoon. In-datacenter performance analysis of a tensor processing unit. In *Proceedings of the 44th Annual International Symposium on Computer Architecture*, ISCA '17, pages 1–12, New York, NY, USA, 2017. ACM.

[203] Julia. https://docs.julialang.org/en/v1/manual/types/index.html.

[204] Svilen Kanev, Juan Pablo Darago, Kim Hazelwood, Parthasarathy Ranganathan, Tipp Moseley, Gu-Yeon Wei, and David Brooks. Profiling a warehouse-scale computer. In *Proceedings of the 42nd Annual International Symposium on Computer Architecture*, ISCA '15, pages 158–169, New York, NY, USA, 2015. ACM.

[205] Svilen Kanev, Sam Likun Xi, Gu-Yeon Wei, and David Brooks. Mallacc: Accelerating memory allocation. *SIGOPS Oper. Syst. Rev.*, 51(2):33–45, April 2017.

[206] Kalapriya Kannan, Suparna Bhattacharya, Kumar Raj, Muthukumar Murugan, and Doug Voigt. Seesaw - similarity exploiting storage for accelerating analytics workflows. In *8th USENIX Workshop on Hot Topics in Storage and File Systems HotStorage 16)*, Denver, CO, 2016. USENIX Association.

[207] Anna R. Karlin, Mark S. Manasse, Lyle A. McGeoch, and Susan Owicki. Competitive randomized algorithms for non-uniform problems. In *Proceedings of the First Annual ACM-SIAM Symposium on Discrete Algorithms*, SODA '90, pages 301–309, Philadelphia, PA, USA, 1990. Society for Industrial and Applied Mathematics.

[208] Andrej Karpathy. Software 2.0. https://medium.com/@karpathy/software-2-0-a64152b37c35.

[209] Christian Kästner. *Virtual Separation of Concerns*. PhD thesis, University of Magdeburg, May 2010.

[210] Christian Kästner and Sven Apel. Virtual separation of concerns – a second chance for preprocessors. In *Journal of Object Technology*, 2009.

[211] Kiyokuni Kawachiya, Kazunori Ogata, and Tamiya Onodera. Analysis and reduction of memory inefficiencies in java strings. In *Proceedings of the 23rd ACM SIGPLAN Conference on Object-oriented Programming Systems Languages and Applications*, OOPSLA '08, pages 385–402, New York, NY, USA, 2008. ACM.

[212] Andy Kellens, Kim Mens, Paolo Tonella, and Département D'ingénierie Informatique. A survey of automated code-level aspect mining techniques. In *In Transactions on Aspect Oriented Software Development*, pages 145–164, 2007.

[213] Juliette Kennedy and Kurt Gödel. In Edward N. Zalta, editor, *The Stanford Encyclopedia of Philosophy*. Metaphysics Research Lab, Stanford University, winter 2018 edition, 2018.

[214] Linux kernel documentation. Direct access for files.

[215] S. Keshav. *An Engineering Approach To Computer Networking: Atm Networks, The Internet, And The Telephone Network.* Pearson Education India, 1997.

[216] Gregor Kiczales, John Lamping, Anurag Mendhekar, Chris Maeda, Cristina Lopes, Jean-Marc Loingtier, and John Irwin. Aspect-oriented programming. In *Proceedings of the European Conference on Object-Oriented Programming, ECOOP.* Springer-Verlag, 1997.

[217] Chang Hwan Peter Kim, Christian Kästner, and Don S. Batory. On the modularity of feature interactions. In *Proceedings of the International Conference on Generative Programming and Component Engineering, GPCE*, 2008.

[218] Mijung Kim, Jun Li, Haris Volos, Manish Marwah, Alexander Ulanov, Kimberly Keeton, Joseph Tucek, Lucy Cherkasova, Le Xu, and Pradeep Fernando. Sparkle: Optimizing spark for large memory machines and analytics. In *Proceedings of the 2017 Symposium on Cloud Computing*, SoCC '17, pages 656–656, New York, NY, USA, 2017. ACM.

[219] Youngsik Kim, Sungjoo Yoo, and Sunggu Lee. Improving write performance by controlling target resistance distributions in MLC PRAM. *ACM Trans. Des. Autom. Electron. Syst.*, 21(2):23:1–23:27, January 2016.

[220] Michael Kircher and Prashant Jain. Pooling. 2002.

[221] Avi Kivity, Dor Laor, Glauber Costa, Pekka Enberg, Nadav Har'El, Don Marti, and Vlad Zolotarov. OS: Optimizing the operating system for virtual machines. In *Proceedings of the 2014 USENIX Conference on USENIX Annual Technical Conference*, USENIX ATC'14, pages 61–72, Berkeley, CA, USA, 2014. USENIX Association.

[222] Donald E. Knuth. Structured programming with go to statements. *ACM Comput. Surv.*, 6(4):261–301, December 1974.

[223] R. Kohavi, T. Crook, R. Longbotham, B. Frasca, R. Henne, J. L. Ferres, and T. Melamed. Online experimentation at Microsoft. *Data Mining Case Studies 11*, 2009.

[224] Jack Kosaian, K. V. Rashmi, and Shivaram Venkataraman. Learning a code: Machine learning for approximate non-linear coded computation. *CoRR*, abs/1806.01259, 2018.

[225] Elias Koutsoupias and Christos Papadimitriou. Worst-case equilibria. *Comput. Sci. Rev.*, 3(2):65–69, May 2009.

[226] Edward M. Kraft. After 40 years why hasn't the computer replaced the wind tunnel? *ITEA Journal 2010*, 31: 329–346.

[227] Tim Kraska, Alex Beutel, Ed H. Chi, Jeffrey Dean, and Neoklis Polyzotis. The case for learned index structures. In *Proceedings of the 2018 International Conference on Management of Data*, SIGMOD '18, pages 489–504, New York, NY, USA, 2018. ACM.

[228] Arvind S. Krishna, Aniruddha S. Gokhale, and Douglas C. Schmidt. Context-specific middleware specialization techniques for optimizing software product-line architectures. In *Proceedings of the 1st ACM SIGOPS/EuroSys European Conference on Computer Systems 2006*, EuroSys '06, pages 205–218, New York, NY, USA, 2006. ACM.

[229] Sameer G. Kulkarni, Wei Zhang, Jinho Hwang, Shriram Rajagopalan, K. K. Ramakrishnan, Timothy Wood, Mayutan Arumaithurai, and Xiaoming Fu. NFVnice: Dynamic backpressure and scheduling for NFV service chains. In *Proceedings of the Conference of the ACM Special Interest Group on Data Communication, SIGCOMM 2017, Los Angeles, CA, USA, August 21-25, 2017*, pages 71–84, 2017.

[230] Sandeep Kumar, Sindhu Padakandla, Chandrashekar L, Priyank Parihar, Gopinath K, and Shalabh Bhatnagar. Scalable performance tuning of Hadoop MapReduce: A noisy gradient approach. *2017 IEEE 10th International Conference on Cloud Computing (CLOUD)*, 25-30 June 2017.

[231] Amlan Kusum, Keval Vora, Rajiv Gupta, and Iulian Neamtiu. Efficient processing of large graphs via input reduction. In *Proceedings of the 25th ACM International Symposium on High-Performance Parallel and Distributed Computing*, HPDC '16, pages 245–257, New York, NY, USA, 2016. ACM.

[232] Yongin Kwon, Sangmin Lee, Hayoon Yi, Donghyun Kwon, Seungjun Yang, Byung-Gon Chun, Ling Huang, Petros Maniatis, Mayur Naik, and Yunheung Paek. Mantis: Automatic performance prediction for smartphone applications. In *Presented as part of the 2013 USENIX Annual Technical Conference (USENIX ATC 13)*, pages 297–308, San Jose, CA, 2013. USENIX.

[233] Pacific Northwest National Laboratory. High performance data analytics (hpda). https://www.pnnl.gov/computing/HPDA/.

[234] Leslie Lamport. *Specifying Systems: The TLA+ Language and Tools for Hardware and Software Engineers*. Addison-Wesley Longman Publishing Co., Inc., Boston, MA, USA, 2002.

[235] James Larus and Galen Hunt. The singularity system. *Commun. ACM*, 53(8):72–79, August 2010.

[236] Nancy Leveson. *Software and the Challenge of Flight Control (in Space Shuttle Legacy: How We Did It and What We Learned, ed. Roger D. Launius, John Krige, and James I. Craig)*. AIAA, Reston, VA, 2013.

[237] Amit Levy, Michael P. Andersen, Bradford Campbell, David Culler, Prabal Dutta, Branden Ghena, Philip Levis, and Pat Pannuto. Ownership is theft: Experiences building an embedded OS in Rust. In *Proceedings of the 8th Workshop on Programming Languages and Operating Systems (PLOS 2015)*, October 2015.

[238] Songze Li, Mingchao Yu, Salman Avestimehr, Sreeram Kannan, and Pramod Viswanath. Polyshard: Coded sharding achieves linearly scaling efficiency and security simultaneously. *CoRR*, abs/1809.10361, 2018.

[239] Ben Liblit, Alex Aiken, Alice X. Zheng, and Michael I. Jordan. Bug isolation via remote program sampling. In *Proceedings of the ACM SIGPLAN 2003 Conference on Programming Language Design and Implementation*, PLDI '03, pages 141–154, New York, NY, USA, 2003. ACM.

[240] Jin Tack Lim, Christoffer Dall, Shih-Wei Li, Jason Nieh, and Marc Zyngier. Neve: Nested virtualization extensions for arm. In *Proceedings of the 26th Symposium on Operating Systems Principles*, SOSP '17, pages 201–217, New York, NY, USA, 2017. ACM.

[241] Xing Lin, Yair Rivenson, Nezih T. Yardimci, Muhammed Veli, Mona Jarrahi, and Aydogan Ozcan. All-optical machine learning using diffractive deep neural networks. *Science*, 26 July 2018.

[242] Jia Liu, Don Batory, and Christian Lengauer. Feature oriented refactoring of legacy applications. In *Proceedings of the International Conference on Software Engineering, ICSE*, 2006.

[243] Lixia Liu and Silvius Rus. Perflint: A context sensitive performance advisor for C++ programs. In *Proceedings of the 7th Annual IEEE/ACM International Symposium on Code Generation and Optimization*, CGO '09, pages 265–274, Washington, DC, USA, 2009. IEEE Computer Society.

[244] Vincent Liu, Aaron Parks, Vamsi Talla, Shyamnath Gollakota, David Wetherall, and Joshua R. Smith. Ambient backscatter: Wireless communication out of thin air. *SIGCOMM Comput. Commun. Rev.*, 43(4):39–50, August 2013.

[245] Robert Love. *Linux System Programming*. OReilly, 2007.

[246] Liang Luo, Suman Nath, Lenin Ravindranath Sivalingam, Madan Musuvathi, and Luis Ceze. Troubleshooting transiently-recurring errors in production systems with blame-proportional logging. In *2018 USENIX Annual Technical Conference (USENIX ATC 18)*, pages 321–334, Boston, MA, 2018. USENIX Association.

[247] Loi Luu, Viswesh Narayanan, Chaodong Zheng, Kunal Baweja, Seth Gilbert, and Prateek Saxena. A secure sharding protocol for open blockchains. In *Proceedings of the 2016 ACM SIGSAC Conference on Computer and Communications Security*, CCS '16, pages 17–30, New York, NY, USA, 2016. ACM.

[248] Anil Madhavapeddy, Richard Mortier, Charalampos Rotsos, David Scott, Balraj Singh, Thomas Gazagnaire, Steven Smith, Steven Hand, and Jon Crowcroft. Unikernels: Library operating systems for the cloud. *SIGPLAN Not.*, 48(4):461–472, March 2013.

[249] Martina Maggio and Henry Hoffmann. ARPE: A tool to build equation models of computing systems. In *Presented as part of the 8th International Workshop on Feedback Computing*, San Jose, CA, 2013. USENIX.

[250] John Maheswaran, Daniel Jackowitz, Ennan Zhai, David Isaac Wolinsky, and Bryan Ford. Building privacy-preserving cryptographic credentials from federated online identities. In *Proceedings of the Sixth ACM Conference on Data and Application Security and Privacy*, CODASPY '16, pages 3–13, New York, NY, USA, 2016. ACM.

[251] Heiko Mantel, Alexandra Weber, and Boris Kopf. A systematic study of cache side channels across AES implementations. *LNCS 10379*, 2017.

[252] Artemiy Margaritovy, Dmitrii Ustiugovz, Edouard Bugnionz, and Boris Groty. Virtual address translation via learned page table indexes. In *32nd Conference on Neural Information Processing Systems*, NIPS 2018.

[253] Marius Marin, Arie Van Deursen, and Leon Moonen. Identifying crosscutting concerns using fan-in analysis. *TOSEM*, 17:3:1–3:37, December 2007.

[254] Robert N. Mayo and Parthasarathy Ranganathan. Energy consumption in mobile devices: Why future systems need requirements-aware energy scale-down. In *Proceedings of the Workshop on Power Aware Computing and Systems*, page 26–40, 2003.

[255] Lawrence McAfee and Kunle Olukotun. EMEURO: A framework for generating multi-purpose accelerators via deep learning. *2015 IEEE/ACM International Symposium on Code Generation and Optimization (CGO)*, 7-11 Feb. 2015.

[256] Stephen McCamant and Michael D. Ernst. Quantitative information-flow tracking for C and related languages. MIT-CSAIL-TR-2006-076, 2006.

[257] Paul E. McKenney. *Exploiting Deferred Destruction: An Analysis of Read-Copy-Update Techniques in Operating System Kernels*. PhD thesis, OGI School of Science and Engineering at Oregon Health and

Sciences University, 2004. Available:http://www.rdrop.com/users/
paulmck/RCU/RCUdissertation.2004.07.14e1.pdf [Viewed October
15, 2004].

[258] Paul E. McKenney. Structured deferral: Synchronization via procrasti-
nation. *Queue*, 11(5):20:20–20:39, May 2013.

[259] Paul E. McKenney and Jonathan Walpole. What is RCU, fundamentally? Available: http://lwn.net/Articles/262464/ [Viewed December 27, 2007], December 2007.

[260] Sasa Misailovic, Daniel M. Roy, and Martin C. Rinard. Probabilistically
accurate program transformations. In *Proceedings of the 18th International Conference on Static Analysis*, SAS'11, pages 316–333, Berlin,
Heidelberg, 2011. Springer-Verlag.

[261] Nikita Mishra, Connor Imes, John D. Lafferty, and Henry Hoffmann.
CALOREE: Learning control for predictable latency and low energy.
In *Proceedings of the Twenty-Third International Conference on Archi-
tectural Support for Programming Languages and Operating Systems*,
ASPLOS '18, pages 184–198, New York, NY, USA, 2018. ACM.

[262] John C. Mitchell, Rahul Sharma, Deian Stefan, and Joe Zimmerman.
Information-flow control for programming on encrypted data. In *Proceedings of the 2012 IEEE 25th Computer Security Foundations Symposium*, CSF '12, pages 45–60, Washington, DC, USA, 2012. IEEE Computer Society.

[263] Nick Mitchell, Edith Schonberg, and Gary Sevitsky. Making sense of
large heaps. In *European Conference on Object-Oriented Programming,
ECOOP*, 2009.

[264] Nick Mitchell, Edith Schonberg, and Gary Sevitsky. Four trends leading
to java runtime bloat. *IEEE Software*, 27(1):56–63, 2010.

[265] Nick Mitchell and Gary Sevitsky. Building Memory Efficient Java Applications: Practices and Challenges. *PLDI 2009 Tutorial*.

[266] Nick Mitchell and Gary Sevitsky. The causes of bloat, the limits
of health. In *Object-Oriented Programming, Systems, Languages and
Applications, OOPSLA*, 2007.

[267] Nick Mitchell and Gary Sevitsky. Building memory efficient java applications. In *Tutorial, International Conference on Software Engineering,
ICSE*, 2008.

[268] Nick Mitchell, Gary Sevitsky, and Harini Srinivasan. Modeling runtime
behaviour in framework based applications. In *European Conference on
Object-Oriented Programming, ECOOP*, 2006.

[269] Sparsh Mittal. A survey of techniques for approximate computing. *ACM Comput. Surv.*, 48(4):62:1–62:33, March 2016.

[270] Sparsh Mittal and Jeffrey S. Vetter. A Survey of Software Techniques for Using Non-Volatile Memories for Storage and Main Memory Systems. *IEEE Transactions on Parallel and Distributed Systems*, 27(5):1537–1550, 2011.

[271] Rajeev Motwani and Prabhakar Raghavan. *Randomized Algorithms*. Cambridge University Press, New York, NY, USA, 1995.

[272] Mozilla. Speculative execution side-channel attack ("spectre"). *Mozilla Foundation Security Advisory 2018-01*, Jan 2018.

[273] Anurag Mukkara, Nathan Beckmann, and Daniel Sanchez. Whirlpool: Improving dynamic cache management with static data classification. *SIGOPS Oper. Syst. Rev.*, 50(2):113–127, March 2016.

[274] Onur Mutlu. The rowhammer problem and other issues we may face as memory becomes denser. In *Proceedings of the Conference on Design, Automation & Test in Europe*, DATE '17, pages 1116–1121, 3001 Leuven, Belgium, 2017. European Design and Automation Association.

[275] Todd Mytkowicz, Amer Diwan, Matthias Hauswirth, and Peter F. Sweeney. Evaluating the accuracy of java profilers. In *Proceedings of the 31st ACM SIGPLAN Conference on Programming Language Design and Implementation*, PLDI '10, pages 187–197, New York, NY, USA, 2010. ACM.

[276] Nachiappan Chidambaram Nachiappan, Haibo Zhang, Jihyun Ryoo, Niranjan Soundararajan, Anand Sivasubramaniam, Mahmut T. Kandemir, Ravishankar Iyer, and Chita R. Das. VIP: virtualizing IP chains on handheld platforms. *2015 ACM/IEEE 42nd Annual International Symposium on Computer Architecture (ISCA)*.

[277] Hoda Naghibijouybari, Khaled N. Khasawneh, and Nael Abu-Ghazaleh. Constructing and characterizing covert channels on GPGPUs. In *Proceedings of the 50th Annual IEEE/ACM International Symposium on Microarchitecture*, MICRO-50 '17, pages 354–366, New York, NY, USA, 2017. ACM.

[278] R. Nair and S. Gupta. Wildfire: Approximate synchronization of parameters in distributed deep learning. *IBM J. Res. Dev.*, 61(4-5):7:1–7:9, July 2017.

[279] Ravi Nair. Models for energy-efficient approximate computing. In *Proceedings of the 16th ACM/IEEE International Symposium on Low Power Electronics and Design*, ISLPED '10, pages 359–360, New York, NY, USA, 2010. ACM.

[280] George C. Necula. Proof-carrying code. In *Proceedings of the 24th ACM SIGPLAN-SIGACT Symposium on Principles of Programming Languages*, POPL '97, pages 106–119, New York, NY, USA, 1997. ACM.

[281] Hai Duc Nguyen, Andrew A. Chien, and Chaojie Zhang. Zero-carbon cloud: Research challenges for datacenters as supply-following loads. Technical Report TR-2019-08, U. Chicago, 2019.

[282] Khanh Nguyen, Lu Fang, Christian Navasca, Guoqing Xu, Brian Demsky, and Shan Lu. Skyway: Connecting managed heaps in distributed big data systems. In *Proceedings of the Twenty-Third International Conference on Architectural Support for Programming Languages and Operating Systems*, ASPLOS '18, pages 56–69, New York, NY, USA, 2018. ACM.

[283] Khanh Nguyen, Lu Fang, Guoqing Xu, Brian Demsky, Shan Lu, Sanazsadat Alamian, and Onur Mutlu. Yak: A high-performance big-data-friendly garbage collector. In *12th USENIX Symposium on Operating Systems Design and Implementation (OSDI 16)*, pages 349–365, Savannah, GA, 2016. USENIX Association.

[284] Khanh Nguyen, Kai Wang, Yingyi Bu, Lu Fang, Jianfei Hu, and Guoqing Xu. Facade: A compiler and runtime for (almost) object-bounded big data applications. In *Proceedings of the Twentieth International Conference on Architectural Support for Programming Languages and Operating Systems*, ASPLOS '15, pages 675–690, New York, NY, USA, 2015. ACM.

[285] Khanh Nguyen and Guoqing Xu. Cachetor: Detecting cacheable data to remove bloat. In *Proceedings of the 2013 9th Joint Meeting on Foundations of Software Engineering*, ESEC/FSE 2013, pages 268–278, New York, NY, USA, 2013. ACM.

[286] Kathleen Nichols and Van Jacobson. Controlling queue delay. *Queue*, 10(5):20:20–20:34, May 2012.

[287] Anant Vithal Nori, Jayesh Gaur, Siddharth Rai, Sreenivas Subramoney, and Hong Wang. Criticality aware tiered cache hierarchy: A fundamental relook at multi-level cache hierarchies. In *Proceedings of the 45th Annual International Symposium on Computer Architecture*, ISCA '18, pages 96–109, Piscataway, NJ, USA, 2018. IEEE Press.

[288] Rei Odaira and Toshio Nakatani. Continuous object access profiling and optimizations to overcome the memory wall and bloat. In *Proceedings of the Seventeenth International Conference on Architectural Support for Programming Languages and Operating Systems*, ASPLOS '12, pages 147–158, New York, NY, USA, 2012. ACM.

[289] Kazunori Ogata, Dai Mikurube, Kiyokuni Kawachiya, Scott Trent, and Tamiya Onodera. A study of java's non-java memory. In *Proceedings of the ACM International Conference on Object Oriented Programming Systems Languages and Applications*, OOPSLA '10, pages 191–204, New York, NY, USA, 2010. ACM.

[290] T.P. Oliveira, J.S. Barbar, and A.S. Soares. Computer network traffic prediction: a comparison between traditional and deep learning neural networks. *Int. J. Big Data Intelligence*, 3(1):28–37, 2016.

[291] John Ousterhout, Arjun Gopalan, Ashish Gupta, Ankita Kejriwal, Collin Lee, Behnam Montazeri, Diego Ongaro, Seo Jin Park, Henry Qin, Mendel Rosenblum, Stephen Rumble, Ryan Stutsman, and Stephen Yang. The ramcloud storage system. *ACM Trans. Comput. Syst.*, 33(3):7:1–7:55, August 2015.

[292] Anand Padmanabha Iyer, Li Erran Li, Mosharaf Chowdhury, and Ion Stoica. Mitigating the latency-accuracy trade-off in mobile data analytics systems. In *Proceedings of the 24th Annual International Conference on Mobile Computing and Networking*, MobiCom '18, pages 513–528, New York, NY, USA, 2018. ACM.

[293] Kostas Pagiamtzis and Ali Sheikholeslami. Content Addressable Memory (CAM) Circuits and Architectures: A Tutorial and Survey. *IEEE Journal of Solid State Circuits*, 41(3), March 2006.

[294] Krishna V. Palem. Energy aware algorithm design via probabilistic computing: From algorithms and models to Moore's law and novel (semiconductor) devices. In *Proceedings of the 2003 International Conference on Compilers, Architecture and Synthesis for Embedded Systems*, CASES '03, pages 113–116, New York, NY, USA, 2003. ACM.

[295] Ashish Panwar, Sorav Bansal, and K. Gopinath. Hawkeye: Efficient fine-grained OS support for huge pages. In *Proceedings of the Twenty-Fourth International Conference on Architectural Support for Programming Languages and Operating Systems*, ASPLOS '19, New York, NY, USA, 2019. ACM.

[296] Ashish Panwar and K. Gopinath. Towards practical page placement for a green memory manager. In *22nd IEEE International Conference on High Performance Computing, HiPC 2015, Bengaluru, India, December 16-19, 2015*, pages 155–164, 2015.

[297] Ashish Panwar, Aravinda Prasad, and K. Gopinath. Making huge pages actually useful. In *Proceedings of the Twenty-Third International Conference on Architectural Support for Programming Languages and Operating Systems*, ASPLOS '18, pages 679–692, New York, NY, USA, 2018. ACM.

[298] Jongse Park, Hadi Esmaeilzadeh, Xin Zhang, Mayur Naik, and William Harris. Flexjava: Language support for safe and modular approximate programming. In *Proceedings of the 2015 10th Joint Meeting on Foundations of Software Engineering*, ESEC/FSE 2015, pages 745–757, New York, NY, USA, 2015. ACM.

[299] David L. Parnas. Designing software for ease of extension and contraction. In *Proceedings of the 3rd International Conference on Software Engineering*, ICSE '78, pages 264–277, Piscataway, NJ, USA, 1978. IEEE Press.

[300] Trevor Parsons and John Murthy. Detecting performance antipatterns in component based enterprise systems. In *Journal of Object Technology*, pages 55–90, March-April, 2008.

[301] Ioannis Patiniotakis, Yiannis Verginadis, and Gregoris Mentzas. Preference-based cloud service recommendation as a brokerage service. In *Proceedings of the 2nd International Workshop on CrossCloud Systems*, CCB '14, pages 5:1–5:6, New York, NY, USA, 2014. ACM.

[302] Edwin Pednault et al., Leveraging Secondary Storage to Simulate Deep 54 qubit Sycamore Circuits. arXiv:1910.09534v2, 22 Oct 2019.

[303] Kexin Pei, Yinzhi Cao, Junfeng Yang, and Suman Jana. Deepxplore: Automated whitebox testing of deep learning systems. *CoRR*, abs/1705.06640, 2017.

[304] Nathan Pemberton. Exploring the disaggregated memory interface design space. In *The First Workshop on Resource Disaggregation (WORD 2019)*, 2019.

[305] Nathan Pemberton. Exploring the disaggregated memory interface design space. *The First Workshop on Resource Disaggregation*, 04 2019.

[306] Nathan Pemberton and Johann Schleier-Smith. The serverless data center: Hardware disaggregation meets serverless computing. *The First Workshop on Resource Disaggregation*, 04 2019.

[307] Phys.org. https://phys.org/news/2018-05-proof-reveals-fundamental-limits-scientific.html.

[308] Pankaj Pipada, Achintya Kundu, K. Gopinath, Chiranjib Bhattacharyya, Sai Susarla, and Nagesh P. C. Loadiq: Online learning to label program phases using storage traces. *HotStorage*, Jun 2012.

[309] J. S. Plank. Erasure codes for storage systems: A brief primer. *login: the Usenix magazine*, 38(6), December 2013.

[310] Klaus Pohl, Günter Böckle, and Frank J. van der Linden. *Software Product Line Engineering: Foundations, Principles and Techniques.* Springer-Verlag New York, Inc., Secaucus, NJ, USA, 2005.

[311] Raghavendra Pradyumna Pothukuchi, Sweta Yamini Pothukuchi, Petros Voulgaris, and Josep Torrellas. Yukta: Multilayer resource controllers to maximize efficiency. In *Proceedings of the 45th Annual International Symposium on Computer Architecture,* ISCA '18, pages 505–518, Piscataway, NJ, USA, 2018. IEEE Press.

[312] Benôit Pradelle, Benôit Meister, Muthu Baskaran, Jonathan Springer, and Richard Lethin. Polyhedral optimization of tensorflow computation graphs. *Lecture Notes in Computer Science, vol 11027,* 2019.

[313] Aravinda Prasad and K. Gopinath. Prudent memory reclamation in procrastination-based synchronization. In *Proceedings of the Twenty-First International Conference on Architectural Support for Programming Languages and Operating Systems,* ASPLOS '16, pages 99–112, New York, NY, USA, 2016. ACM.

[314] Aravinda Prasad and K. Gopinath. A frugal approach to reduce RCU grace period overhead. In *Proceedings of the Thirteenth EuroSys Conference,* EuroSys '18, pages 41:1–41:15, New York, NY, USA, 2018. ACM.

[315] Aravinda Prasad, K. Gopinath, and Paul E. McKenney. The RCU-reader preemption problem in VMs. In *Proceedings of the 2017 USENIX Conference on Usenix Annual Technical Conference,* USENIX ATC '17, pages 265–270, Berkeley, CA, USA, 2017. USENIX Association.

[316] Christian Prehofer. Feature-oriented programming: A fresh look at objects. In *Proceedings of the European Conference on Object-Oriented Programming, ECOOP,* 1997.

[317] Google Github project. Flat Buffers.

[318] SBE Github project. Simple Binary Encoding Wiki.

[319] Anh Quach, Rukayat Erinfolami, David Demicco, and Aravind Prakash. A multi-OS cross-layer study of bloating in user programs, kernel and managed execution environments. In *Proceedings of the 2017 Workshop on Forming an Ecosystem Around Software Transformation,* FEAST '17, pages 65–70, New York, NY, USA, 2017. ACM.

[320] Anh Quach, Aravind Prakash, and Lok Yan. Debloating software through piece-wise compilation and loading. In *27th USENIX Security Symposium (USENIX Security 18),* pages 869–886, Baltimore, MD, 2018. USENIX Association.

[321] Quantamagazine. Quantum vs classical algorithms. https://www.quan
      tamagazine.org/quantum-computers-struggle-against-classical-algorith
      ms-20180201/.

[322] Arnab Raha and Vijay Raghunathan.   Towards full-system energy-
      accuracy tradeoffs: A case study of an approximate smart camera sys-
      tem. In *Proceedings of the 54th Annual Design Automation Conference
      2017*, DAC '17, pages 74:1–74:6, New York, NY, USA, 2017. ACM.

[323] K. Gopinath and Rahul Simha. Insurable storage services: Creating a
      marketplace for long-term document archival. *2006 International Con-
      ference on Computational Science (3), Reading, UK*, 2006.

[324] B. Rajendran, R. Cheek, L. Lastras, M. Franceschini, M. Breitwisch,
      A. Schrott, J. Li, R. Montoye, Chang L., and C. Lam. Demonstration
      of CAM and TCAM using Phase Change Devices. *IEEE International
      Memory Workshop*, 2011.

[325] Ajinkya Rajput and K. Gopinath.   Towards a more secure Aadhaar.
      ICISS2017, 2017.

[326] Dušan Ramljak, Deepak Abraham Tom, Doug Voigt, and Krishna Kant.
      Modular Framework for Data Prefetching and Replacement at the Edge,
      pages 18–33. 06 2018.

[327] Vaibhav Rastogi, Drew Davidson, Lorenzo De Carli, Somesh Jha, and
      Patrick McDaniel. Cimplifier: Automatically debloating containers. In
      *Proceedings of the 2017 11th Joint Meeting on Foundations of Software
      Engineering*, ESEC/FSE 2017, pages 476–486, New York, NY, USA,
      2017. ACM.

[328] Spring Java/J2EE Application Framework Reference. Aspect Oriented
      Programming with Spring.

[329] Jack Regula. Overcoming latency in PCie systems. 2007.

[330] Lakshminarayanan Renganarayana, Vijayalakshmi Srinivasan, Ravi
      Nair, and Daniel Prener. Programming with relaxed synchronization.
      In *Proceedings of the 2012 ACM Workshop on Relaxing Synchronization
      for Multicore and Manycore Scalability*, RACES '12, pages 41–50, New
      York, NY, USA, 2012. ACM.

[331] Meghan Revelle, Bogdan Dit, and Denys Poshyvanyk. Using data fusion
      and web mining to support feature location in softwareusing data fusion
      and web mining to support feature location in software. In *Proceedings of
      the IEEE International Conference on Program Comprehension, ICPC*,
      2010.

[332] Chris Richardson. *Microservices Patterns*. Manning Publications, NY,
      2019.

[333] Martin Rinard, Cristian Cadar, Daniel Dumitran, Daniel M. Roy, Tudor Leu, and William S. Beebee, Jr. Enhancing server availability and security through failure-oblivious computing. In *Proceedings of the 6th Conference on Symposium on Operating Systems Design & Implementation - Volume 6*, OSDI'04, pages 21–21, Berkeley, CA, USA, 2004. USENIX Association.

[334] Martin Rinard, Henry Hoffmann, Sasa Misailovic, and Stelios Sidiroglou. Patterns and statistical analysis for understanding reduced resource computing. In *Proceedings of the ACM International Conference on Object Oriented Programming Systems Languages and Applications*, OOPSLA '10, pages 806–821, New York, NY, USA, 2010. ACM.

[335] Martin P. Robillard and Gail C. Murphy. Representing concerns in source code. *ACM Trans. Softw. Eng. Methodol.*, 16(1), February 2007.

[336] Ian Robinson, Jim Webber, and Emil Eifrem. *Graph Databases: New Opportunities for Connected Data*. O'Reilly Media, Inc., 2nd edition, 2015.

[337] Ohad Rodeh. Deferred reference counters for copy-on-write B-trees. *Technical Report rj10464, IBM*, 2010.

[338] Pooja Roy, Rajarshi Ray, Chundong Wang, and Weng Fai Wong. ASAC: Automatic sensitivity analysis for approximate computing. *SIGPLAN Not.*, 49(5):95–104, June 2014.

[339] Andy Rudoff. Persistent memory development kit. 2017.

[340] J. M. Rushby. Design and verification of secure systems. In *Proceedings of the Eighth ACM Symposium on Operating Systems Principles*, SOSP '81, pages 12–21, New York, NY, USA, 1981. ACM.

[341] Adrian Sampson, Werner Dietl, Emily Fortuna, Danushen Gnanapragasam, Luis Ceze, and Dan Grossman. EnerJ: Approximate data types for safe and general low-power computation. In *Proceedings of the 32nd ACM SIGPLAN Conference on Programming Language Design and Implementation*, PLDI '11, pages 164–174, New York, NY, USA, 2011. ACM.

[342] Dipankar Sarma and Paul E. McKenney. Making RCU safe for deep sub-millisecond response realtime applications. In *Proceedings of the Annual Conference on USENIX Annual Technical Conference*, ATEC '04, pages 32–32, Berkeley, CA, USA, 2004. USENIX Association.

[343] T. Savage, M. Revelle, and D. Poshyvanyk. FLAT3: Feature location and textual tracing tool. In *Proceedings of the International Conference on Software Engineering, ICSE*, 2010.

[344] L. Sha, R. Rajkumar, and J. P. Lehoczky. Priority inheritance protocols: An approach to real-time synchronization. *IEEE Trans. Comput.*, 39(9):1175–1185, September 1990.

[345] Ajeet Shankar, Matthew Arnold, and Rastislav Bodik. JOLT: lightweight dynamic analysis and removal of object churn. In *Object-Oriented Programming, Systems, Languages and Applications, OOPSLA*, 2008.

[346] Hashim Sharif, Muhammad Abubakar, Ashish Gehani, and Fareed Zaffar. Trimmer: Application specialization for code debloating. In *Proceedings of the 33rd ACM/IEEE International Conference on Automated Software Engineering*, ASE 2018, pages 329–339, New York, NY, USA, 2018. ACM.

[347] Prateek Sharma, Stephen Lee, Tian Guo, David Irwin, and Prashant Shenoy. SpotCheck: Designing a derivative IaaS cloud on the spot market. In *Proceedings of the Tenth European Conference on Computer Systems*, EuroSys '15, pages 16:1–16:15, New York, NY, USA, 2015. ACM.

[348] Robert Sheldon. Why an edge computing platform benefits from hci. https://searchconvergedinfrastructure.techtarget.com/tip/Why-an-edge-computing-platform-benefits-from-HCI.

[349] K. Shen, C. Li, and C. Ding. Quantifying the cost of context switch. *Proceedings of the 2007 Workshop on Experimental Computer Science*, 2007.

[350] David Shepherd, Zachary P. Fry, Emily Hill, Lori Pollock, and K. Vijay-Shanker. Using natural language program analysis to locate and understand action-oriented concerns. In *AOSD*, 2007.

[351] Jack Shirazi. Java performance tuning. O'Reilly, 2003.

[352] Smitha Shivshankar and Abbas Jamalipour. Effect of node neighborhood on the evolution of cooperation using public goods game in vehicular networks. In *WCNC*, pages 3242–3247. IEEE, 2014.

[353] Julian Shun, Guy E. Blelloch, Jeremy T. Fineman, Phillip B. Gibbons, Aapo Kyrola, Harsha Vardhan Simhadri, and Kanat Tangwongsan. Brief announcement: The problem based benchmark suite. In *Proceedings of the Twenty-fourth Annual ACM Symposium on Parallelism in Algorithms and Architectures*, SPAA '12, pages 68–70, New York, NY, USA, 2012. ACM.

[354] Stelios Sidiroglou-Douskos, Sasa Misailovic, Henry Hoffmann, and Martin Rinard. Managing performance vs. accuracy trade-offs with loop perforation. In *Proceedings of the 19th ACM SIGSOFT Symposium and*

the 13th European Conference on Foundations of Software Engineering, ESEC/FSE '11, pages 124–134, New York, NY, USA, 2011. ACM.

[355] Benjamin H. Sigelman, Luiz Andre Barroso, Mike Burrows, Pat Stephenson, Manoj Plakal, Donald Beaver, Saul Jaspan, and Chandan Shanbhag. Dapper, a large-scale distributed systems tracing. *Infrastructure Google Technical Report dapper-2010-1*, April 2010.

[356] Dwight Silverman. Your smartphone is light years ahead of NASA computers that guided Apollo moon landings. Available at: `https://www.houstonchronicle.com/local/space/mission-moon/article/Your-smartphone-is-light-years-ahead-of-NASA-13757565.php` Accessed April, 2019.

[357] Hyogi Sim, Youngjae Kim, Sudharshan S. Vazhkudai, Devesh Tiwari, Ali Anwar, Ali R. Butt, and Lavanya Ramakrishnan. AnalyzeThis: An analysis workflow-aware storage system. In *Proceedings of the International Conference for High Performance Computing, Networking, Storage and Analysis*, SC '15, pages 20:1–20:12, New York, NY, USA, 2015. ACM.

[358] Aaron Smith, Jon Gibson, Bertrand Maher, Nick Nethercote, Bill Yoder, Doug Burger, Kathryn S. McKinley, and Jim Burrill. Compiling for edge architectures. In *Proceedings of the International Symposium on Code Generation and Optimization*, CGO '06, pages 185–195, Washington, DC, USA, 2006. IEEE Computer Society.

[359] SNIA. Cloud data management interface version 1.1.1. 2015.

[360] SNIA. Swordfish scalable storage management API specification version 1.0.6. 2018.

[361] Open source. Using the block translation table for sector atomicity. *http://pmem.io/2014/09/23/btt.html*, 2014.

[362] J. C. Spall. Multivariate stochastic approximation using a simultaneous perturbation gradient approximation. *IEEE Transactions on Automatic Control*, 1992.

[363] Apache Spark. Lightning-fast cluster computing. `http://spark.apache.org/`.

[364] Joel Spolsky. The law of leaky abstractions. https://www.joelonsoftware.com/2002/11/11/the-law-of-leaky-abstractions/.

[365] Akshitha Sriraman and Thomas F. Wenisch. μtune: Auto-tuned threading for OLDI microservices. In *13th USENIX Symposium on Operating Systems Design and Implementation, OSDI 2018, Carlsbad, CA, USA, October 8-10, 2018*, pages 177–194, 2018.

[366] Renee St. Amant, Amir Yazdanbakhsh, Jongse Park, Bradley Thwaites, Hadi Esmaeilzadeh, Arjang Hassibi, Luis Ceze, and Doug Burger. General-purpose code acceleration with limited-precision analog computation. In *Proceeding of the 41st Annual International Symposium on Computer Architecuture*, ISCA '14, pages 505–516, Piscataway, NJ, USA, 2014. IEEE Press.

[367] Dylan Stamat. Microcontainers – tiny, portable docker containers. https://blog.iron.io/microcontainers-tiny-portable-containers/.

[368] Standard Performance Evaluation Corporation. http://www.specbench.org.

[369] Phillip Stanley-Marbell, Virginia Estellers, and Martin Rinard. Crayon: Saving power through shape and color approximation on next-generation displays. In *Proceedings of the Eleventh European Conference on Computer Systems*, EuroSys '16, pages 11:1–11:17, New York, NY, USA, 2016. ACM.

[370] Phillip Stanley-Marbell and Martin Rinard. A hardware platform for efficient multi-modal sensing with adaptive approximation. arXiv:1804.09241 [physics.app-ph].

[371] Bjarne Stroustrup. *The Design and Evolution of C++*. ACM Press/Addison-Wesley Publishing Co., New York, NY, USA, 1994.

[372] Emma Strubell, Ananya Ganesh, and Andrew McCallum. Energy and policy considerations for deep learning in NLP. arXiv:1906.02243 [cs.CL].

[373] Ryan Stutsman, Collin Lee, and John Ousterhout. Experience with rules-based programming for distributed, concurrent, fault-tolerant code. In *2015 USENIX Annual Technical Conference (USENIX ATC 15)*, pages 17–30, Santa Clara, CA, 2015. USENIX Association.

[374] S. Swanson and A. M. Caulfield. Refactor, reduce, recycle: Restructuring the I/O stack for the future of storage. *Computer*, 46(8):52–59, August 2013.

[375] Andrew S. Tanenbaum. *Modern Operating Systems*. Prentice Hall Press, Upper Saddle River, NJ, USA, 3rd edition, 2007.

[376] Ewin Tang. A quantum-inspired classical algorithm for recommendation systems. *CoRR*, abs/1807.04271, 2018.

[377] Yang Tang and Junfeng Yang. Secure deduplication of general computations. In *2015 USENIX Annual Technical Conference (USENIX ATC 15)*, Santa Clara, CA, 2015.

[378] J. Tate, R.V. Dias, I. Dikanarov, J. Kelly, P. Mescher, and IBM Redbooks. *IBM System Storage SAN Volume Controller and Storwize V7000 Replication Family Services*. IBM redbooks. IBM Redbooks, 2017.

[379] TensorFlow.org. https://www.tensorflow.org/lite/performance/post_training_quantization.

[380] Sai Deep Tetali, Mohsen Lesani, Rupak Majumdar, and Todd Millstein. Mrcrypt: Static analysis for secure cloud computations. *SIGPLAN Not.*, 48(10):271–286, October 2013.

[381] Martin Thompson. Simple Binary Encoding.

[382] Niraj Tolia, Zhikui Wang, Manish Marwah, Cullen Bash, Parthasarathy Ranganathan, and Xiaoyun Zhu. Delivering energy proportionality with non energy-proportional systems – optimizing the ensemble. In *Proceedings of the Workshop on Power Aware Computing and Systems, Hot-Power'08*, 2008.

[383] Paul Tschirhart, Jim Stevens, Zeshan Chishti, Shih-Lien Lu, and Bruce Jacob. Bringing modern hierarchical memory systems into focus: A study of architecture and workload factors on system performance. *Proceedings of the 2015 International Symposium on Memory Systems*, pages 179–190, 2015.

[384] Uresh Vahalia. *UNIX Internals: The New Frontiers*. Prentice Hall Press, Upper Saddle River, NJ, USA, 1996.

[385] Amin Vahdat and Thomas Anderson. Transparent result caching. In *Proceedings of the Annual Conference on USENIX Annual Technical Conference*, ATEC '98, 1998.

[386] Jelle van den Hooff, David Lazar, Matei Zaharia, and Nickolai Zeldovich. Vuvuzela: Scalable private messaging resistant to traffic analysis. In *Proceedings of the 25th Symposium on Operating Systems Principles*, SOSP '15, pages 137–152, New York, NY, USA, 2015. ACM.

[387] George Varghese and Mahesh Jayaram. The fault span of crash failures. *J. ACM*, 47(2):244–293, March 2000.

[388] Arthur H. Veen. Dataflow machine architecture. *ACM Comput. Surv.*, 18(4):365–396, December 1986.

[389] Akshat Verma, Ricardo Koller, Luis Useche, and Raju Rangaswami. SRCMap: energy proportional storage using dynamic consolidation. In *Proceedings of the 8th USENIX Conference on File and Storage Technologies*, FAST'10, Berkeley, CA, USA, 2010. USENIX Association.

[390] Doug Voigt. NVM PM remote access for high availability. 2016.

[391] Doug Voigt. Persistent memory atomics and transactions. 2017.

[392] Haris Volos, Kimberly Keeton, Yupu Zhang, Milind Chabbi, Se Kwon Lee, Mark Lillibridge, Yuvraj Patel, and Wei Zhang. Memory-oriented distributed computing at rack scale. In *Proceedings of the ACM Symposium on Cloud Computing*, SoCC '18, pages 529–529, New York, NY, USA, 2018. ACM.

[393] Haris Volos, Andres Jaan Tack, and Michael M. Swift. Mnemosyne: Lightweight persistent memory. *SIGPLAN Not.*, 47(4):91–104, March 2011.

[394] Keval Vora, Guoqing Xu, and Rajiv Gupta. Load the edges you need: A generic I/O optimization for disk-based graph processing. In *2016 USENIX Annual Technical Conference (USENIX ATC 16)*, pages 507–522, Denver, CO, 2016. USENIX Association.

[395] Mythili Vutukuru, Hari Balakrishnan, and Kyle Jamieson. Cross-layer wireless bit rate adaptation. In *Proceedings of the ACM SIGCOMM 2009 Conference on Data Communication*, SIGCOMM '09, pages 3–14, New York, NY, USA, 2009. ACM.

[396] Gregor Wagner. *Domain Specific Memory Management in a Modern Web Browser*. PhD thesis, California State University at Long Beach, Long Beach, CA, USA, 2011. AAI3487080.

[397] Kai Wang, Guoqing Xu, Zhendong Su, and Yu David Liu. GraphQ: Graph query processing with abstraction refinement—scalable and programmable analytics over very large graphs on a single PC. In *2015 USENIX Annual Technical Conference (USENIX ATC 15)*, pages 387–401, Santa Clara, CA, 2015. USENIX Association.

[398] Shu Wang, Chi Li, Henry Hoffmann, Shan Lu, William Sentosa, and Achmad Imam Kistijantoro. Understanding and auto-adjusting performance-sensitive configurations. In *Proceedings of the Twenty-Third International Conference on Architectural Support for Programming Languages and Operating Systems*, ASPLOS '18, pages 154–168, New York, NY, USA, 2018. ACM.

[399] Weihu Wang and Gang Huang. Pattern-driven performance optimization at runtime: experiment on JEE systems. In *Proceedings of the 9th International Workshop on Adaptive and Reflective Middleware*, ARM '10, pages 39–45, New York, NY, USA, 2010. ACM.

[400] Eric W. Weisstein. Blum's speed-up theorem. http://mathworld. wolfram.com/BlumsSpeed-UpTheorem.html.

[401] KA Wetterstrand. DNA sequencing costs: Data from the nhgri genome sequencing program (gsp). Available at: `www.genome.gov/sequencingcostsdata` Accessed April, 2019.

[402] Anna Wiedemann, Nicole Forsgren, Manuel Wiesche, Heiko Gewald, and Helmut Krcmar. The DevOps phenomenon. *Queue*, 17(2):40:93–40:112, April 2019.

[403] Wikipedia.

[404] Wikipedia. Feature (machine learning). `https://en.wikipedia.org/wiki/Feature_(machine_learning)`.

[405] Wikipedia. Separation of Concerns.

[406] Wikipedia. Software bloat. `http://en.wikipedia.org/wiki/Softwarebloat`.

[407] Samuel Williams, Andrew Waterman, and David Patterson. Roofline: An insightful visual performance model for multicore architectures. *Commun. ACM*, 52(4):65–76, April 2009.

[408] Christian Wimmer and Hanspeter Mössenbösck. Automatic feedback-directed object fusing. *ACM Transactions on Architecture and Code Optimization*, 7(2):7:1–7:35, October 2010.

[409] Terry Winograd and Fernando Flores, editors. *Understanding Computers and Cognition*. Ablex Publishing Corp., Norwood, NJ, USA, 1985.

[410] Niklaus Wirth. A plea for lean software. *Computer*, 28(2):64–68, February 1995.

[411] David H. Wolpert. Constraints on physical reality arising from a formalization of knowledge. arXiv:1711.03499v3, 28 Jun 2018.

[412] Qiang Wu et al. Dynamic-compiler-driven control for microprocessor energy and performance. In *IEEE Micro*, 2006.

[413] Chenning Xie, Rong Chen, Haibing Guan, Binyu Zang, and Haibo Chen. Sync or async: Time to fuse for distributed graph-parallel computation. In *Proceedings of the 20th ACM SIGPLAN Symposium on Principles and Practice of Parallel Programming*, PPoPP 2015, pages 194–204, New York, NY, USA, 2015. ACM.

[414] Guoqing Xu. Finding reusable data structures. In *Proceedings of the ACM International Conference on Object Oriented Programming Systems Languages and Applications*, OOPSLA '12, pages 1017–1034, New York, NY, USA, 2012. ACM.

[415] Guoqing Xu. Coco: Sound and adaptive replacement of java collections. In *Proceedings of the 27th European Conference on Object-Oriented Programming*, ECOOP'13, pages 1–26, Berlin, Heidelberg, 2013. Springer-Verlag.

[416] Guoqing Xu. Resurrector: A tunable object lifetime profiling technique for optimizing real-world programs. *SIGPLAN Not.*, 48(10):111–130, October 2013.

[417] Guoqing Xu, Matthew Arnold, Nick Mitchell, Atanas Rountev, and Gary Sevitsky. Go with the flow: profiling copies to find runtime bloat. In *Programming Language Design and Implementation (PLDI)*, 2010.

[418] Guoqing Xu et al. Finding low-utility data structures. In *Programming Language Design and Implementation (PLDI)*, 2010.

[419] Guoqing Xu, Nick Mitchell, Matthew Arnold, Atanas Rountev, Edith Schonberg, and Gary Sevitsky. Scalable runtime bloat detection using abstract dynamic slicing. *ACM Trans. Softw. Eng. Methodol.*, 23(3):23:1–23:50, June 2014.

[420] Guoqing Xu, Nick Mitchell, Matthew Arnold, Atanas Rountev, and Gary Sevitsky. Software bloat analysis: Finding, removing, and preventing performance problems in modern large-scale object-oriented applications. In *Future of Software Engineering Research*, 2010.

[421] Guoqing Xu and Atanas Rountev. Detecting inefficiently-used containers to avoid bloat. In *Programming Language Design and Implementation (PLDI)*, 2010.

[422] Guoqing Xu, Dacong Yan, and Atanas Rountev. Static detection of loop-invariant data structures. In *Proceedings of the European Conference on Object-Oriented Programming, ECOOP*, pages 738–763, 2012.

[423] Neeraja J. Yadwadkar, Chiranjib Bhattacharyya, K. Gopinath, Thirumale Niranjan, and Sai Susarla. Discovery of application workloads from network file traces. In *Proceedings of the 8th USENIX Conference on File and Storage Technologies*, FAST'10, pages 14–14, Berkeley, CA, USA, 2010. USENIX Association.

[424] Neeraja Jayant Yadwadkar, Bharath Hariharan, Joseph E. Gonzalez, and Randy Howard Katz. Multi-task learning for straggler avoiding predictive job scheduling. *Journal of Machine Learning Research*, 17:106:1–106:37, 2016.

[425] Da Yan, Bu Yingyi, Yuanyuan Tian, and Amol Deshpande. *Big Graph Analytics Platforms*, 2017.

[426] X. Yao, J. Cafaro, A.J. McLaughlin, F.R. Postma, D.L. Paul, G. Awatramani, and G.D. Field. Gap junctions contribute to differential light adaptation across direction-selective retinal ganglion cells. *Neuron*, 2018.

[427] Young Jin Yu, Dong In Shin, Woong Shin, Nae Young Song, Jae Woo Choi, Hyeong Seog Kim, Hyeonsang Eom, and Heon Young Yeom. Optimizing the block I/O subsystem for fast storage devices. *ACM Trans. Comput. Syst.*, 32(2):6:1–6:48, June 2014.

[428] Eleni Tzirita Zacharatou, Harish Doraiswamy, Anastasia Ailamaki, Cláudio T. Silva, and Juliana Freiref. GPU rasterization for real-time spatial aggregation over arbitrary polygons. *Proc. VLDB Endow.*, 11(3):352–365, November 2017.

[429] Matei Zaharia, Mosharaf Chowdhury, Tathagata Das, Ankur Dave, Justin Ma, Murphy McCauley, Michael J. Franklin, Scott Shenker, and Ion Stoica. Resilient distributed datasets: A fault-tolerant abstraction for in-memory cluster computing. In *Proceedings of the 9th USENIX Conference on Networked Systems Design and Implementation*, NSDI'12, pages 2–2, Berkeley, CA, USA, 2012. USENIX Association.

[430] Charles Zhang and Hans-Arno Jacobsen. Efficiently mining crosscutting concerns through random walks. In *Proceedings of the 6th International Conference on Aspect-oriented Software Development*, AOSD '07, pages 226–238, New York, NY, USA, 2007. ACM.

[431] Jiaxing Zhang, Ying Yan, Liang Jeff Chen, Minjie Wang, Thomas Moscibroda, and Zheng Zhang. Impression store: Compressive sensing-based storage for big data analytics. In *Proceedings of USENIX HotCloud'14*, 2014.

[432] Mingxing Zhang, Yongwei Wu, Youwei Zhuo, Xuehai Qian, Chengying Huan, and Kang Chen. Wonderland: A novel abstraction-based out-of-core graph processing system. *SIGPLAN Not.*, 53(2):608–621, March 2018.

[433] Yiming Zhang, Jon Crowcroft, Dongsheng Li, Chengfen Zhang, Huiba Li, Yaozheng Wang, Kai Yu, Yongqiang Xiong, and Guihai Chen. Kylinx: A dynamic library operating system for simplified and efficient cloud virtualization. In *2018 USENIX Annual Technical Conference (USENIX ATC 18)*, pages 173–186, Boston, MA, 2018. USENIX Association.

[434] W. Zhao, L. Zhang, Y. Liu, J. Sun, and F. Yang. SNIAFL: Towards a static non-interactive approach to feature location. In *ACM Transactions on Software Engineering and Methodologies*, volume 15, pages 195–226, April 2006.

[435] Yi Zhao, Jin Shi, Kai Zheng, Haichuan Wang, Haibo Lin, and Ling Shao. Allocation wall: a limiting factor of java applications on emerging multicore platforms. In *Object-Oriented Programming, Systems, Languages and Applications, OOPSLA*, 2009.

[436] Yang Zheng, Cong Xu, and Yuan Xie. Modeling framework for cross-point resistive memory design emphasizing reliability and variability issues. *20th Asia and South Pacific Design Automation Conference, ASP-DAC 2015*, pages 112–117, 03 2015.

[437] Nadav Amit, Fred Jacobs, Michael Wei, JumpSwitches: Restoring the performance of Indirect Branches In the era of Spectre, USENIX ATC, 2019.

Index

Index